ULTRA-FAST
FIBER LASERS

Principles and Applications
with MATLAB® Models

ULTRA-FAST
FIBER LASERS

Principles and Applications
with MATLAB® Models

Le Nguyen Binh
Nam Quoc Ngo

CRC Press
Taylor & Francis Group
Boca Raton London New York

CRC Press is an imprint of the
Taylor & Francis Group, an **informa** business

CRC Press
Taylor & Francis Group
6000 Broken Sound Parkway NW, Suite 300
Boca Raton, FL 33487-2742

First issued in paperback 2018

© 2011 by Taylor and Francis Group, LLC
CRC Press is an imprint of Taylor & Francis Group, an Informa business

No claim to original U.S. Government works

ISBN-13: 978-1-4398-1128-3 (hbk)
ISBN-13: 978-1-138-37417-1 (pbk)

Library of Congress Cataloging-in-Publication Data

Binh, Le Nguyen.
   Ultra-fast fiber lasers : principles and applications with MATLAB models / authors, Le Nguyen Binh, Nam Quoc Ngo.
      p. cm.
   Includes bibliographical references and index.
   ISBN 978-1-4398-1128-3
   1. Fiber optics. 2. Laser pulses, Ultrashort. I. Ngo, Nam Quoc. II. Title.

TA1800.B56 2010
621.382'7--dc22   2010006603

Visit the Taylor & Francis Web site at
http://www.taylorandfrancis.com

and the CRC Press Web site at
http://www.crcpress.com

Contents

Preface

Ultrashort pulses in mode-locked lasers are a topic of extensive research due to their wide range of applications from optical clock technology [1] to measurements of the fundamental constants of nature [2] and ultrahigh-speed optical communications [3,4].

Ultrashort pulses are especially important for the next generation of ultrahigh-speed optical systems and networks operating at 100 Gbps per carrier. Pulse sequences with pulse width on the order of a few picoseconds to less than one picosecond are considered to be short and are probable for the generation of multi-Gbps optical-carrier data. The most practical short pulse sources must be in the form of guided wave photonic structures in order to minimize the alignment and losses and ease of coupling into fiber transmission systems. Fiber ring lasers operating in active mode are suitable ultrashort pulse sources because they meet these requirements.

This book is written as a stand-alone reference book for professional engineers, scientists, and graduate students in the fields of applied photonics and optical communications. Theoretical and experimental results are presented, and MATLAB® files are included in order to provide a basic grounding in the simulation of the generation of short pulses and the propagation or circulation around nonlinear fiber rings.

The principal objectives of the book are as follows: (1) To describe the fundamental principles of the generation of ultrashort pulses employing fiber ring lasers that incorporate active optical modulators of amplitude or phase types. (2) To present experimental techniques for the generation, detection, and characterization of ultrashort pulse sequences. Several schemes are described by detuning the excitation frequency of modulation of the optical modulator embedded in the ring. The birefringence of the guided medium ring that influences the locking and polarization multiplexed sequences is considered. (3) To describe the multiplication of ultrashort pulse sequences using the Talbot diffraction effects in the time domain via the use of highly dispersive media. (4) To develop the theoretical and experimental developments of multiple short pulses in the form of solitons binding together by phase states. (5) To describe the generation of short pulse sequences and multiple wavelength channels from a single fiber laser. This book consists of up-to-date research materials from the authors and researchers working in the field.

Le Nguyen Binh
Monash University
Melbourne, Australia

Nam Quoc Ngo
Nanyang Technological University
Singapore

MATLAB® is a registered trademark of The MathWorks, Inc. For product information, please contact:

The MathWorks, Inc.
3 Apple Hill Drive
Natick, MA 01760-2098 USA
Tel: 508 647 7000
Fax: 508-647-7001
E-mail: info@mathworks.com
Web: www.mathworks.com

References

1. J. Ye, S. Cundiff, *Femtosecond Optical Frequency Comb: Technology, Principles, Operation and Application*, Springer, New York, 2005.
2. M. Fischer, N. Kolachevsky, M. Zimmermann, R. Holzwarth, T. Udem, T. Hänsch, M. Abgrall et al., New limits on the drift of fundamental constants from laboratory measurements, *Phys. Rev. Lett.*, 92, 230802, 2004.
3. H. Haus, Mode-locking of lasers, *IEEE J. Sel. Top. Quant. Electron.*, 6, 1173–1185, 2000.
4. L. J. Tong, Future networks, Plenary paper, *International Conference on Optical Communications Networks*, Beijing, September 2009.

Acknowledgments

We are grateful to the Faculty of Engineering at Monash University, Melbourne, Australia, for allowing us to use their laboratories and facilities to demonstrate the fiber ring laser experiments presented in this book. Furthermore, a number of fiber ring lasers presented have also been carried out at the Network Technology Research Center (NTRC) and at the Photonics Research Center of Nanyang Technological University of Singapore.

We are grateful to Professor Ping Shum for his advice on the use of the facilities at NTRC to construct and demonstrate the generation of ultrashort pulse sequences. We are also indebted to our doctoral students, Dr. WenJing Lai, Dr. Lam Quoc Huy, Gary Teo, and Nguyen Duc Nhan, for discussions on the development and research of mode-locked fiber lasers. We would also like to acknowledge Dr. Shilong Pan and Dr. Caiyun Lou from Tsinghua University, Beijing, China, and the Institute of Electrical and Electronic Engineering for permission to use the figures in Chapter 11.

Last but not least, Dr. Binh would like to thank his family, Phuong and Lam, who have put up with his long hours of writing this book and have supported him over the years; he is also grateful to his parents who have inculcated in him the value of *learning for life*.

Le Nguyen Binh
Monash University
Melbourne, Australia

Nam Quoc Ngo
Nanyang Technological University
Singapore

Authors

Le Nguyen Binh received his BE (Hons) and PhD degrees in electronic engineering and integrated photonics in 1975 and 1980, respectively, from the University of Western Australia, Nedlands, Western Australia, Australia. In 1980, he joined the Department of Electrical Engineering at Monash University, Clayton, Victoria, Australia, after a three-year period with Commonwealth Scientific and Industrial Research Organisation (CSIRO), Clayton, Victoria, Australia, as a research scientist.

In 1995, he was appointed as reader at Monash University. He has worked in the Department of Optical Communications of Siemens AG Central Research Laboratories in Munich, Germany, and in the Advanced Technology Centre of Nortel Networks at Harlow, United Kingdom. He has also served as a visiting professor of the Faculty of Engineering of Christian Albrechts University of Kiel, Kiel, Germany.

Dr. Binh has published more than 250 papers in leading journals and refereed conferences, and two books in the field of photonic signal processing and digital optical communications: the first is *Photonic Signal Processing* and the second is *Digital Optical Communications* (both published by CRC Press, Boca Raton, Florida). His current research interests are in advanced modulation formats for long haul optical transmission, electronic equalization techniques for optical transmission systems, ultrashort pulse lasers, and photonic signal processing.

Nam Quoc Ngo received his BE and PhD degrees in electrical and computer systems engineering from Monash University, Melbourne, Victoria, Australia, in 1992 and 1998, respectively. From July 1997 to July 2000, he was a lecturer at Griffith University, Brisbane, Queensland, Australia. Since July 2000, he has been with the School of Electrical and Electronic Engineering (EEE), Nanyang Technological University, Singapore, where he is presently an associate professor. Since March 2009, he has been the deputy director of the Photonics Research Centre at the School of EEE. Among his other significant contributions, he has pioneered the development of the theoretical foundations of arbitrary-order temporal optical differentiators and arbitrary-order temporal optical integrators, which resulted in the creation of these two new research areas. He has also pioneered the development of a general theory of the Newton–Cotes digital integrators, from which he has designed a wideband integrator and a wideband differentiator known as the Ngo integrator and the Ngo differentiator, respectively, in the literature. His current research interests are on the design and development of fiber-based and waveguide-based devices for application in optical communication systems and optical sensors. He has published more than 110 international journal papers and over 60 conference papers in these areas. He received two awards for outstanding contributions in his PhD dissertation. He is a senior member of IEEE.

1

Introduction

1.1 Ultrahigh Capacity Demands and Short Pulse Lasers

1.1.1 Demands

Four decades ago, G.R. Moore made a prediction, which is now the famous Moore's law. He predicted that the number of integrated transistors on a chip would be doubled every 18 months. This is true even now, and is expected to be valid for another decade at least. Moore's law is applied not only to study the growth of the semiconductor industry but also to predict the growth of data traffic. Data traffic is projected to grow exponentially, as shown in Figure 1.1a [1]. This is mainly driven by the emergence of high-bandwidth-consuming applications and services such as Internet protocol television (IPTV), file sharing, and high-definition television (HDTV). Besides, the expansion of major telecommunications infrastructure in the Far East (China, India) and in developing countries also results in huge data traffic.

The rapid deployment of fiber to the x (FTTx) (see Figure 1.1b) further pushes the demand for network capacity for delivering the data traffic. Recently, the Federal Government of Australia initiated a project to design, build, and operate the active infrastructure of Australia's next-generation national broadband network [2]. Similarly, in Southeast Asia, Singapore has also introduced its national broadband networks [3–5] for the twenty-first century. Both projects aim at providing broadband access links, which can scale up to 100 Mbps, to 50% of residents in Australia and Singapore by 2012, and then to 1 Gbps soon after that period. This development is considered as the digital economy of the twenty-first century. The delivery of information at this rate is unheard of in the history of human communication. This would not be possible if optical fibers, especially the single-mode fibers, were not invented and exploited over the last three decades. For this invention, Dr. Charles Kao was awarded the Nobel Prize for physics in 2009. Even at this data rate to the home, the bandwidth usage is only a tiny fraction of the huge fiber bandwidth of ~25 THz provided by a single-mode fiber. The principal point is how to deliver the services effectively and economically both in the core systems/networks and the last-mile distribution networks. In addition, societies live in an age of creativity and the trend of using video transmission in global communities has risen tremendously (see Figure 1.2), indicating that the bit rate or capacity of the backbone networks must be increased significantly to

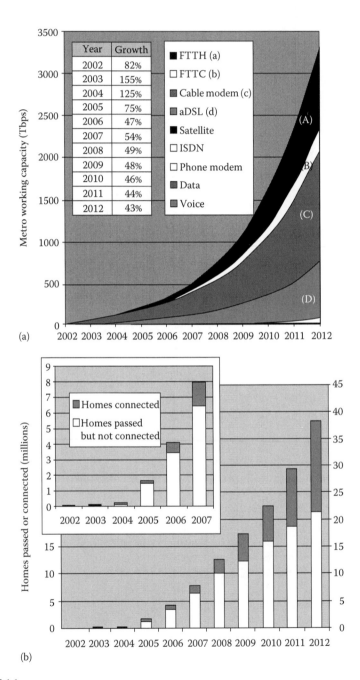

FIGURE 1.1
Metro Internet traffic data growth over time and in the future.

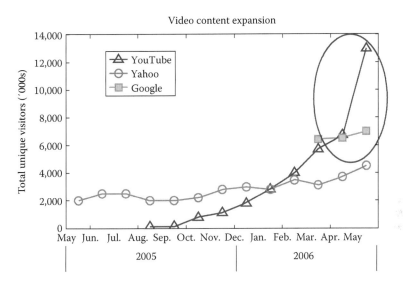

FIGURE 1.2
Demands on video transmission.

respond to these demands. The video of the future will include super-high-definition television, high-definition holographic television, and image transmission whose compressed bit rate is on the order of Gbps. The increase in the capacity–distance product in one single-mode fiber over the years is shown in Figure 1.3, indicating a 100-fold increase in every 4 years.

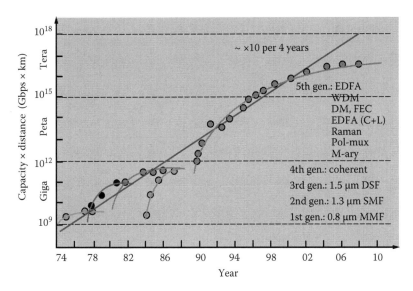

FIGURE 1.3
Capacity–distance product in Gbps achieved over the years in one single-mode fiber.

The development of optical communications after the invention of the optical fiber has fulfilled the demand in the past. But in order to keep up with the exponential growth of data traffic in the near and not-so-near future, more and more hardware components and transmission technologies have to be developed. Wavelength-division multiplexing (WDM), the technology of combining a number of wavelengths into the same fiber, successfully increases the fiber capacity by 32 times or even hundreds of times if its advanced version, dense wavelength-division multiplexing (DWDM), is employed. However, the bottleneck is at the routers and switches where hundreds of channels must be demultiplexed for O/E (optical-to-electrical conversion), routing, E/O (electrical-to-optical conversion), and then multiplexing [6,7]. The whole network speed can further be significantly increased if the optical signal can be routed/switched directly in the optical domain without the need for O/E. All-optical switching, signal processing, and optical time-division multiplexing (OTDM) are promising solutions [5,8–13].

In OTDM systems, data is encoded using ultrashort optical pulses occupying N time slots in the OTDM time frame. Each channel is assigned a time slot, and its data can be accessed with the aid of an optical clock pulse train corresponding to that time slot. Hence, the generation of ultrashort optical pulses with multiple-gigabits repetition rate is critical for ultrahigh bit rate optical communications, particularly for the next generation of Tbps optical fiber systems. The mode-locked fiber laser offers a potential source of such a pulse train. Furthermore, the field of optical packet switching has gained recognition in recent years and requires ultrashort and high-peak power pulse generators to provide all-optical switching [14–20].

Although the generation of ultrashort pulses by mode locking of a multimodal ring laser is well known, the applications of such short pulse trains in multi-gigabits/s optical communications challenge designers on their stability and spectral properties, as well as the multiplication of the repetition rate. Recent reports on the generation of short pulse trains at repetition rates on the order of 40 Gbps [21,22], and possibly higher in the near future, have motivated us to design, experiment, and model these sources in order to evaluate whether they can be employed in practical optical communications and all-optical switching systems. Furthermore, the development of multiplexed transmission at 160 Gbps and higher in the foreseeable future requires us to experiment with an optical pulse source with a short pulse duration and high repetition rates.

1.1.2 Ultrashort Pulse Lasers

Since the invention of optical maser or laser in 1958 [23], several research fields have bloomed in optics and photonics, especially ultrafast optics that began in the mid-1960s, with the production of narrow pulses by the mode locking in a laser cavity [24,25]. Currently, ultrafast pulse generation remains an active research field. As optical communication technologies reach ultrahigh-speed transmission and greater sensitivity, the performance of fiber optic devices and electronic systems has begun to approach their fundamental physical limits.

Transmission using short pulses is a fundamental technology for a high-speed, long-haul transmission system. Among several optical transmission pulse formats, optical soliton [26,27], which is a very stable optical pulse formed from balancing between the anomalous dispersion and the fiber's self-phase modulation (SPM) effect, offers a great potential to realize such a system. The optical soliton was first observed experimentally by Mollenauer et al. in 1980 [28], and the first soliton laser was constructed later [29]. Since then, the soliton has attracted tremendous research interest in the optical communication community. Hasegawa [30] has proposed the use of a soliton for transoceanic transmission, compensating the fiber loss by Raman gain, with no pulse regeneration over the entire reach. This proposal could only be proven in practice with the availability of erbium-doped fiber amplifiers (EDFAs) [31]. Several published works have explored and demonstrated ways to boost the information-carrying capacity of fibers such as by increasing the modulation frequency, decreasing the channel spacing, and utilizing efficient modulation schemes, so that it will be brought closer to the predicted Shannon's law for optical fiber [32].

Recent progress on soliton technology has pushed the transmission limit into the terahertz range by means of time division multiplexing (TDM), WDM, or polarization division multiplexing (PDM). Nakazawa and coworkers [33] demonstrated a 1.28 Tbps transmission over 70 km by OTDM 128 channels at 10 Gbps. Sotobayashi et al. [34] also showed a 3.24 Tbps transmission capacity by 81 wavelength channels at 40 Gbps with the carrier-suppressed return-to-zero (CSRZ) format. More recently, Bigo et al. [35] successfully transmitted a 10.2 Tbps (2×128 WDM channels $\times$ 42.7 Gbps) signal over 100 km by using the PDM/WDM technique. Technology advancements over the last two decades have led to fourth-generation communication systems referred to as coherent optical communication systems, and finally to fifth-generation systems, rather known as soliton communication systems.

1.2 Principal Objectives of the Book

The principal objectives of the book are as follows: (1) To describe the fundamental principles of the generation of ultrashort pulses employing fiber ring lasers incorporating active optical modulators of amplitude or phase types. (2) To present experimental techniques for the generation, detection, and characterization of ultrashort pulse sequences. Several schemes are described by detuning the excitation frequency of modulation of the optical modulator imbedded in the ring. The birefringence of the guided medium ring influencing the locking and polarization multiplexed sequences is considered. (3) To describe the multiplication of ultrashort pulse sequences using the Talbot diffraction effects in the time domain via the use of highly dispersive media. (4) To develop the theoretical and experimental developments of multiple short pulses in the form of solitons binding together by phase states. (5) To describe the generation of multiple wavelength channels of short pulse sequence.

1.3 Organization of the Book Chapters

Chapter 2 presents the fundamentals of actively mode-locked fiber lasers and corresponding analytical techniques. A review on the development of mode-locked lasers is presented and discussed. A mathematical description of the principles of mode locking is derived. Two techniques for mode locking, passive mode locking and active mode locking, are discussed. Passive mode locking is suitable for generation of ultrashort pulses and high-peak power, while active mode locking is preferred for the generation of high-repetition-rate pulses for high-speed telecommunication systems. The repetition rate of actively mode-locked fiber lasers can even be increased by using the rational harmonic mode-locking technique where the modulation frequency is deliberately detuned by a fraction of the fundamental frequency. Alternatively, an intra-cavity Fabry–Perot etalon filter can be used to increase the pulses' repetition rate by suppressing the unwanted modes. In addition, supermode noise causes the pulse-amplitude fluctuation and can be suppressed by filtering or the SPM effect. The change of temperature and environment can make the modulation frequency different to the harmonic of the laser fundamental frequency and hence causes the laser to become unstable. Monitoring the repetition rate and using a piezoelectric transducer (PZT) for dynamical adjustment of the cavity length can stabilize the lasers. Furthermore, a series approach to obtain an analytical solution for mode locking is described. Instead of assuming a Gaussian pulse solution for the mode-locking equation and applying an invariant condition to obtain the pulse steady-state parameters, we propose using a mathematical series to trace the pulse evolution in the cavity and derive the pulse parameters directly from the series.

Chapter 3 then presents techniques and methodology for experimental setup and measurements of mode-locked fiber lasers. The corroboration of the experimental results and theoretical analyses is given.

In Chapter 4, we present the modeling techniques for studying the actively mode-locked lasers under the influence of amplified stimulated emission (ASE) noise. Detuning of the laser is also investigated and the locking range is derived with the consideration of ASE noise. We show that ASE noise reduces the locking range of the laser and hence the optical amplifier noise should be minimized in order to improve the laser performance.

Dispersion and nonlinear effects on the performance of actively mode-locked fiber lasers are presented in Chapter 5, both by analytical and experimental techniques. A comprehensive laser model, which includes the dispersion and nonlinear effects, is presented. It is demonstrated that nonlinearity assists in shortening the pulse and plays an important role in soliton shaping of the laser pulse. Moreover, dispersion improves the stability of the laser against the modulation detuning through shifting the lasing wavelength to compensate for delay/advance caused by the detuning. We show that there is a trade-off between pulse shortening and stability of the laser. Cavity dispersion and nonlinearity should be optimized to obtain the shortest pulse while the laser is still stable within a certain amount of detuning.

Chapter 5 presents an ultrafast, electrically wavelength-tunable, actively mode-locked fiber laser. Based on a Lyot birefringence filter, we introduce a phase modulator to shift the phase of the optical field so that the transmission peak wavelength of the filter can be tuned by an external electrical voltage. The filter is then applied to the laser cavity design. Moreover, the phase modulator of the filter is also used for mode locking and thus the cost can be reduced. The polarization dynamics of actively mode-locked fiber lasers are presented in this chapter. We investigate the behaviors of the lasers when the birefringence is introduced into the cavity. Pulses with dual polarization states are generated, and polarization switching of the mode-locked pulses is observed. Pulse dropout and sub-harmonic mode locking due to mode competition is also reported.

Chapter 6 presents the structures and operations of multiwavelength mode-locked fiber lasers for the generation of ultrashort pulse sequence and in several wavelength regions.

The remaining chapters of the book introduce detailed investigations on the system behaviors of several fiber ring laser structures, namely, active-mode-locked erbium-doped fiber ring laser, regenerative active-mode-locked erbium-doped fiber ring laser, fractional-temporal-Talbot-based repetition-rate-multiplication system, parametric-amplifier-based fiber ring laser, regenerative parametric-amplifier-based fiber ring laser to generate a terahertz-repetition-rate pulse sequence (Chapter 8), and nonlinear bidirectional propagating NOLM-NALM fiber ring laser.

Although some of the laser structures are common in literature, Chapter 7 tackles the issue from different perspectives, hence arriving at novel system observations and analyses such as phase plane analysis on the system stability issue, Gaussian modulating signal in mode-locked laser systems, and frequency detuning in parametric-amplifier-based fiber ring laser. The phase plane analysis, which is a subsidiary of the nonlinear control engineering, is used for the first time in the laser system behavior analysis. We use it to analyze the transient and steady-state behaviors of the fractional-temporal-Talbot-based repetition-rate-multiplication system and rational harmonic detuning in active-mode-locked erbium-doped fiber ring lasers.

Conventionally, the modulating signal of a mode-locked laser system is a co-sinusoidal signal. However, with a change in the pulse shape and duty cycle of the modulating signal, we achieve a record high-order rational harmonic mode-locked laser system. An analytical model for this Gaussian modulating signal in a mode-locked laser system has been developed and validated by experiments. The phase-matching condition is an essential criterion in parametric amplification systems. However, by applying some frequency detuning to the system, some interesting phenomenon can be observed. We study the frequency detuning behavior in the parametric-amplifier-based fiber ring laser system, and ultrahigh-repetition-rate operation is observed. We believe this ultrahigh-speed operation results from the combination of rational harmonic detuning and modulation instability.

In Chapter 8, the modulating signal integrated in a conventional actively mode-locked laser system is modified. The stability of the generated pulses is studied. The novel fiber laser structure in generating terahertz operation is also described with its theoretical model to concur with the experimental demonstration.

With the NOLM-NALM fiber ring laser described in Chapter 9, we describe the bidirectional lightwaves' propagation behaviors. In addition, with our developed model, different operation regimes are obtained numerically, namely, single operation, period-doubling operation, and chaotic operation. Unfortunately, due to the hardware limitations, only the first two operations are observed experimentally. We believe that this laser structure will have a good potential in various photonics applications due to its peculiar operating characteristics.

Chapter 10 introduces the generation of bound solitons from actively mode-locked fiber lasers in which optical phase modulation under chirp operations are used. Groups of bound soliton states of up to the sixth order can be generated. The propagation of these bound solitons is also described.

Finally, Chapter 11 presents the simulation and experimental results of a particular type of an actively mode-locked multiwavelength erbium-doped fiber laser (AMLM-EDFL) operating in the 1550 nm communication window. Different central wavelength channels of short pulse sequences are generated. The chapter begins with an overview of the various important applications of the AMLM-EDFLs and their unique performance and cost advantages over the competing technologies. The chapter then discusses the numerical model used in the analysis of the AMLM-EDFL through the use of the nonlinear Schrödinger equation, which includes the loss, and dispersive and nonlinear effects of the fiber laser cavity. The effectiveness of the numerical model is demonstrated by simulating two lasing wavelengths at 10 Gbps.

The experimental results of a dual-wavelength fiber laser are presented to verify the numerical model and the simulation results. Several suggestions for the future are also given.

1.4 Historical Overview of Ultrashort Pulse Fiber Lasers

1.4.1 Overview

This section gives an overview of the operational principles of mode-locked fiber lasers, especially when an active optical component such as an optical modulator is integrated within the ring cavity. A detailed analysis of the evolution of the pulse sequence in the laser is described.

In 1964, DiDomenico predicted that small signal modulating of cavity loss in a laser at a frequency equal to multiples of axial mode spacing causes mode coupling with well-defined amplitude and phase [36]. This phenomenon was later called mode locking of the laser. In the same year, the first mode-locked laser was demonstrated by Hargrove et al. [37]. The authors incorporated an acousto-optic modulator inside a He–Ne laser to internal loss modulate the laser and obtained mode locking. The mode-locked amplitude was fivefold over the average unlocked one, and the rapidly fluctuating optical spectrum changed to a stationary one. The repetition rate and pulse width are 56 MHz and 2.5 ns, respectively.

In 1970, the analytic theory of active mode locking was firmly established by Kuizenga and Siegman [38]. The authors studied the process in the time domain and derived mode-locking equations for both amplitude modulation (AM) and frequency modulation (FM). In the paper, the authors assumed a Gaussian pulse solution, and approximations of the line shape and modulation function to keep the pulse Gaussian were applied. However, the intracavity filter, which is normally used to stabilize the laser, was not included in the analysis.

In 1975, Haus presented a frequency domain approach for the mode-locking theory [39]. Using a circuit model, which is similar to that of Ref. [40], the author analyzed the evolution of the pulse in the frequency domain and derived differential equations for forced mode locking. The equations had a Hermite–Gaussian solution, and the mode-locked pulse characteristic was obtained. The author also derived differential equations in the time domain, which follow the form of the Schrödinger equation. It is noted that a co-sinusoidal modulation was used to acquire injection locking. Furthermore, continuum spectrum approximation of a discrete frequency spectrum and the assumption of a dense spectrum were made to derive the equations.

In 1989, the first report of mode-locked operation in an Yb–Er-doped fiber laser was presented [41]. The authors used an AM modulator to achieve mode locking. Pulses with a repetition rate of 100 MHz and a pulse width of 70 ps were generated in a linear cavity.

One year later, Takada and Miyazawa applied harmonic mode locking with a ring configuration to generate pulses with a repetition rate as high as 30 GHz by using a high-speed LiNO$_3$ modulator [42]. Instead of modulating the signal at the cavity's fundamental frequency, the authors applied a high-frequency microwave signal at a high-order harmonic of the fundamental frequency to the modulator and obtained the pulse train at the modulating rate.

The major problem in active mode-locked lasers is the stability. Several research works have been conducted to overcome this problem. In 1993, Harvey and Mollenauer used a subcavity with a free spectral range that matches the modulation frequency to stabilize the pulse energy [43]. The coupling of a portion of the energy of each individual pulse to successive pulses causes injection locking the pulse train optically and hence equalizes the pulse energy. The disadvantage of this method is that it requires strict matching between the fundamental frequency, the free spectral range of the subcavity, and the modulation frequency.

Another approach is the additive pulse limiting (APL) technique [44], in which the high-energy pulse is rotated more than the low-energy and experiences more loss. By properly adjusting the polarization bias, the pulse energy can be clamped to a specific level.

The SPM effect was also used to stabilize the pulse energy in 1996 [45]. SPM causes more intense pulse and generates a wider spectrum. With a spectral filter placed inside the cavity, the high-intensity pulse experiences more loss and hence energy stability is obtained.

A variation of the fiber length due to temperature fluctuation may cause the pulse to lose synchronism with the modulation. Several techniques have been proposed to stabilize the laser against temperature. A common technique is to adjust the cavity length to compensate for the variation such as in Refs. [46,47]. The authors used a piezoelectric drum driven by an error signal to adjust the fiber length. The phase mismatch between the output pulse and the modulation signal was used as the error signal. Alternatively, the modulation frequency can be adjusted so that it is kept in synchronization with the fiber. This technique is called regenerative feedback since the modulation frequency is derived directly from the pulse change [48,49].

However, the maximum pulse repetition rate is limited by the bandwidth of the LiNbO$_3$ modulator. The repetition rate can be increased by using the rational harmonic mode-locking technique [50–52]. The rational harmonic mode-locked fiber laser (RHMLFL) configuration is the same as that of the harmonic mode-locked fiber laser (HMLFL), except that the modulating frequency f_m is not an integer number of the fundamental frequency. The modulating frequency f_m is now detuned by a fraction of the fundamental frequency f_R, which means $f_m = (p + 1/k)f_R$. With that detuning, the output pulse train has a repetition rate of $(kp + 1)f_R$, k times the modulating frequency.

The disadvantage of RHMLFL is that the pulse amplitudes vary from pulse to pulse due to the different losses the pulses experienced when passing through the modulator. In 2002, Gupta et al. proposed to use a Fabry–Perot filter (FFP) inserted into the ring to equalize the pulse amplitudes in the RHMLFL [53]. However, this method requires that the FFP's free spectral range (FSR) be equal to the repetition rate and additional circuit to stabilize the FFP.

1.4.2 Mode-Locking Mechanism in Fiber Ring Resonators

The technique examined in this section is the generation of repetitive pulse sequences from locking into one of the longitudinal modes of a ring laser, whether it be a fiber-based or a semiconductor-based structure [54]. A fundamental schematic diagram of mode-locked lasers is shown in Figure 1.4, which consists of a nonlinear waveguide section, an amplifying device to compensate for the energy loss as well as to provide sufficient gain to induce the nonlinear effects, a tuning section to generate the locking condition of the lightwave energy to a particular harmonic, an input and output coupling section for tapping the laser source, and an optical modulator to generate the repetition rate in association with the locking mechanism. All these optical components are interconnected by a ring of single-mode optical fiber. The chromatic dispersion of the single-mode fiber is used to achieve good balancing interplay with the nonlinear effects induced by

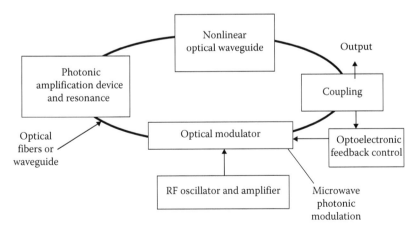

FIGURE 1.4

Schematic diagram of a generic mode-locked laser for generation of a multi-gigahertz-repetition-rate pulse sequence. Note that the optical modulator is incorporated in the ring → the repetition rate is limited by the bandwidth of the modulator.

the nonlinear waveguide. The length of the optical waveguide is used to determine the frequency spacing between adjacent longitudinal modes of the ring laser.

The pulse repetition rate generated by the above convention technique is often limited by the operating frequency of the optical modulator imbedded in the ring resonator. Some techniques have been proposed to increase the repetition rate of the mode-locked system using temporal Talbot effects (temporal diffraction [55,56]), rational harmonic detuning [57], optical division multiplexing [58], etc.

The generation of an extremely high-repetition-rate pulse sequence in the terahertz range is possible by using the nonlinear parametric amplification and the degenerate four-wave mixing (FWM) phenomenon in a special optical waveguide and the mechanism of rational frequency detuning, as well as the interference between the modulated pump signals via an optical modulator.

1.4.2.1 Amplifying Medium and Laser System

The fundamental structure of a fiber ring laser is the amplifying resonance ring. There are certain discrete energy levels in a particle that electrons can occupy, as shown in Figure 1.5. The electrons can jump between those levels depending on whether they receive or release energy. When receiving energy from other sources such as an electronic or optical source, they jump from a lower level to higher levels. The electrons at high-level states are excited electrons, and they can randomly jump back to a lower energy state without any stimulating source. When changing from high-energy level E_i to lower-energy E_j, the electrons release the energy in the form of a photon whose frequency can be calculated from

$$v_{ij} = \frac{E_i - E_j}{h} \tag{1.1}$$

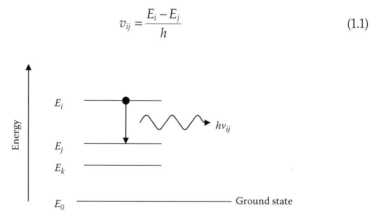

FIGURE 1.5
Energy level diagram of a particle and photon emission.

where

v_{ij} is the released photon's frequency
h is Planck's constant
E_i and E_j are the energies of the electron at level i and j, respectively

The emission of photons through this random jumping process is called spontaneous emission, and the phases of the released photons are random. By contrast, when a photon enters a medium with excited electrons, as shown in Figure 1.6, it will cause a stimulated emission when the electron transits to a lower level and releases a new photon. The photons released through this emission have the same energy and phase as those of the stimulating photon. They then stimulate new emission and hence more and more photons are generated. Therefore, the optical signal is amplified, and the medium with electrons pumped into excited states is called an amplifying medium.

The amplifying medium is now put inside a cavity formed by two mirrors to form a typical laser, as shown in Figure 1.7. One of the mirrors totally reflects the light, while the other reflects part of the light and transmits the rest to the output. The photons initially emitted from the spontaneous emission process are amplified when traveling through the amplifying medium via the stimulated emission. The photons are reflected back and forth between the mirrors and grow exponentially if their frequencies or wavelengths satisfy the phase condition

$$\omega = \frac{N\pi c}{nL} \tag{1.2}$$

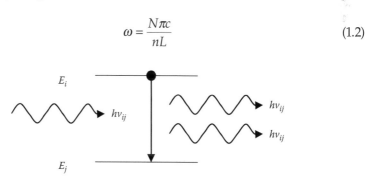

FIGURE 1.6
Stimulated emission and energy levels in an optically amplifying medium.

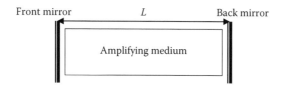

FIGURE 1.7
Schematic diagram of a basic laser system.

where
  L is the cavity length
  n is the refractive index of the medium or the effective refractive index of
    the guided mode if the medium is an optical waveguide
  N is an integer number, $c = 3 \times 10^8$ m/s is the speed of light in vacuum

It can be seen from (1.2) that the laser can generate lightwaves in a wide spectral range of wavelengths that satisfy the cavity resonance conditions. These are the longitudinal modes. In several published works, experiments have been conducted to force the laser to resonate at only one particular mode so as to narrow the laser linewidth [59–61].

Alternatively, the laser can be designed to oscillate at multiple longitudinal modes, but those modes are controlled so that sequences of very short pulses can be generated. Such pulses find important applications in photonic switching and photonic communication systems [62–64]. A typical technique to generate a short pulse laser is the Q-switching [55,56,65–67] in which the quality factor Q of the optical resonator can be lowered to prevent the laser from oscillating during the energy-pumping period. The gain/population inversion can therefore build up to a higher value than that obtained for the normal pumping case. When the inversion population reaches its peak, the quality factor Q is switched to the high level. Therefore, the threshold for oscillating is restored to a level much below the gain. This causes an exponential growth of the intra-cavity intensity, and the laser delivers a pulse of high intensity and short duration. However, the Q-switching technique can just generate pulses no shorter than a few nanoseconds. For generating picosecond pulses, mode-locking techniques are commonly employed [68,69].

1.4.2.2 Active Modulation in Laser Cavity

Active modulation in the fiber-laser cavity assists the concentration of the energy,circulating in the ring into specific pulses distributed evenly in time. The active modulation can be amplitude, phase, or frequency (continuous phase). For amplitude modulation, the intensity of the lightwaves circulating in the ring would be at the minimum or the maximum in periodic time instants. Thus the resonance happens at these periodic instants due to the sufficient energy provided by the pulses for further amplification and compensation of the loss of the ring cavity.

On the other hand, phase modulation forces the resonant conditions at the instants of the applied electrical signals. Thus, only at these instants the resonant conditions could be satisfied, and hence the pulse trains would be generated.

For frequency or continuous phase modulation, the phase conditions for resonance could be satisfied not only at the periodic frequency, but also at a very wide range of frequencies when the pulses are shortened further.

This is very useful when very short pulses are required such as the bound solitons. These phenomena will be described in Chapter 10.

1.4.2.3 Techniques for Generation of Terahertz-Repetition-Rate Pulse Trains

Alternatively, a terahertz-repetition-rate laser can be implemented using parametric amplification in a section of a special optical waveguide, which can be a highly nonlinear dispersion-flattened and dispersion-shifted fiber optical waveguide [70] or an integrated optical waveguide. Parametric amplification is achieved by mixing four waves at three different frequencies based on the nonlinear intensity-dependent change of the waveguide refractive index.

Under a perfect phase-matching condition, a signal at the frequency f_s and a pump source at the frequency f_p mix and modulate the refractive index of the nonlinear optical waveguide section in such a way that a third lightwave, also at f_p, creates sidebands at $f_p \pm (f_s - f_p)$, and this process results in the generation of an idler with gain.

In an active ring laser, the repetition frequency, which is generated by locking at a harmonic, is determined by the modulation frequency of the modulator. The optical modulator is shown in both Figures 1.4 and 1.8. The modulation frequency f_m is $f_m = q f_c$, where f_c is the fundamental resonant frequency of the waveguide ring determined by the length of the laser cavity (the ring length) and q is the order of the harmonics of the cavity's longitudinal modes.

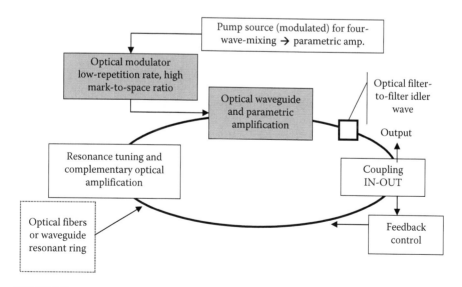

FIGURE 1.8

Structure of a mode-locked laser using parametric amplification for the generation of sidebands and, hence, a terahertz repetitive pulse sequence.

By applying a small deviation of the modulation frequency $\Delta f = f_c/N$ (N = large integer), the modulation frequency becomes $f_m = qf_c + f_c/N$. This leads to an N times increase of the repetition rate, with the repetition rate expressed as $f_R = Nf_m$.

In conventional high-repetition-rate mode-locked lasers [71], the modulation of the circulating lightwave for mode locking is implemented by integrating an optical modulator (MZM) within the resonance cavity as shown in Figure 1.8. In this system, one can modulate the pump signal for mode locking. This locking, in association with the FWM effect (or four-photon interaction) under a phase-matching condition, will transform a low modulation rate into an extremely high repetition rate by the detuning of the locking frequency.

Figure 1.8 shows the schematic diagram of the proposed laser system. The experimental setup will be described in detail in later chapters. The modulation of the pump source can be implemented using a step recovery diode (SRD) which would produce a very short pulse and a low repetition rate to modulate the amplitude of the pump lightwave via an optical modulator of moderately wide bandwidth. This is the principal novelty of our scheme. That is, the bandwidth limitation of the modulator in conventional mode-locked lasers can be overcome. Hence, it is possible to push the repetition rate to the extremely high region, in the few tens of terahertz range. An optical filter is placed at the output of the parametric amplifier to filter the generated idler wave from the FWM process.

1.4.2.4 Necessity of Highly Nonlinear Optical Waveguide Section for Ultrahigh-Speed Modulation

Parametric amplification is achieved by the interaction of the four photons of the signals and the pump waves. The nondegenerative process starts with two photonic waves at different frequencies that co-propagate through the guided waveguide. As they propagate, they are beating with each other. The intensity-modulated group wave at the different frequency will modulate the refractive index (RI) of the guided medium via the SPM effect (Kerr's effect).

When a third wave is injected, it becomes phase modulated at this beating frequency due to the modulated RI. There will then be a phase-modulation to amplitude conversion, and sidebands would be developed whose amplitudes are proportional to the signal amplitude. This happens for both the signal and the pump lightwaves. This process enhances the gain of the signals, and hence parametric amplification. This process is not possible if there is no nonlinear change of the refractive index of the guided medium. Therefore, a highly nonlinear optical waveguide, either in the form of an optical fiber or an integrated photonic waveguide having highly nonlinear core materials such as polymers or doped silica or diffused $LiNbO_3$, can be employed. Furthermore, the optical waveguide must have zero dispersion characteristics at the signal frequency for phase matching.

Highly nonlinear optical fibers with a nonlinear coefficient of around 10 W/km, zero dispersion in the region 1540–1550 nm, and a dispersion slope of about 0.035 ps/(nm)²/km can be used.

References

1. R.E. Wagner, J.R. Igel, R. Whitman, M.D. Vaughn, A.B. Ruffin, and S. Bickharn, Fiber-based broadband-access deployment in the United States, *J. Lightwave Technol.*, 24, 4526–4540, 2006.

2. Department of Broadband, Communications and Digital Economy, Government of Australia. http://www.dbcde.gov.au/communications/national_broadband_network. Access date, June 2009.

3. H.Y. Khoong, Next Generation National Broadband Network for Singapore (Next Gen NBN), http://www.ida.gov.sg/doc/News%20and%20Events/News_and_Events_Level2/20080407164702/OpCoRFP7Apr08.pdf, Last access: May 2008.

4. Singapore IDA, IDA To Pre-Qualify Interested Parties To Be Operating Company For Singapore's Next Generation National Broadband Network, 2008, http://www.ida.gov.sg/News%20and%20Events/20080303140126.aspx, Last access, May 2008.

5. H. Kogelnik, Perspectives on optical communications, in *Proceedings of the Optical Fiber Communication Conference and Exposition and the National Fiber Optic Engineers Conference (OFC/NFOEC), 2008, OFC 2008*, San Diego, CA, 2008.

6. A. Bogoni, L. Poti, P. Ghelfi, M. Scaffardi, C. Porzi, F. Ponzini, G. Meloni, G. Berrettini, A. Malacarne, and G. Prati, OTDM-based optical communications networks at 160 Gbit/s and beyond, *Opt. Fiber Technol.*, 13, 1–12, 2007.

7. K. Vlachos, N. Pleros, C. Bintjas, G. Theophilopoulos, and H. Avramopoulos, Ultrafast time-domain technology and its application in all-optical signal processing, *J. Lightwave Technol.*, 21, 1857–1868, 2003.

8. M.M. Mosso, W.R. Ruziscka, F.S. da Silva, and C.F.C. da Silva, OTDM quasi-all-optical demultiplexing techniques comparative analysis, in *1997 SBMO/IEEE Mtts-S—International Microwave and Optoelectronics Conference, Proceedings*, Vols. 1 and 2, Natal, Brazil, 1997, pp. 692–697.

9. A. Bogoni, L. Poti, R. Proietti, G. Meloni, E. Ponzini, and P. Ghelfi, Regenerative and reconfigurable all-optical logic gates for ultra-fast applications, *Electron. Lett.*, 41, 435–436, 2005.

10. H.J.S. Dorren, A.K. Mishra, Z.G. Li, H.K. Ju, H. de Waardt, G.D. Khoe, T. Simoyama, H. Ishikawa, H. Kawashima, and T. Hasama, All-optical logic based on ultrafast gain and index dynamics in a semiconductor optical amplifier, *IEEE J. Select. Top. Quantum Electron.*, 10, 1079–1092, 2004.

11. M. Scaffardi, N. Andriolli, G. Meloni, G. Berrettini, F. Fresi, P. Castoldi, L. Poti, and A. Bogoni, Photonic combinatorial network for contention management in 160 Gb/s-interconnection networks based on all-optical 2 × 2 switching elements, *IEEE J. Select. Top. Quantum Electron.*, 13, 1531–1539, 2007.

12. A. Bogoni, P. Ghelfi, M. Scaffardi, C. Porzi, F. Ponzini, and L. Poti, Demonstration of feasibility of a complete 160 Gbit/s OTDM system including all-optical 3R, *Opt. Commun.*, 260, 136–139, 2006.

13. T. Houbavlis, K.E. Zoiros, M. Kalyvas, G. Theophilopoulos, C. Bintjas, K. Yiannopoulos, N. Pleros et al., All-optical signal processing and applications within the ESPRIT project DO_ALL, *J. Lightwave Technol.*, 23, 781–801, 2005.

14. J. Herrera, O. Raz, E. Tangdiongga, Y. Liu, H.C.H. Mulvad, F. Ramos, J. Marti et al., 160-Gb/s all-optical packet switching over a 110-km field installed optical fiber link, *J. Lightwave Technol.*, 26, 176–182, 2008.

15. V. Eramo, M. Listanti, and A. Germoni, Cost evaluation of optical packet switches equipped with limited-range and full-range converters for contention resolution, *J. Lightwave Technol.*, 26, 390–407, 2008.

16. S.N. Fu, P. Shum, N.Q. Ngo, C.Q. Wu, Y.J. Li, and C.C. Chan, An enhanced SOA-based double-loop optical buffer for storage of variable-length packet, *J. Lightwave Technol.*, 26, 425–431, 2008.

17. D. Klonidis, C.T. Politi, R. Nejabati, M.J. O'Mahony, and D. Simeonidou, OPSnet: Design and demonstration of an asynchronous high-speed optical packet switch, *J. Lightwave Technol.*, 23, 2914–2925, 2005.

18. P. Zhou and O. Yang, How practical is optical packet switching in core networks?, in *Proceedings of the Global Telecommunications Conference, 2003 (GLOBECOM '03), IEEE Global Telecommunications Conference, 2003 (GLOBECOM '03, IEEE)*, Vol. 5, Anchorage, AK, 2003.

19. M.J. O'Mahony, D. Simeonidou, D.K. Hunter, and A. Tzanakaki, The application of optical packet switching in future communication networks, *Commun. Mag. IEEE*, 39, 128–135, 2001.

20. T.S. El-Bawab and J.-D. Shin, Optical packet switching in core networks: Between vision and reality, *Commun. Mag. IEEE*, 40, 60–65, 2002.

21. B. Bakhski and P.A. Andrekson, 40 GHz actively modelocked polarisation-maintaining erbium fibre ring laser, *Electron. Lett.*, 36, 411–413, 2000.

22. M. Nakazawa and E. Yoshida, A 40-GHz 850-fs regeneratively FM mode-locked polarization-maintaining erbium fiber ring laser, *IEEE Photon. Technol. Lett.*, 12(12), 1613–1615, 2000.

23. A.L. Schawlow and C.H. Townes, Infrared and optical masers, *Phys. Rev.*, 112(6), 1940–1949, 1958.

24. K. Gürs and R. Müller, Breitband-modulation durch Steuerung der emission eines optischen masers (Auskopple-modulation), *Phys. Lett.*, 5, 179–181, 1963.

25. H. Statz and C.L. Tang, Zeeman effect and nonlinear interactions between oscillating modes, in *Quantum Electronics III*, P. Grivet and N. Bloembergen (eds.), Columbia University Press, New York, 1964, pp. 469–498.

26. A. Hasegawa and F.D. Tappert, Transmission of stationary nonlinear optical pulses in dispersive dielectric fibers—I: Anomalous dispersion, *Appl. Phys. Lett.*, 23(3), 142–144, 1973.

27. A. Hasegawa and F.D. Tappert, Transmission of stationary nonlinear optical pulses in dispersive dielectric fibers—II: Normal dispersion, *Appl. Phys. Lett.*, 23(4), 171–173, 1973.

28. L.F. Mollenauer, R.H. Stolen, and J.P. Gordon, Experimental observation of picosecond pulse narrowing and solitons in optical fibers, *Phys. Rev. Lett.*, 45(13), 1095–1098, 1980.

29. L.F. Mollenauer and R.H. Stolen, The soliton laser, *Opt. Lett.*, 9(1), 13–15, 1984.

30. A. Hasegawa, Amplification and reshaping of optical solitons in a glass fiber—IV: Use of the stimulated Raman process, *Opt. Lett.*, 8(12), 650–652, 1983.

31. M. Nakazawa, Y. Kimura, K. Suzuki, and H. Kubota, Wavelength multiple soliton amplification and transmission with an Er^{3+} doped optical fiber, *J. Appl. Phys.*, 66(7), 2803–2812, 1989.

32. N. Le Binh, *Digital Optical Communications*, CRC Press, Taylor & Francis Group, Boca Raton, FL, 2009.

33. M. Nakazawa, T. Yamamoto, and K.R. Tumaura, 1.28 Tbit/s – 70 km OTDM transmission using third and fourth order simultaneous dispersion compensation with a phase modulator, *Electron. Lett.*, 36(24), 2027–2029, 2000.

34. H. Sotobayashi, A. Konishi, W. Chujo, and T. Ozeki, Simultaneously generated 3.24Tbit/s (81 WDM × 40 Gbit/s) carrier suppresses RZ transmission using a single supercontinuum source, in *European Conference on Optical Communications (ECOC 2001)*, Vol. 1, Copenhagen, Denmark, 2001, pp. 56–57.

35. S. Bigo, Y. Frignac, G. Charlet, W. Idler, S. Borne, H. Gross, R. Dischler et al., 10.2 Tbit/s (256 42.7 Gbit/s PDM/WDM) transmission over 100 km TeraLight/sup TM/fiber with 1.28 bit/s/Hz spectral efficiency, in *Optical Fiber Communication Conference and Exhibit, OFC 2001*, Anaheim, CA, Vol. 4, PD25, 2001, pp. P1–3.

36. M. DiDomenico, Small-signal analysis of internal (coupling-type) modulation of lasers, *J. Appl. Phys.*, 35, 2870–2876, 1964.

37. L.E. Hargrove, R.L. Fork, and M.A. Pollack, Locking of He-Ne laser modes induced by synchronous intracavity modulation, *Appl. Phys. Lett.*, 5, 4–5, 1964.

38. D. Kuizenga and A. Siegman, FM and AM mode locking of the homogeneous laser—Part I: Theory, *IEEE J. Quantum Electron.*, 6, 694–708, 1970.

39. H. Haus, A theory of forced mode locking, *IEEE J. Quantum Electron.*, 11, 323–330, 1975.

40. J. Fontana, Theory of spontaneous mode locking in lasers using a circuit model, *IEEE J. Quantum Electron.*, 8, 699–703, 1972.

41. D.C. Hanna, A. Kazer, M.W. Phillips, D.P. Shepherd, and P.J. Suni, Active mode-locking of an Yb:Er fibre laser, *Electron. Lett.*, 25, 95–96, 1989.

42. A. Takada and H. Miyazawa, 30 GHz picosecond pulse generation from actively mode-locked erbium-doped fibre laser, *Electron. Lett.*, 26, 216–217, 1990.

43. G.T. Harvey and L.F. Mollenauer, Harmonically mode-locked fiber ring laser with an internal Fabry–Perot stabilizer for soliton transmission, *Opt. Lett.*, 18, 107, 1993.

44. C.R. Doerr, H.A. Haus, E.P. Ippen, L.E. Nelson, M. Shirasaki, and K. Tamura, Additive-pulse limiting, *Opt. Lett.*, 19, 31, 1994.

45. M. Nakazawa, K. Tamura, and E. Yoshida, Supermode noise suppression in a harmonically modelocked fibre laser by selfphase modulation and spe, *Electron. Lett.*, 32, 461, 1996.

46. X. Shan, D. Cleland, and A. Ellis, Stabilising Er fibre soliton laser with pulse phase locking, *Electron. Lett.*, 28, 182–184, 1992.

47. M. Yoshida, T. Hirayama, M. Yakabe, M. Nakazawa, Y. Koga, and K. Hagimoto, An ultrastable PLL mode-locked fiber laser with a hydrogen maser clock, in *Proceedings of the Lasers and Electro-Optics, 2005. CLEO/Pacific Rim 2005. Pacific Rim Conference*, Tokyo, Japan, 2005.

48. M. Nakazawa, E. Yoshida, and Y. Kimura, Ultrastable harmonically and regeneratively modelocked polarisation-maintaining erbium fibre ring laser, *Electron. Lett.*, 30, 1603–1605, 1994.

49. G. Zhu, Q. Wang, H. Chen, H. Dong, and N.K. Dutta, High-quality optical pulse train generation at 80 Gb/s using a modified regenerative-type mode-locked fiber laser, *IEEE J. Quantum Electron.*, 40, 721–725, 2004.

50. G. Lin, J. Wu, and Y. Chang, 40 GHz rational harmonic mode-locking of erbium-doped fiber laser with optical pulse injection, in *Optical Fiber Communications Conference, OFC 2003*, Atlanta, GA, 2003.

51. D.L.A. Seixas and M.C.R. Carvalho, 50 GHz fiber ring laser using rational harmonic mode-locking, in *Microwave and Optoelectronics Conference, 2001. IMOC 2001. Proceedings of the 2001 SBMO/IEEE MTT-S International*, Belem, PA, Brazil, 2001.

52. E. Yoshida and M. Nakazawa, 80~200 GHz erbium doped fibre laser using a rational harmonic mode-locking technique, *Electron. Lett.*, 32, 1370–1372, 1996.

53. K.K. Gupta, N. Onodera, and M. Hyodo, Equal amplitude optical pulse generation at higher-order repetition frequency in fibre ring lasers using intra-cavity Fabry-Perot etalon and rational harmonic mode-locking, in *Microwave Photonics, 2001, MWP'01, 2001 International Topical Meeting*, Long Beach, CA, 2002.

54. H.A. Haus, Mode-locking of lasers, *IEEE J. Select. Top. Quantum Electron.*, 6, 1173–1185, 2000.

55. T.R. Schibli, U. Morgner, and F.X. Kartner, Control of Q-switched mode locking by active feedback, in *Lasers and Electro-Optics, 2000, (CLEO 2000), Conference*, Bethesda, MD, 2000.

56. J.M. Sousa and O.G. Okhotnikov, Multiple wavelength Q-switched fiber laser, *IEEE Photon. Technol. Lett.*, 11, 1117–1119, 1999.

57. W.J. Lai, P. Shum, and L.N. Binh, Stability and transient analyses of temporal Talbot-effect-based repetition-rate multiplication mode-locked laser systems, *IEEE Photon. Technol. Lett.*, 16, 437–439, 2004.

58. G.R. Lin, Y.C. Chang, and J.R. Wu, Rational harmonic mode-locking of erbium-doped fiber laser at 40 GHz using a loss-modulated Fabry-Perot laser diode, *IEEE Photon. Technol. Lett.*, 16, 1810–1812, 2004.

59. P.D. Humphrey and J.E. Bowers, Fiber-birefringence tuning technique for an erbium-doped fiber ring laser, *IEEE Photon. Technol. Lett.*, 5, 32–34, 1993.

60. J.L. Zyskind, J.W. Sulhoff, Y. Sun, J. Stone, L.W. Stulz, G.T. Harvey, D.J. Digiovanni, et al., Singlemode diode-pumped tunable erbium-doped fibre laser with linewidth less than 5.5 kHz, *Electron. Lett.*, 27, 2148–2149, 1991.

61. R. Wyatt, High-power broadly tunable erbium-doped silica fibre laser, *Electron. Lett.*, 25, 1498–1499, 1989.

62. L.E. Nelson, D.J. Jones, K. Tamura, H.A. Haus, and E.P. Ippen, Ultrashort-pulse fiber ring lasers, *Appl. Phys. B (Lasers and Optics)*, B65, 277–294, 1997.

63. K. Smith and J.K. Lucek, Modelocked fiber lasers promise high-speed data networks, *Laser Focus World*, 29, 85, 1993.

64. K.R. Tamura, Short pulse lasers and their applications to optical communications, in *Lasers and Electro-Optics Society 1999 12th Annual Meeting, LEOS '99, IEEE*, San Francisco, CA, 1999.

65. D.-W. Huang, W.-F. Liu, and C.C. Yang, Actively Q-switched all-fiber laser with a fiber Bragg grating of variable reflectivity, in *Optical Fiber Communication Conference, 2000*, Baltimore, MD, 2000.

66. J. Liu, D. Shen, S.-C. Tam, and Y.-L. Lam, Modeling pulse shape of Q-switched lasers, *IEEE J. Quantum Electron.*, 37, 888–896, 2001.

67. P.R. Morkel, K.P. Jedrzejewski, E.R. Taylor, and D.N. Payne, Short-pulse, high-power Q-switched fiber laser, *IEEE Photon. Technol. Lett.*, 4, 545–547, 1992.
68. A. Yariv, *Optical Electronics in Modern Communications*, 5th edition, Oxford University Press, New York, 1997.
69. G.P. Agrawal, *Applications of Nonlinear Fiber Optics*, 3rd edition, Academic Press, New York, 2001.
70. K.K. Gupta, N. Onodera, K.S. Abedin, and M. Hyodo, Pulse repetition frequency multiplication via intracavity optical filtering in AM mode-locked fiber ring lasers, *IEEE Photon. Technol. Lett.*, 14, 284–286, 2002.
71. J. Hansryd, P.A. Andrekson, M. Westlund, J. Li, and P.-O. Hedekvist, Fibre-based optical parametric amplifiers and their applications, *IEEE J. Select. Top. Quantum Electron.*, 8, 506–519, 2002.

2

Principles and Analysis of Mode-Locked Fiber Lasers

This chapter deals with the principles of mode locking in fiber ring lasers and then the analyses and some simulation results of the generation of ultrashort pulses from mode-locked fiber lasers (MLFLs), which form the bases for the following chapters on the experimental generation of pulse sequences.

2.1 Principles of Mode Locking

Recalling the generic structure of the fiber laser in a ring cavity given in Chapter 1, a laser can oscillate at a number of resonant frequencies whose spacing is equal to the fundamental frequency ω_R of the cavity:

$$\omega_i - \omega_{i-1} = \frac{\pi c}{L} = \omega_R \tag{2.1}$$

where
L is the cavity length
c is the speed of light in vacuum

The output electric field of the generated lightwave in the temporal domain is the summation of all the oscillating modes given as

$$e(t) = \sum_n E_n e^{j[(\omega_0 + n\omega_R)t + \Phi_n]} \tag{2.2}$$

where
ω_0 is the referenced center oscillating frequency
E_n and Φ_n are the amplitude and phase of the nth mode, respectively
$j = (-1)^{1/2}$

When the laser is in the free oscillating state, E_n and Φ_n can take any value without any bound leading to the generation of a continuous wave (CW) source. Figure 2.1 shows the typical electric fields and pulse energy as a

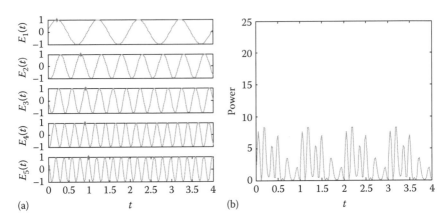

(a)   (b)

FIGURE 2.1

Plot of (a) electric field amplitudes of five individual modes of randomly distributed phases and (b) power of the total signal of a multi-longitudinal mode laser. Please note that the unit of t in both plots and the unit of power in plot (b) are in arbitrary units.

function of time generated from this source in which electrical fields of five modes with random phase and the power of the total signal distributed in a random fashion are illustrated.

On the other hand, when the modes are forced to lock together, that means all the modes N are either in phase or different in a multiple number of 2π, E_n and Φ_n will be constants. The simplest case is when $E_n = 1$, and $\Phi_n = 0$, using (2.2) leads to

$$e(t) = \cos \omega_0 t \, \frac{\sin(N\omega_R t/2)}{\sin(\omega_R t/2)} \tag{2.3}$$

This is an oscillation at frequency ω_0 modulated with the *sinc* envelope function

$$f(t) = \frac{\sin(N\omega_R t/2)}{\sin(\omega_R t/2)} \tag{2.4}$$

The average power is thus given by

$$P(t) \propto \frac{\sin^2(N\omega_R t/2)}{\sin^2(\omega_R t/2)} \tag{2.5}$$

This represents a periodic train of pulses that have the following properties: (1) the pulse period is $T = 2\pi/\omega_R$; (2) the peak power is N times the

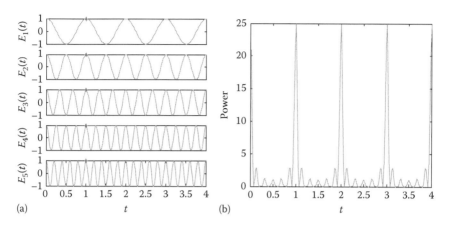

FIGURE 2.2
(a) Electrical field amplitudes of five in-phase individual modes and (b) the total power of a periodic pulse train. Please note that the unit of t in both plots and the unit of power in plot (b) are not given.

average power; (3) the peak field amplitude is N times the amplitude of a single mode; and (4) the pulse width, defined as the time from the peak to the first zero, is $\tau = T/N$, which shortens as N increases. An example of such a pulse sequence is plotted in Figure 2.2.

2.2 Mode-Locking Techniques

There are several techniques for mode locking and they are divided into two classifications: passive mode locking and active mode locking.

2.2.1 Passive Mode Locking

Passive mode locking technique refers to those locking without any external radio frequency (RF) signal for seeding. The pulses are formed passively through the internal structure of the laser that gives more advantages (less loss, high gain) to signal if it travels in the pulse form rather than in a CW [1–6]. The simplest method is to insert a saturable absorber into the cavity, as shown in Figure 2.3. The saturable absorber is a nonlinear optical component whose absorption coefficient decreases when the optical intensity increases. Thus, the pulse train with high peak intensity would pass through the absorber with much less loss as compared to that in a CW laser with several modes as the energy available is concentrated in the periodic pulses, thus a stable pulse train can be formed in the cavity.

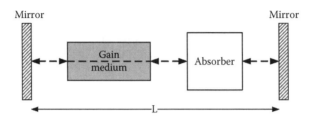

FIGURE 2.3
Schematic of a passive MLL integrated with a saturable absorber in the cavity.

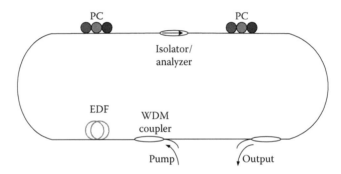

FIGURE 2.4
Schematic of an NPR passive MLFL.

Instead of using the saturable absorber, a nonlinear polarization rotation (NPR) technique can be used for passive mode locking, as shown in Figure 2.4 [7]. In this system, a nonlinear phase shift is imposed on the signal with high peak power and thus rotates its polarization state to align with the analyzer axis. The signal thus passes through the analyzer with minimum loss. On the other hand, the CW signal with a low average power experiences zero nonlinear phase shift and thus its polarization state is not rotated to align with the analyzer axis. It is then blocked by the analyzer. Therefore, pulses with high peak power are formed in the cavity instead of the CW with a low average power.

A passive mode-locked laser (MLL) has the potential to produce a very short laser pulse, down to femtoseconds. The shortest pulse generated in the communication wavelength window to date is 5 fs [8,9]. Moreover, passive mode-locking is a self-locking phenomenon, it means that the pulse is formed without any external modulating signal. However, the spacing between two pulses is varied from pulse to pulse since there is no control mechanism to force the pulses to be equally spaced. This is the disadvantage that makes passive MLLs not suitable for high-speed optical transmission systems, in which precise timing from pulse to pulse is required.

2.2.2 Active Mode Locking by Amplitude Modulation

In actively MLLs, mode locking can be induced either by amplitude modulation (AM), phase modulation (PM), or frequency modulation (FM). When amplitude is modulated by an external RF signal with a frequency of f_m, the oscillating mode at the optical frequency f_0 shades its energy to the two side bands located at f_0+f_m and f_0-f_m, as shown in Figure 2.5. If the modulating frequency f_m is chosen so that $f_m=f_R$, the side bands are coincident with the adjacent modes of the laser and hence the energy from ith mode is transferred to its adjacent $(i+1)$th and $(i-1)$th modes. In other words, the energies of the modes in the laser are transferred from one to the other. This causes the phase of the modes locked together and hence a mode-locked pulse train is formed.

In actively MLLs, mode locking is induced by modulating the gain or loss of the cavity with an external signal at the fundamental frequency f_R.

A common method of active mode locking is inserting an acousto-optic modulator (AOM) into the cavity, as shown in Figure 2.6. This is the first method used in the experiment for the observation of mode-locked pulses

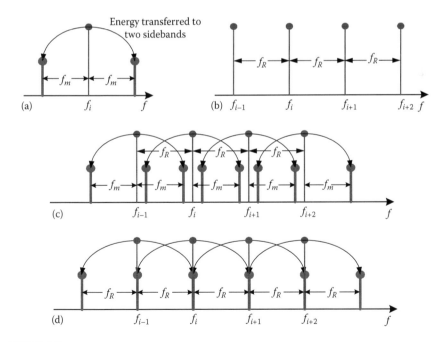

FIGURE 2.5
Active mode locking by modulation: (a) energy transferred to two sidebands when modulating, (b) oscillating modes in a cavity with fundamental frequency of f_R, (c) energy distribution when modulating the oscillating modes with frequency of f_m, and (d) energy transferring between the modes when $f_m=f_R$.

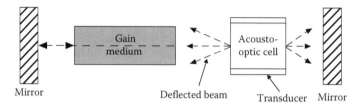

FIGURE 2.6
MLL with an intra-cavity acousto-optic cell/deflector. (Adapted from Davis, C.C., *Laser and Electro-optics: Fundamentals and Engineering*, Cambridge University Press, Cambridge, U.K., 1996.)

in a helium–neon laser [10]. The AOM is used to deflect light energy from the cavity by varying the refractive index of the medium in a standing wave profile. Thus, the undeflected beam is amplitude modulated. AOM is widely used in several MLLs as it offers a very low insertion loss [11–15]. However, the operating frequency of the AOM is limited and far below the pulse rate required for modern optical communication, typically, from a few hundred megahertz to a few gigahertz depending on the acoustic velocity of the acousto-optic material.

Due to the limited bandwidth of AOMs [16–18], they are normally replaced by LiNbO$_3$ intensity or phase modulators whose bandwidth may reach 40 GHz or higher in high repetition rate actively MLLs. In addition, all-fiber cavity is also used in those lasers to take advantage of fiber lasers such as low threshold, single mode, and high beam quality. Moreover, high optical intensity in a small cross section of the fiber enables one to obtain nonlinear effects for short pulses with even a low pump power.

Another method of mode locking is to modulate the gain of the amplifying medium as in Refs. [19,20]. In those experiments, the gain of the semiconductor amplifier is modulated by an external modulating signal. The lightwaves therefore experience a periodical gain and form the mode-locked pulses in the cavity. A semiconductor laser has the potential to generate very short pulses due to its wide spectral gain region.

2.2.3 Active Medium and Pump Source

In order to compensate for the losses within the laser cavity, an amplifying element must be included, and an EDFA is commonly used in this regard. The main transition of interest in EDFA is the high gain $^4I_{13/2} \rightarrow {}^4I_{15/2}$ transition centered around the 1550 nm spectral region, because the $^4I_{13/2}$ is the only metastable state for common oxide glasses at room temperature and gain is only available for these materials at 1550 nm $^4I_{13/2} \rightarrow {}^4I_{15/2}$ emission band. The peak absorption and emission wavelengths occur around 1530 nm; therefore, the operating wavelength must be chosen in the band that the emission cross section is much greater than the absorption cross section, which is around 1550 nm, the third optical communication window (Figure 2.7).

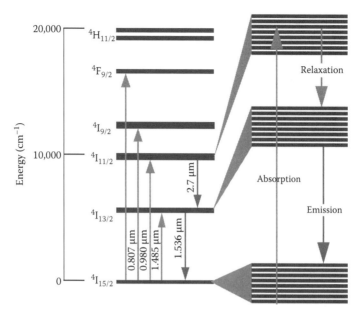

FIGURE 2.7
Erbium transition diagram.

With current technology, EDFA pumping with laser diode is possible around 810, 980, and 1480 nm. However, the pump wavelength of 810 nm suffers from strong excited state absorption (ESA), which causes an undesirable waste of pump photons. As a result, 980 and 1480 nm are more commonly used as the pumping wavelengths. Nine hundred and eighty nanometer (InGaAs/GaAs-based lasers) pumping wavelength provides higher gain efficiency and signal-to-noise ratio (SNR) for small signal amplifiers [21]. It also gives better noise figures (NFs) and quantum conversion efficiencies for power amplifiers. Quantum conversion efficiency is defined as number of photons added to the signal for every pump photon coupled into the fiber. The power conversion efficiency for power amplifier is higher for 1480 nm (InGaAsP/InP-based lasers) pumping because the energy per pump photon is lower, based on $E_{ph} = hc/\lambda$, where E_{ph} is the photon energy, h is Planck's constant, c is the speed of light in vacuum, and λ is the wavelength.

Most pump power is consumed within the first portion from the ends, leaving the middle part of the fiber present with signal and amplified spontaneous emission (ASE). At this part of the EDF, the ASE acts as a pump source on the signal. This is possible only when the signal wavelength at 1550 nm is on the long wavelength side of the ASE peak at 1530 nm. Thus, the bidirectional pumping scheme may be viewed as three serially connected amplifiers, with the middle amplifier being pumped by forward- and backward-propagating ASE generated in the surrounding amplifiers. The major part of the gain improvement is indirectly achieved through the strong ASE attenuation in the middle part of the EDF [22]. This causes fewer pump photons to be used to amplify the spontaneous emission, thereby increasing the pump efficiency.

2.2.4 Filter Design

By introducing a filter in the mode-locked fiber ring laser (MLFRL), the noise generated in the system could be reduced, especially the ASE generated by EDFA. It also provides stabilization against the energy fluctuation of the pulses propagating through the cavity. When the pulse energy increases beyond the average designed value, it will shorten the pulse width and broaden the spectrum due to strong nonlinear effect. Pulses with a broader spectrum experience excess loss and, thus, energy increases are reduced by the filtering. The reverse is true for the decrease of energy. This energy equalization by filtering acts against the gain variation over the erbium bandwidth. However, filtering is associated with the noise penalty. The pulses generated by the laser require higher gain to compensate for the insertion loss of the filter, and the noise at the center frequency is not affected by the filter and sees the excess gain. This noise eventually affects the pulse generated.

Besides the noise reduction capability, the filter is also used for mode selection. The tunability of the filter facilitates the determination of the center operating wavelength of the laser. The wavelength tuning range is chosen in such a way that it falls within the erbium bandwidth, which is in the conventional band (C-band) of the communication systems. With 3 dB bandwidth of 1 nm ($\approx$125 GHz @ 1550 nm), there will be 12 longitudinal modes locked in the laser if a 10 GHz repetition pulse train is desired. However, this number will reduce to 3 if a 40 GHz pulse train is required. Based on the mode-locking equations shown in Section 2.1, the more longitudinal modes involved in the mode locking, N, the narrower the width and the higher the intensity of the pulse generated. As a result, this filter may not be suitable for 40 GHz operation. By changing to a wider bandwidth filter, a narrower pulse train could be obtained, however, with an increase in the system noise level. This is the trade-off between the pulse width and SNR of the system. Alternatively, the 40 GHz operation can be achieved by using the linear repetition rate multiplication, such as direct multiplication by fractional temporal Talbot effect, which will be evaluated in Chapter 3.

2.2.5 Modulator Design

Modulator is the essential component in the MLFRL. Amplitude modulator is selected due to its ability to generate a more stable pulse as compared to phase modulator, where the pulses have the tendency to switch randomly between two maxima phases of the modulation signal. Structurally, there are two main types of modulators, namely, bulk and integrated-optic. Bulk modulators are made out of discrete pieces of a nonlinear optical crystal. They feature very low insertion losses and high power handling capability. Integrated-optic modulators use waveguide technology to lower the drive voltage. They are fiber pigtailed and compact. Due to their small size and compatibility with single-mode fiber, they have been chosen in our design of an active MLFL [23].

Both X-cut and Z-cut $LiNbO_3$ are commonly used for fabrication of optical modulators in which in-diffusion of titanium are employed to form waveguides. The other process for patterning single-mode optical waveguide on $LiNbO_3$ can be annealed proton exchange (APE). In Z-cut waveguide structures, the microwave coplanar traveling wave electrode is positioned on top of the optical waveguide. It is possible to reverse the polarity of the modulating electric field by offsetting the electrode, forming a phase reversal electrode structure. Velocity mismatch between the microwave and optical fields in the traveling wave device can be overcome over a large modulation frequency range by suitably choosing a pattern of phase reversals. In X-cut devices, the phase reversal designs are not possible since the waveguide must be placed in the gap between the electrodes to utilize the strong r_{33} electro-optic coefficient. To achieve fast operation, the device must be made short to overcome the velocity mismatch between the optical guided waves and the microwave traveling waves.

High-frequency response is limited mainly by the microwave-optical velocity mismatch. The response at high frequencies can be improved by using a SiO_2 buffer layer between the electrode and the waveguide. The low dielectric constant of the buffer layer increases the microwave velocity nearer to the optical velocity, which reduces the velocity mismatch. Also, the buffer layer reduces the optical insertion loss by shielding the optical mode from the conductive electrodes [24].

At lower frequency (<40 GHz), the X-cut structure is preferred due to its better modulation response. It is because the X-cut structure offers better overlapping area between the applied electric and the optical fields and the absence of the phase reversals. However, when the operating frequency approaches 40 GHz, the Z-cut device is more efficient than the X-cut one.

The optical transfer function of a $LiNbO_3$ electro-optic modulator can be expressed by [25]

$$T(t) = 1 + \cos\left(\pi \frac{V_b + V_m \sin(\omega_m t)}{V_\pi}\right)$$ (2.6)

where
V_b is the bias voltage
V_m is the modulation voltage amplitude
V_π is the half wavelength voltage that creates a π phase shift

For the most linear operation, the device must be biased half-on, i.e., in quadrature. Variation in the bias point with changing environmental conditions is a serious problem in LiNbO$_3$ devices.

Three main types of bias drift are generally reported: bias variation due to temperature changes via the pyroelectric effect, long-term drift after applying an external electrical signal to set the bias point, and optical damage to the device due to the photorefractive effect causing bias point changes [26]. Pyroelectric effect is a change of electric polarization in a crystal due to the change of temperature, it is much more severe in Z-cut than in X-cut LiNbO$_3$. Over-coating Z-cut devices with a conductive layer to bleed off surface charges significantly reduces the pyroelectric bias drift. However, the drift in X-cut devices is very small; hence, no temperature control is required. Electrically biasing a modulator produces a slow drift to the final bias value. This drift depends on the bias voltage applied, the time over which the bias is applied and the temperature, which is proportional to the time and temperature. Finally, the photorefractive effect caused by the high optical power may contribute to the bias point drift too.

2.2.6 Active Mode Locking by Phase Modulation

Mode locked lasers can be classified into two common types in terms of the gain medium: solid-state mode locked lasers and MLFLs. Although the solid-state mode locked lasers have high reliability and long-term stability, these lasers still require a very stable lab-like environment and high power consumption as well because they are bulky and not highly adaptive with the fiber communication systems. Hence, the MLFLs become an attractive optical source for optical fiber communication systems with its simplicity and broad gain bandwidth. The applications in the telecommunication field have driven the developments of these lasers; however, they are currently also developed for variety of other applications such as optical coherence tomography, medical applications, and terahertz generation.

The mode locking of a laser refers to the locking of the phase relation between longitudinal modes of a cavity. Traditionally, mechanisms existing for MLFLs fall into one of two kinds: passive or active. Passive mode locking in MLFLs is based on the exploitation of an optical effect in a material without any time-varying intervention such as saturated absorption (SA), NPR, and Kerr lens mode locking. On the contrary, active MLFLs perform mode locking with the help of externally modulated devices such as an amplitude modulator (AM) or phase modulator (PM) as a mode-locking element. These two mode-locking methods have different applications. Passive MLFL can produce ultrashort pulses with high pulse energy, but its repetition rate of the output pulse sequence is limited and not sufficient for high-speed communication systems. The optical signal processing and transmission systems based on soliton-like short pulses preferably use the

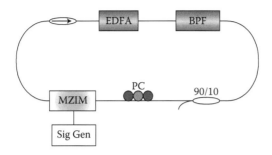

FIGURE 2.8
Schematic of an active harmonic MLFL. The component with an arrow is an isolator.

active MLFLs with high and stable repetition rate of the order of tens of gigabytes per second.

Although phase modulator has been used in the development of active MLFLs, amplitude modulator has been employed more often in the actively MLFLs. Hence, an actively phase-modulated or frequency-modulated MLFL remains to be exploited for different and novel phenomena. The structure of a PM or FM MLFL is similar to the AM one (see Figure 2.8) but replacing the optical amplitude modulator by a phase modulator. An optical phase modulator is usually a straight optical guide integrated with a pair of electrodes placed across its propagation path. An applied voltage creates an electric field across the section of the waveguide, thereby inducing a change of the refractive index of the optical waveguide due to the electro-optic effect. This induced change of the refractive index slows down and changes the phase of the guide wave.

There have been many various structures of FM MLFLs, which have been developed to improve the performance of output pulse trains and to increase the repetition rate since the first FM MLFL was reported in 1988. In 1994, a 2.5 GHz FM MLFL with a pulse width of 9.6 ps was experimentally demonstrated. By using rational harmonic mode locking, a record repetition rate has been established to date in an FM MLFL operating at 80–200 GHz. Another technique to multiply repetition rate in an FM MLFL was demonstrated by using a high-finesse Fabry–Perot filter in the fiber ring. The performance of FM MLFLs can meet some requirements of short pulse sources used in soliton transmission systems.

One problem of FM MLFLs is their high sensitivity to the fluctuations of the surrounding environment that prevents them from generating a stable short pulse sequence for transmission purpose. A practical common solution is to use regenerative mode locking technique besides using the polarization maintaining fiber inside the cavity. This technique will be presented in Chapters 3 and 4. Some other approaches such as a high-finesse Fabry–Perot etalon and the sigma configuration can also be applied to stabilize the output pulse sequence.

In theoretical treatment, most of the research efforts have concentrated on AM MLFLs rather than FM MLFLs, which are due to the difficulty of solving the master of mode locking in the presence of PM in the ring cavity. In a pioneering work, Kuizenga and Siegman described active mode locking in the linear operating regime [56]. For a FM mode locking, the predicted output is a train of chirped Gaussian-shaped pulses, which is only valid when the nonlinear Kerr effects are negligible in the fiber cavity. To investigate a full model of mode locking, the equation known as the master equation of mode locking including all linear and nonlinear effects in the fiber cavity such as dispersion, nonlinearity, and active mode locker need to be used. In case the perturbation of active modulation is small, the master equation can be reduced to the Ginzburg–Landau equation, which can be solved to give an approximate solution. The studies on active MLFLs show that when the power in the cavity changes, the pulses can evolve through three different operating regimes besides the linear regime mentioned above. As the intracavity power increases, the Kerr effect such as self-PM becomes important and results in pulse shortening, but a large amount of pulse dropouts may occur. The stable operating regime is attained when the power exceeds a certain threshold. If the power is further increased, multi-pulsing may occur in some time slots that is often considered as an instable operating state. However, these studies on operating regimes were inferred from the investigation of the actively AM MLFLs.

It should be noted that AM and FM mode-locking schemes differ in some aspects. A PM mode locker interacts with the Kerr effects in a direct manner. Additionally, phase modulators do not require biasing while amplitude modulators would in the AM MLFL. Hence, there are some significant differences between AM and FM MLFLs. FM MLFLs can jump between two degenerate states where the mode-locked pulses would locate themselves at the two extremes of the PM profile. However, dispersion in the fiber cavity suppresses this tendency of MLFL to switch between states. Both theoretical and experimental analyses of timing jitter show that the optimum dispersion would minimize the jitter for FM MLFLs and the quantum-limited jitter for FM MLFLs can be less than that for AM MLFLs. Rational harmonic mode locking in an FM MLFL has recently been demonstrated by the contributions of the harmonics of the modulation frequency in the amplified electrical driving signal, so the output pulses do not experience the problem of unequal amplitudes due to the unique property of the PM. Thus, the operating states of FM MLFLs may not be the same as those of the AM type, especially at high intracavity powers where the nonlinearity has a strong effect on mode-locked pulses. Hence, there are still many interesting research issues of the FM MLFLs in new operation conditions.

Besides the conventional use of the phase modulator in FM MLFRLs, there is another type of active MLFL using a phase modulator, which is based on the intensity modulator using a phase-modulated Sagnac loop. The intensity modulator consists of a traveling phase modulator inside a fiber Sagnac interferometer, which converts FM to AM. This structure was also applied to measure dispersion of fibers and to generate negative coefficients in photonic signal processors. In an MLFL, the phase modulated Sagnac loop can be considered as an amplitude modulator or mode locker. The advantages of this intensity modulator are bidirectional operation without bias, bias drift free, and polarization independent. The bidirectional operation in the interferometer forms a dual output modulator with double the modulation frequency, so it can be used to multiply the repetition rate in an actively MLFL. However, a little residual chirp by PM can still remain and accumulate during the circulation of pulses in the loop; this influences not only the repetition rate multiplication in the fiber laser but also possibly the shape and width of pulses. The output transmittance of the Sagnac loop can be changed by changing the PM depth and the length mismatch of the loop mirror. With the control of amplitude through PM, the phase modulated Sagnac loop can promise new characteristics in actively MLFLs; however, the influence of the operating parameters that affect the phase matching condition in the loop mirror are investigated in Chapter 10. Experimental and simulation works of PM MLFLs are given in Chapters 3 and 4, respectively.

In recent times, bound soliton states have become a new interesting topic in nonlinear fiber systems although they were theoretically predicted in some papers over a decade ago about the stable existence of bound solitons in nonlinear optical fiber systems. The bound states of solitons have recently been experimentally observed in mostly passive MLFLs. The researches on bound solitons have been mostly focused on passive mode locking regime because it can easily generate short pulses like solitons (see Chapter 10). Different techniques such as NPR, additive mode locking and figure-8 fiber lasers have been used to investigate the bound states. The bound soliton pairs were experimentally observed for the first time in the NPR fiber lasers, and then they were also obtained in the figure-8 structure. The mechanism of a bound state formation has been explained by the multi-pulsing operation at high gain cavity, which refers to the soliton energy quantization effect and direct interactions between pulses, so the bound soliton pairs show complicated behavior in the output pulse sequence, which could depend on the relative phase difference in the bound soliton pairs. The multi-soliton bound states, period-doubling bifurcations, and period doubling route to chaos of bound solitons have also been experimentally observed in passively dispersion-managed mode-locked fiber. One limitation of passive bound soliton fiber lasers is the variation of the relative position of bound soliton pairs in time and power, which are limited in some potential applications. Similar to the single soliton passive MLLs, these bound soliton lasers operate at low frequency; even the harmonic mode locking has been used. In addition, there are few reports on bound solitons in active MLFLs. The observation of bound soliton pairs was first reported in a hybrid FM MLFL. Chapter 10 describes the formation and generation of multiple bound soliton states.

2.3 Actively Mode-Locked Fiber Lasers

2.3.1 Principle of Actively Mode-Locked Fiber Lasers

Figure 2.9 shows the schematic diagram of an actively MLFL. The laser has a ring configuration incorporating an isolator for unidirectional lasing. An EDFA is used to provide the gain for the lightwave traveling in the ring. The lasing wavelength is selected by using an optical bandpass filter (BPF). Mode locking is obtained by inserting into the ring a Mach–Zehnder intensity modulator (MZIM) that periodically modulates the loss of the lightwaves traveling around the ring. A polarization controller (PC) is also employed to maximize the coupling of lightwaves from the fiber to the diffused waveguides of the MZIM modulator. Output pulse trains are extracted via a 90:10 fiber coupler.

The fundamental frequency of the ring is

$$f_R = \frac{c}{nL} \tag{2.7}$$

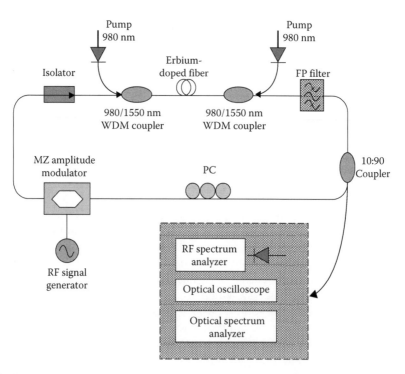

FIGURE 2.9
An RHMLFL. (Adapted from Yoshida, E. and Nakazawa, M., *Electron. Lett.*, 32, 1370, 1996.)

where

　n is effective refractive index of the single-mode fiber
　$c = 3 \times 10^8$ m/s speed of light in vacuum
　L is the total length of the ring

Since the total length of the ring L is usually very long, from a few meters to hundreds of meters, the fundamental frequency f_R is low and hence the repetition rate of the output pulse train is also low. To increase the repetition rate, harmonic mode locking is used. In an HMLFL, the modulator is driven by a signal with frequency equal to a high harmonic order of the fundamental frequency. With this arrangement, the laser is mode-locked to the modulating frequency, which can be up to few tens of gigahertz, instead of the low fundamental frequency f_R, depending on the bandwidth of the optical modulator.

2.3.2 Multiplication of Repetition Rate

Although high repetition rate pulse trains can be obtained by an driving actively MLFL at the harmonic of the fundamental frequency of the ring cavity, the maximum pulse repetition rate is still limited by the bandwidth of the modulator. To increase the repetition rate, rational harmonic mode-locking technique can be used [20,27–33].

The RHMLFL configuration is the same as that of the harmonic MLFL except that the modulation frequency f_m is not an integer number of the fundamental frequency. The modulation frequency f_m is now detuned by a fraction of fundamental frequency f_R, that means $f_m = (p + 1/k)f_R$ where p and k are integers. With that detuning, the output pulse train has a repetition rate of $(kp + 1)f_R$, k times the modulation frequency.

A rational harmonic MLL with a repetition rate of 200 GHz when modulating at 40 GHz has been reported in [30] and its setup is shown in Figure 2.9. A multiplication factor of 5 is achieved. The disadvantage of RHMLFL is that the pulse amplitude is varied from pulse to pulse, as seen in Figure 2.10, due to the different losses the pulses experience when passing through the modulator.

Alternatively, the multiplication of the repetition rate can be obtained by using an intra-cavity optical filter [34,35]. Figure 2.11 shows the schematic of an HMLFL with pulse repetition rate multiplication by using a Fabry–Perot filter (FFP). The schematic is similar to that of the RHMLFL with an FFP inserted into the ring. The FSR of the FFP is K times the modulation signal frequency, where K is the multiplication factor. The FFP is used to select higher-order longitudinal modes that correspond to their transmission peaks, while filtering out the intermediate longitudinal modes. The condition for this laser is that the modulation frequency f_m is not only a harmonic multiple of the fundamental frequency f_R but also a sub-harmonic multiple of the FSR of the FFP:

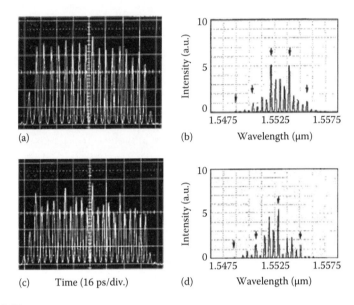

FIGURE 2.10
Output pulse train of the RHMLFL: (a) autocorrelation waveform at 160 GHz, (b) spectral profile at 160 GHz, (c) autocorrelation waveform at 200 GHz, and (d) spectral profile at 200 GHz. (Adapted from Yoshida, E. and Nakazawa, M., *Electron. Lett.*, 32, 1370, 1996.)

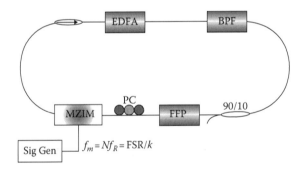

FIGURE 2.11
An RHMLFL. The component with an arrow is an isolator.

$$f_m = Nf_R \tag{2.8}$$

and

$$f_m = \frac{\text{FSR}}{k} \tag{2.9}$$

where N and k are integers.

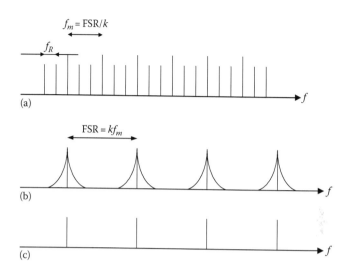

FIGURE 2.12
Frequency domain depiction of pulse repetition frequency multiplication in HMLFL via intra-cavity optical filtering: (a) spectral structure of the pulse train without the FFP, (b) the FFP transmission characteristic, and (c) spectral structure of the pulse train filtered by the FFP.

When those conditions are satisfied, only those longitudinal modes equal to the peak frequencies of the FFP are selected while the other intermediate longitudinal modes are suppressed, as shown in Figure 2.12. This results in a pulse train whose repetition rate is equal to the FSR. The pulse train generated using this technique has equal amplitude but the repetition rate is fixed to the FSR.

2.3.3 Equalizing and Stabilizing Pulses in Rational HMLFL

The FFP can also be used to equalize the pulse amplitudes in RHMLFLs [36–39]. As mentioned above, RHMLFL can generate pulse trains with high repetition rates, but the pulse trains suffer from pulse amplitude instability that includes both amplitude noise and unequal pulse amplitudes. The amplitude noise is mainly due to supermode noise caused by the unequal energy distribution among the cavity modes. The unequal pulse amplitudes in the optical pulse train are mainly due to the asymmetric cavity loss modulation within the cavity caused by the presence of randomly oscillating intermediate modes that are frequency spaced at f_m and become predominant at a pulse repetition rate greater than $2f_m$.

The RHMLFL setup is shown in Figure 2.13. When the modulation frequency is detuned by f_R/k (k is an integer), that is, $f_m = (p + 1/k)f_R$, the output pulse train repetition frequency is $(kp + 1)f_R$, which is k times the modulation

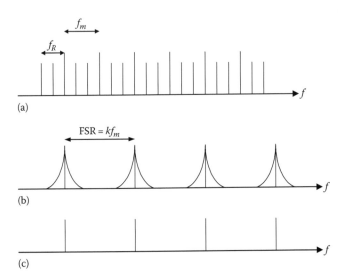

FIGURE 2.13

Frequency domain depiction of intermediate frequency filtering of FFP: (a) spectrum structure of the pulse train without the FFP, (b) the FFP transmission characteristic, and (c) spectrum structure of the pulse train filtered by the FFP.

frequency. However the modulation signal also generates intermediate longitudinal modes at f_m, $2f_m$, $3f_m$ .... These intermediate modes cause the pulse amplitude instability. If the FFP is chosen so that its FSR is equal to kf_m, then those intermediate modes are filtered out, leaving the modes at kf_m, as shown in Figure 2.13. Therefore, the pulse becomes stable at the repetition rate of kf_m.

HMLFL has the ability to generate very short pulses at high repetition rates, but a long-term stable output pulse train can hardly be obtained due to its long cavity, which is easily affected by the change of the environment such as mechanical vibration or temperature variation [40–43]. Therefore, there is a need to stabilize the laser so that the output pulse train can be used as a practical source for transmission of ultrahigh speed data.

The length of the ring in the HMLFL can be varied when the temperature changes, which in turn changes the fundamental frequency of the ring and hence the oscillating longitudinal modes are out of sync with the modulating signal. The length of the ring can be controlled by wounding the fiber on a PZT and using a feedback circuit to control the PZT and hence stabilize the laser performance.

Figure 2.14 shows a schematic of HMLFL using the PZT to stabilize the laser. The fiber is wound on a PZT so that the length of the fiber can be controlled by applying a DC voltage on the PZT. The output of the laser is converted to an electrical pulse by a photodetector (PD) and filtered by a BPF. The output is then compared with the modulation signal that is extracted

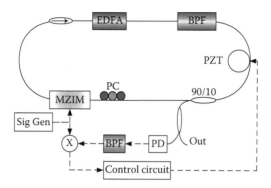

FIGURE 2.14
Schematic of an HMLFL using the PZT to control the length of the cavity. The component with an arrow is an optical isolator. What is the splitting ratio of the output coupler that extracts the output signal?

from a signal generator. The error is used for controlling the PZT through a control circuit.

Alternately, laser stabilization can also be achieved by using a feedback scheme, as shown in Figure 2.15. This configuration can be called a regenerative harmonic MLFL [41,43,44]. The electrical output pulses after detection by the PD are fed into the clock recovery circuitry to extract the repetition frequency of the pulses. The output signal is used to control the modulator after being amplified and phase shifted. The phase between the pulses and the modulation signal is adjusted by the phase shifter so that the pulses always experience the optimum loss when they pass through the modulator. Long-term stability using this stabilization technique has been demonstrated by Nakazawa et al. [41].

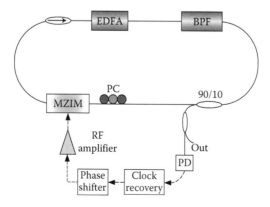

FIGURE 2.15
Schematic of a regenerative harmonic MLFL. The component with an arrow is an optical isolator. What is the splitting ratio of the output coupler that extracts the output signal?

The instability of the fiber laser is also caused by the birefringence of the fiber. The random change of the polarization state of the light traveling in the fiber causes fluctuation of the pulse amplitudes. A polarization maintaining fiber can be used to preserve the polarization of the light and hence help to stabilize the laser [45–49]. Lasers using PM fibers have been reported to produce stable pulse trains with repetition rates of up to 40 GHz and at wavelengths tunable over the whole C-band (from 1530 to 1565 nm) [50,51].

2.4 Analysis of Actively Mode-Locked Lasers

2.4.1 Introduction

As discussed in the previous sections, actively MLLs have potential applications in future high-speed optical networks due to their ability to generate very short pulses at high repetition rates with low timing jitters. Since their first demonstration in 1964 by Hargrove et al. [10], actively MLLs have been improved over the years with numerous proposed designs [52–55]. Besides this, there were extensive theoretical developments on mode locking to study the MLL, and to serve as a guideline for laser design. In 1970, an analytic theory of active mode locking was firmly established by Kuizenga and Siegman [56]. They studied the process in the time domain and derived mode locking equations for both AM and FM. In their paper, a Gaussian pulse solution was assumed and the approximations of the line, shape, and modulation function to keep the pulse Gaussian were applied.

In 1975, Haus presented a frequency domain approach for mode locking theory [57]. Using a circuit model, which is similar to that of [58], the author analyzed the evolution of pulses in the frequency domain and derived differential equations for forced mode locking. The equations had a Hermite–Gaussian solution and the mode-locked pulse characteristics were obtained. In addition, the analysis of the detuning effects in the actively MLL was reported by Li et al. with the use of consistence condition and the assumption of the Gaussian pulse shape [59]. This analysis was based on the invariant condition of the Gaussian pulse shape after traveling a round trip. Pulse parameters in steady state and locking range were derived. The dispersion effect on the detuning properties of actively MLFLs was recently reported by Zhu and Dutta [60]. The authors found that with a strong cavity dispersion effect, the lasing wavelength was shifted from the central wavelength of the filter when the laser was detuned.

However, the analyses were based on the assumption that the pulse shape is Gaussian in the steady state or the consistence condition held to obtain the mode-locked pulse parameters. Moreover, ASE noise, which cannot be avoided in the optical gain medium of the laser, has not been considered in the models. Thus, there is a need to study the evolution of the pulse shape in MLLs under the existence of optically amplified noises. In this chapter, we employ a mathematical series instead of the self-consistence condition to trace the evolution of a white-noise pulse traveling inside the cavity. The shape of the noise pulse is featured by eight series: S_n, Q_n, R_n, a_n, b_n, c_n, r_n, and P_m, which can be easily calculated from the laser parameters. Using this method, we observe that the noise pulse is gradually shaped to the steady-state pulse under the shaping effect of the filter and modulator for every round trip. We obtain the steady-state pulse through transient processes without the need of a presumed Gaussian pulse shape. Equations for determining the steady-state pulse parameters and SNR are also derived using this mathematical series. These equations can be used to study the effects of the laser parameters and the detuning condition on the output pulse characteristics, and to determine the locking range for a specific requirement of the laser SNR.

The analysis of actively MLLs using the self-consistence condition with a presumed Gaussian pulse shape is presented in Section 2.4.2 and series approach is presented in Section 2.4.3. Following is the description of our series analysis method for actively MLLs. The applications of our method in the analysis of the MLLs with and without detuning are presented in Section 2.4.4. The results are then confirmed with the simulation results. And finally, conclusions are given.

2.4.2 Analysis Using Self-Consistence Condition with Gaussian Pulse Shape

In this section, we present the analysis of an actively MLL by using the self-consistence condition method with a presumed Gaussian pulse shape [61]. Figure 2.16 shows the laser model that consists of an in-line optical amplifier, a bandpass optical filter, and an intensity modulator. Since the line width of the gain medium is normally much larger than the filter bandwidth for lasers with long gain-recovery time constant, e.g., in an erbium-doped gain medium, the gain can be considered flat over the filter spectrum with a saturated value of G_0.

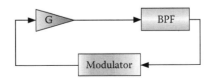

FIGURE 2.16
Actively MLL model.

The filter transmittance function $T(\omega)$ and the Mach–Zehnder modulator transmission function $M(t)$ can be written as

$$T(\omega) = \exp\left(-\beta\omega^2\right) \tag{2.10}$$

$$M(t) = \cos\left[\frac{\pi}{2}(1-b) - \frac{\pi}{4}m\cos(\omega_m t)\right] \tag{2.11}$$

where

$\beta = 2\ln 2/\Delta\omega^2$, $\Delta\omega$ is the filter 3 dB bandwidth
b accounts for the bias point of the modulator
$b = 0.5$ for the quadrature bias point
m is the modulation index
ω_m is the modulation frequency

If the pulses pass the modulator at the time near the peak of the transmission function, $t = 0$, we can expand (2.11) using the Taylor series, and hence $M(t)$ can be approximated by

$$M(t) = \Gamma\exp\left(-\delta_m t^2\right) \tag{2.12}$$

in which $\delta_m = m\omega_m^2/4$, and Γ accounts for the total loss of the laser cavity, including the insertion loss, modulator bias loss, and coupler loss. Suppose the pulse entering the gain medium after the nth round trip is

$$A_n(t) = A\exp\left[-\alpha(t-t_s)^2\right] \tag{2.13}$$

where

A is the pulse amplitude
α is the Gaussian pulse parameter
t_s is the position of the pulse with reference to the transmission peak of the modulator

Hence the signal after the optical filter, in the time domain, is

$$A_{nl}(t) = F^{-1}\left\{G_0 T(\omega)F\left[A_n(t)\right]\right\} \tag{2.14}$$

where F and F^{-1} denote the Fourier and its inverse transform, respectively. Substituting (2.10) and (2.13) into (2.14) we obtain

$$A_{nl}(t) = \frac{G_0 A}{\sqrt{1+4\alpha\beta}} \exp\left[\frac{-(t-t_s)^2}{(1+4\alpha\beta)}\right] \qquad (2.15)$$

If the modulation frequency is detuned by Δf_m from its ideal Nth order harmonic mode-locking frequency f_{m0}: $\Delta f_m = f_m - f_{m0}$, the pulse arrives at the modulator with the delay of

$$t_d = N\left(\frac{1}{f_{m0}} - \frac{1}{f_m}\right) \approx \frac{N\Delta f_m}{f_{m0}^2} \qquad (2.16)$$

After passing through the modulator, the pulse completes one round trip and becomes

$$A_{n+1}(t) = A_{nl}(t)M(t) = \frac{AG_0\Gamma}{\sqrt{1+4\alpha\beta}} \exp\left\{-\left[\frac{4\alpha}{1+4\alpha\beta} + \delta_m\right](t-t_s-t_d)^2\right\} \qquad (2.17)$$

Using the self-consistence condition in the steady state $A_{n+1}(t) = A_n(t)$, and balancing the terms on both sides, we obtain

$$\alpha = \frac{1}{2}\sqrt{\delta_m^2 + \frac{\delta_m}{\beta}} + \frac{\delta_m}{2} \qquad (2.18)$$

$$t_s = \frac{t_d}{4\alpha\beta} \qquad (2.19)$$

$$G_0 = \frac{\sqrt{1+4\alpha\beta}}{\Gamma} \exp\left(\frac{t_d^2}{4\beta}\right) \qquad (2.20)$$

From these analytical results, we observe the following: (1) The pulse width is independent of the detuning and identical to that obtained with exact tuning. (2) Detuning does not change the pulse width of the steady-state pulse but shifts the pulse position away from the transmission peak. The shift of the pulse position from the transmission peak is linearly proportional to the detuning Δf_m. (3) Due to additional loss caused by position shifting, the required gain is increased by a factor of $\exp(t_d^2/4\beta)$.

When exact tuning happens, that is, $\Delta f_m = 0$, $t_d = 0$, we get

$$\alpha = \frac{1}{2}\sqrt{\delta_m^2 + \frac{\delta_m}{\beta}} + \frac{\delta_m}{2} \qquad (2.21)$$

$$t_s = 0 \tag{2.22}$$

$$G_0 = \frac{\sqrt{1+4\alpha\beta}}{\Gamma} \tag{2.23}$$

This confirms the result obtained in [56] if we substitute β and δ_m in the above definitions.

2.4.3 Series Approach Analysis

The ASE noise of the optical amplifier is included in the laser model shown in Figure 2.17. The lasing process is started by the ASE noise, which is modeled as a white noise source ρ_{ASE} with a variance of σ_N^2. For every round trip, the ASE noise is filtered, shaped, and amplified when passing through the optical BPF, modulator, and amplifier in the loop.

In the frequency domain, the ASE signal after passing through the filter is

$$\rho_{1A} = \sigma_N e^{-\beta\omega^2} \tag{2.24}$$

Due to detuning, after one round trip, the pulse arrives at the modulator with the delay of t_d. The delay t_d in the time domain is equivalent to multiplying by $\exp(-j\omega t_d)$ in the frequency domain. Hence, the noise signal in the frequency domain after the first round trip is

$$\rho_1(\omega) = G \times F\left[F^{-1}\left(\sigma_N e^{-\beta\omega^2} e^{-j\omega t_d}\right) \times \Gamma e^{-\delta_m t^2}\right]$$

$$= \Gamma G \sigma_N \times F\left[\frac{1}{2\sqrt{\pi\beta}} e^{-(t-t_d)^2/4\beta} \times e^{-t^2 x/4\beta}\right]$$

$$= \Gamma G \sigma_N \times F\left[\frac{1}{2\sqrt{\pi\beta}} e^{-A_1(t)}\right] \tag{2.25}$$

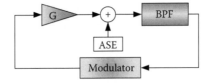

FIGURE 2.17
Actively MLL model with ASE noise included.

where F and F^{-1} denote the Fourier transform and its inverse transform, respectively, $x = 4\beta\delta$ and

$$A_1(t) = \frac{(t - t_d)^2}{4\beta} + \frac{xt^2}{4\beta}$$

$$= \frac{1+x}{4\beta}\left(t - \frac{t_d}{1+x}\right)^2 + \frac{xt_d^2}{(1+x)}$$

$$= \frac{Q_1}{4\beta S_1}\left(t - \frac{S_1}{Q_1}t_d\right)^2 + \frac{R_1}{4\beta Q_1}t_d^2 \qquad (2.26)$$

Therefore,

$$\rho_1(\omega) = \sigma_N \Gamma G \times F\left[\frac{1}{2\sqrt{\pi\beta S_1}}e^{-\frac{Q_1}{4\beta S_1}\left(t - \frac{S_1}{Q_1}t_d\right)^2}e^{-\frac{R_1}{4\beta Q_1}t_d^2}\right]$$

$$= \sigma_N \Gamma G \frac{1}{2\sqrt{\pi\beta S_1}}\sqrt{\frac{\pi 4\beta S_1}{Q_1}}e^{-\frac{S_1}{Q_1}\beta\omega^2}e^{-j\omega\frac{S_1}{Q_1}t_d}e^{-\frac{R_1}{4\beta Q_1}t_d^2}$$

$$= \sigma_N \Gamma G \frac{1}{\sqrt{Q_1}}e^{-\frac{R_1}{4\beta Q_1}t_d^2}e^{-\frac{S_1}{Q_1}\beta\omega^2}e^{-j\omega\frac{S_1}{Q_1}t_d} \qquad (2.27)$$

in which $S_1 = 1, Q_1 = x + 1, R_1 = x$.

Similarly, the noise signal in the frequency domain after the second round trip is

$$\rho_2(\omega) = G \times F\left\{F^{-1}\left[\rho_1(\omega)e^{-\beta\omega^2}e^{-j\omega t_d}\right] \times \Gamma e^{-\delta_m t^2}\right\}$$

$$= \frac{\sigma_N(\Gamma G)^2}{\sqrt{Q_1}} \times F\left\{F^{-1}\left[e^{-\frac{R_1}{4\beta Q_1}t_d^2}e^{-\frac{S_1+Q_1}{Q_1}\beta\omega^2}e^{-j\omega\frac{S_1+Q_1}{Q_1}t_d}\right]e^{-\frac{x}{4\beta}t^2}\right\}$$

$$= \sigma_N(\Gamma G)^2\frac{e^{-\frac{R_1}{4\beta Q_1}t_d^2}}{\sqrt{Q_1}} \times F\left\{\frac{e^{-\frac{Q_1}{4\beta(S_1+Q_1)}\left(t - \frac{S_1+Q_1}{Q_1}t_d\right)^2}e^{-\frac{x}{4\beta}t^2}}{2\sqrt{\pi\beta(S_1+Q_1)/Q_1}}\right\}$$

$$= \sigma_N(\Gamma G)^2\frac{e^{-\frac{R_1}{4\beta Q_1}t_d^2}}{2\sqrt{\pi\beta(S_1+Q_1)}} \times F\left\{e^{-A_2(t)}\right\} \qquad (2.28)$$

where

$$A_2(t) = \frac{Q_1}{4\beta(S_1+Q_1)}\left(t - \frac{S_1+Q_1}{Q_1}t_d\right)^2 + \frac{x}{4\beta}t^2$$

$$= \frac{Q_1 + x(S_1+Q_1)}{4\beta(S_1+Q_1)}\left(t - \frac{S_1+Q_1}{Q_1+x(S_1+Q_1)}t_d\right)^2$$

$$+ \frac{2xS_1+R_1}{4\beta[Q_1+x(S_1+Q_1)]}t_d^2 - \frac{R_1}{4\beta Q_1}t_d^2$$

$$= \frac{Q_2}{4\beta S_2}\left(t - \frac{S_2}{Q_2}t_d\right)^2 + \frac{R_2}{4\beta Q_2}t_d^2 - \frac{R_1}{4\beta Q_1}t_d^2 \qquad (2.29)$$

in which $S_2 = S_1 + Q_1, Q_2 = xS_2 + Q_1, R_2 = 2xS_2 + R_1$. Hence,

$$\rho_2(\omega) = \frac{\sigma_N(\Gamma G)^2}{2\sqrt{\pi\beta S_2}} e^{-\frac{R_1}{4\beta Q_1}t_d^2} \times F\left\{ e^{-\frac{Q_2}{4\beta S_2}\left(t - \frac{S_2}{Q_2}t_d\right)^2} e^{-\frac{R_2}{4\beta Q_2}t_d^2} e^{\frac{R_1}{4\beta Q_1}t_d^2} \right\}$$

$$= \sigma_N(\Gamma G)^2 \frac{1}{2\sqrt{\pi\beta S_2}}\sqrt{\frac{\pi 4\beta S_2}{Q_2}} e^{-\frac{S_2}{Q_2}\beta\omega^2} e^{-j\omega\frac{S_2}{Q_2}t_d} e^{-\frac{R_2}{4\beta Q_2}t_d^2}$$

$$= \sigma_N(\Gamma G)^2 \frac{1}{\sqrt{Q_2}} e^{-\frac{R_2}{4\beta Q_2}t_d^2} e^{-\frac{S_2}{Q_2}\beta\omega^2} e^{-j\omega\frac{S_2}{Q_2}t_d} \qquad (2.30)$$

Thus, in general, the ASE signal in the frequency domain after the *n*th round trip is

$$\rho_n(\omega) = \sigma_N(\Gamma G)^n \frac{1}{\sqrt{Q_n}} e^{-\frac{R_n}{4\beta Q_n}t_d^2} e^{-\frac{S_n}{Q_n}\beta\omega^2} e^{-j\omega\frac{S_n}{Q_n}t_d} \qquad (2.31)$$

where

$$\begin{cases} S_1 = 1, Q_1 = x+1, R_1 = x \\ S_n = S_{n-1} + Q_{n-1}, Q_n = xS_n + Q_{n-1}, R_n = nxS_n + R_{n-1} \end{cases} \qquad (2.32)$$

The total ASE noise signal in the frequency domain in the ring is sum of the 1st, 2nd, ..., ∞th round-trip signals, and is given as

$$\eta(\omega) = \sum_{n=1}^{\infty} \rho_n(\omega) \qquad (2.33a)$$

The total ASE noise signal in the ring in the time domain can be obtained by performing the inverse Fourier transform on (2.33a), and it is given as

$$\eta(t) = F^{-1}\{\eta(\omega)\} = \sum_{n=1}^{\infty} \rho_n(t) \qquad (2.33b)$$

where $\rho_n(t) = F^{-1}\{\rho_n(\omega)\}$ is the ASE signal in the time domain after the nth round trip and it can be easily obtained, using the result in (2.28), by performing the inverse Fourier transform on (2.31) to give

$$\rho_n(t) = \sigma_N \left(\Gamma G\right)^n \frac{1}{\sqrt{Q_n}} e^{-\frac{R_n}{4\beta Q_n} t_d^2} \left[e^{-\frac{Q_n}{4\beta S_n} \left(t - \frac{S_n}{Q_n} t_d\right)^2} \right] \qquad (2.33c)$$

Using (2.33b) and (2.33c), one can observe the total ASE noisy pulse evolution in the time domain. This time-domain information could be useful if one would like to know how much noise is contributed to the signal propagating in the cavity or if one would like to isolate or extract the actual signal from the noisy signal at the output.

2.4.4 Mode Locking

2.4.4.1 Mode Locking without Detuning

First of all, we consider the case of zero detuning in which the modulation frequency is exactly equal to a harmonic of the fundamental frequency of the laser ring. Without detuning, there is no additional delay per round trip, thus we can substitute $t_d = 0$ into (2.31) and then analyze the result in two conditions: noiseless amplifier and amplifier with ASE noise.

2.4.4.1.1 Noiseless Amplifier

To compare our result with those obtained using classical approaches in which the gain medium is modeled as a noiseless amplifier [56], we assume that the amplifier used here is also noiseless. The signal is formed from a white noise traveling the loop infinitely without any ASE added for every round trip. Therefore, the steady-state signal in the frequency domain is

$$\eta(\omega) = \lim_{n \to \infty} \rho_n(\omega) \propto e^{-\lim_{n \to \infty}\left(\frac{S_n}{Q_n}\right)\beta\omega^2} = e^{-\frac{2}{x+\sqrt{4x+x^2}}\beta\omega^2} \qquad (2.34a)$$

Using the inverse Fourier transform, we obtain the Gaussian pulse shape in the time domain as

$$\eta(t) \propto e^{-\frac{x+\sqrt{4x+x^2}}{8\beta}t^2} = e^{-\alpha t^2} \tag{2.34b}$$

where $\alpha = x + \sqrt{4x + x^2}/8\beta$ is the Gaussian pulse parameter. This is exactly the same result as in (2.18).

To calculate the gain required for lasing, we consider the power of the signal in the frequency domain after n round trips:

$$P_n = \frac{1}{2\pi} \int_{-\infty}^{+\infty} |P_n(\omega)|^2 d\omega = \frac{\sigma_N^2}{\sqrt{8\pi\beta}} \frac{(\Gamma G)^{2n}}{\sqrt{S_n Q_n}} \tag{2.35}$$

The requirement is that the power is not reduced to zero or increased to infinity but approaches a constant as $n \to \infty$. It is found that $\lim_{n \to \infty} P_n = \text{const}$ when

$$G = G_{0I} = \frac{\sqrt{1 + 4\alpha\beta}}{\Gamma} \tag{2.36}$$

This is the required gain for lasing and is the same as (2.23), which is obtained using the self-consistence approach, but here we obtain it by using a series approach analysis. Compared with the self-consistence approach, one unique advantage of the series approach analysis is that it provides more insights into the laser dynamics by allowing one to observe and determine the behavior of pulse evolution at every round trip of the cavity.

2.4.4.1.2 Amplifier with ASE Noise

In practice, the noise is continuously added to the signal for every round trip passing through the gain medium. The newly added noise is then shaped closer to the steady-state Gaussian pulse in (2.34b) for every round trip in the loop. Hence, the total signal is the sum of all shaped noises as presented in (2.34b). Since the individual shaped noise converges to the Gaussian pulse in (2.34b), the summation of these individual noise signals also converges to the steady-state Gaussian pulse. It should be noted that if the gain is kept the same as that in the ideal amplifier case, the power of the individual Gaussian pulse is unchanged but the total power increases to infinity due to the contribution of the added noise in every round trip. This cannot happen in practice, thanks to the saturation of the gain medium [62–65]. The increase of power will reduce the gain, and thus the net gain in one round trip is less than the one, which makes the power decrease. The result is that the total power is kept unchanged and the gain is reduced by a factor of r compared to the noiseless amplifier case

$$G = G_{0I}r = \frac{\sqrt{1+4\alpha\beta}}{\Gamma}r \qquad (2.37)$$

where G_{0I} is defined in (2.36). Since the gain is reduced, the power of the individual signal P_n is no longer conserved but exponentially decreased to zero as $n \to \infty$. Using (2.37), the power of the signal after n round trips now becomes

$$P_n = \frac{\sigma_N^2}{\sqrt{8\pi\beta}} \frac{(1+4\alpha\beta)^n}{\sqrt{S_n Q_n}} r^{2n} \qquad (2.38)$$

From the above analysis, the evolution of the steady-state pulse can be described as follows. The spontaneous noise of the amplifying medium is shaped to the steady-state pulse under the shaping effect of the filter and modulator for every round trip. It converges to the steady-state pulse as n approaches infinity. By examining (2.31) and (2.38), we see that there are three series that control the dynamic evolution of the pulse: $a_n = S_n/Q_n$, $b_n = (1+4\alpha\beta)^n/sqrt(S_n Q_n)$, and $r_n = r^{2n}$. The $\{a_n\}$ series controls the evolution of the pulse width, the b_n series controls the evolution of the power, and the r_n series controls the decaying process of the individual signal.

The value of r is determined from balancing the total power:

$$P = \sum_{n=1}^{\infty} P_n = \sum_{n=1}^{\infty} \frac{\sigma_N^2}{\sqrt{8\pi\beta}} \frac{(1+4\alpha\beta)^n}{\sqrt{S_n Q_n}} r^{2n} \approx \sum_{n=1}^{\infty} P_I r^{2n} = \frac{P_I}{1-r^2} \qquad (2.39)$$

where

$$P_I = \lim_{n\to\infty} \frac{\sigma_N^2}{\sqrt{8\pi\beta}} \frac{(1+4\alpha\beta)^n}{\sqrt{S_n Q_n}}$$

with the power related to the saturated gain [66]

$$P = \frac{G^2}{G^2-1} P_{sat} \ln\frac{G_{ss}}{G^2} \approx \frac{r^2}{r^2-1/G_{0I}^2} P_{sat} \ln\frac{G_{ss}}{G_{0I}^2} \qquad (2.40)$$

where
G_{ss} is the small-signal gain
P_{sat} is the saturation power
G_{0I} is defined in (2.36)

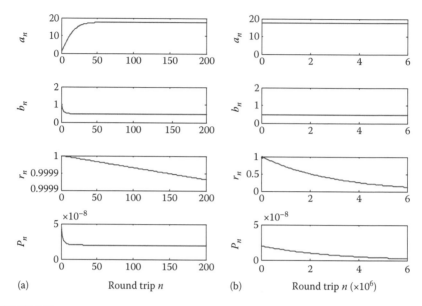

(a)   Round trip n (b)   Round trip n ($\times 10^6$)

FIGURE 2.18
Evolution of the a_n, b_n, r_n, P_n series in an actively MLL: (a) first 200 round-trip evolutions and (b) long-term evolution; the a_n series controlling the evolution of the pulse width and the b_n series controlling the evolution of the power rapidly converge to the steady state after about 50 round trips; the gain reduced factor series r_n plays a negligible role during the first round trips (shaping phase), but is the main factor causing the power series P_n to decrease in the long term (decaying phase).

Figure 2.18 plots the evolution of the a_n, b_n, r_n, P_n series for an MLL with the following parameter values: central wavelength $\lambda = 1550$ nm, filter bandwidth $BW = 1.2$ nm, modulation frequency $f_m = 10$ GHz, modulation index $m = 0.5$, small signal gain $G_{ss} = 20$ dB, $NF = 3$ dB, and total cavity loss $\Gamma = 6$ dB. Figure 2.18a shows the evolution of the shaped noise for the first 200 round trips. During the first round trips (shaping phase), the shaping effects of the filter and modulator are dominant and change the shaped noise after every round trip. The pulse width, which is determined by a_n, quickly converges to the steady-state value and reaches 99% of the value after about 50 round trips. Similarly, the power parameter, b_n, also rapidly converges to the steady-state value and obtains less than 1% difference from the steady-state value after about 20 round trips. The gain reduced factor r is determined to be 0.99999969 and hence has negligible effect compared to the filtering and modulating effects during the first round trips. However, it has a long-term effect on the power of the shaped noise when a_n and b_n already settle to their steady state values. Figure 2.18b shows the long-term evolution of the signal. After obtaining the shaping phase at the steady state, the signal power does not remain constant but decreases for every round trip since the net gain is negative due to the gain reduced factor r. The power decreases at a very slow rate since $r \approx 1$. It can be seen that the power is just reduced by about six times after 5,000,000 round trips.

Given the steady-state pulse signal of $\exp(-\omega^2/4\alpha)$, we can determine the noise contained in $\eta_n(\omega)$ as

$$\eta_n^{noise}(\omega) = \frac{\sigma_n (G_o\Gamma)^n}{\sqrt{Q_n}} \left[e^{-\omega^2\beta a_n} - e^{-\omega^2/4\alpha} \right] \tag{2.41}$$

The noise power is

$$P_n^{noise} = \frac{1}{2\pi} \int_{-\infty}^{+\infty} \left| \rho_n^{noise}(\omega) \right|^2 d\omega$$

$$= P_n \left[1 + 2\sqrt{\alpha\beta a_n} - 4\sqrt{\frac{2\alpha\beta a_n}{1 + 4\alpha\beta a_n}} \right] \tag{2.42}$$

Therefore, the SNR of the laser is

$$SNR = \frac{\sum_{n=1}^{\infty} \left(P_n - P_n^{noise} \right)}{\sum_{n=1}^{\infty} \left(P_n^{noise} \right)}$$

$$= \frac{\sum_{n=1}^{\infty} P_n \left(4\sqrt{\frac{2\alpha\beta a_n}{1 + 4\alpha\beta a_n}} - 2\sqrt{\alpha\beta a_n} \right)}{\sum_{n=1}^{\infty} P_n \left(1 + 2\sqrt{\alpha\beta a_n} - 4\sqrt{\frac{2\alpha\beta a_n}{1 + 4\alpha\beta a_n}} \right)} \tag{2.43}$$

Using (2.43), we can now evaluate the effects of the laser parameters such as the small-signal gain, NF, and cavity loss on the laser performance. Figure 2.19 shows the dependence of the laser's SNR on the NF and cavity loss. The SNR decreases as the cavity loss increases. This can be understood that the required gain for lasing increases as the cavity loss increases. Thus, the optical amplifier must work in the high gain region, which generates more ASE noise than in the saturated region. Higher NF also adds more noise to the signal for every round trip and hence also reduces the signal quality. For the given laser parameters, the SNR is about 53.7 dB. However, it is noted that SNR as high or more than 60 dB can be obtained if the cavity loss and the NF of the amplifier are kept at low values. This result is consistent with the low-noise feature of reported MLLs [67,68].

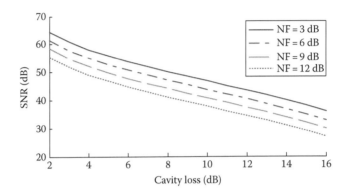

FIGURE 2.19
Dependence of the SNR on the cavity loss and amplifier NF; small cavity loss and small NF help to increase the SNR.

2.4.4.2 Mode Locking by Detuning

We here consider the case of frequency detuning where the modulation frequency is not exactly equal to the harmonic of the fundamental frequency of the laser ring. With frequency detuning, there is an additional delay per round trip and thus $t_d \neq 0$ in (2.31), and, just like Section 2.4.4.1 where $t_d = 0$ in (2.31) for no frequency detuning, we analyze the result in two conditions: noiseless amplifier and amplifier with ASE noise.

2.4.4.2.1 Noiseless Amplifier

The steady-state signal is the limit of $\rho_n(\omega)$ when n approaches infinity and is determined as

$$\eta(\omega) = \lim_{n \to \infty} \rho_n(\omega)$$

$$\propto e^{-\lim_{n \to \infty}\left(\frac{S_n}{Q_n}\right)\beta\omega^2} e^{-j\omega \lim_{n \to \infty}\left(\frac{S_n}{Q_n}\right)t_d} = e^{-\frac{2}{x+\sqrt{4x+x^2}}\beta\omega^2} e^{-j\omega\frac{2}{x+\sqrt{4x+x^2}}t_d} \quad (2.44)$$

Therefore, the pulse shape in the time domain is

$$\eta(t) \propto e^{-\frac{x+\sqrt{4x+x^2}}{8\beta}\left(t - \frac{2}{x+\sqrt{4x+x^2}}t_d\right)^2} = e^{-\alpha\left(t - \frac{t_d}{4\alpha\beta}\right)^2} \quad (2.45)$$

where $\alpha = x + \sqrt{4x+x^2}/8\beta$ is the Gaussian pulse parameter. The pulse width under detuning is the same as that of exact tuning. However, the pulse position does not pass the modulator at the peak of the modulation transfer function but shifted by an amount of $t_s = t_d/4\alpha\beta$, which is proportional to the detuning Δf_m. This is exactly the same result as in Section 2.2.

The power of the signal after n round trips is given as

$$P_n = \frac{1}{2\pi} \int\limits_{-\infty}^{+\infty} \rho_n^2(\omega)d\omega = \frac{\sigma_N^2}{\sqrt{8\pi\beta}} \frac{(\Gamma G)^{2n}}{\sqrt{S_n Q_n}} e^{-\frac{R_n}{2\beta Q_n}t_d^2} \tag{2.46}$$

The gain required for $\lim_{n\to\infty} P_n = \text{const}$ is

$$G = G_I = \frac{\sqrt{1+4\alpha\beta}}{\Gamma} e^{(t_d^2/4\beta)} \tag{2.47}$$

which is the same as (2.20) that is obtained using the classical approach. Compared to the zero-detuning case, the required gain increases by $\exp(t_d^2/4\beta)$ due to the additional loss caused by the frequency shifting.

2.4.4.2.2 Locking Range

Detuning causes the pulse to pass through the modulator not at the transmission peak but at a shifted position t_s. This induces more loss to the pulse and hence requires more gain to compensate for the loss. The larger the detuning, the higher the gain required. However, the maximum gain provided by the amplifying medium is limited to G_{ss}, the small-signal gain. If the required gain exceeds G_{ss}, mode locking is not obtained. Using (2.47), we have

$$\frac{\sqrt{1+4\alpha\beta}}{\Gamma} \exp\left(\frac{t_d^2}{4\beta}\right) < G_{ss} \tag{2.48}$$

Putting t_d from (2.16) into (2.49) gives

$$\Rightarrow |\Delta f_m| < \frac{f_m^2}{N} \sqrt{4\beta \ln\left(\frac{G_{ss}\Gamma}{\sqrt{1+4\alpha\beta}}\right)} \tag{2.49}$$

The normalized locking range is

$$|\delta f_m| \equiv \frac{|\Delta f_m|}{f_m} < \frac{f_m}{N} \sqrt{4\beta \ln\left(\frac{G_{ss}\Gamma}{\sqrt{1+4\alpha\beta}}\right)} \tag{2.50}$$

In addition to the limitation impinged by the required gain, the locking range is also limited by the maximum allowable value of the pulse position shift when detuned. In order to reshape and retime the pulse to keep the pulse unchanged for every round trip, a negative second derivative in the modulation transmission function is required [69]. In the case of the MZIM, the second derivative changes sign at about $T_m/4$, where $T_m = 1/f_m$ and f_m is defined in (2.9). Therefore, the pulse must arrive at the modulator no later than $T_m/4$. Equation 2.45 shows that the steady-state pulse position when detuned is t_s. When it arrives at the modulator in the next round, its position becomes $t_s + t_d$, which must satisfy the condition $|t_s + t_d| < T_m/4$. Using this condition, we obtain the locking range as

$$\left|\delta f_m\right| \equiv \frac{\left|\Delta f_m\right|}{f_m} < \frac{\alpha\beta}{N(1+4\alpha\beta)} \tag{2.51}$$

Combining (2.50) and (2.51), the locking range is determined as

$$\left|\delta f_m\right| < \min\left\{\frac{f_m}{N}\sqrt{4\beta\ln\left(\frac{G_{ss}\Gamma}{\sqrt{1+4\alpha\beta}}\right)}, \frac{\alpha\beta}{N(1+4\alpha\beta)}\right\} \tag{2.52}$$

In most of the cases, the loss of the cavity is kept very low while the amplifying medium gain is large enough that the first term in the RHS of (2.52) is larger than the later one. Hence, the locking range is usually limited by the later term. Using the given laser parameters, we obtain $|\delta f_m| < \min\{2.64 \times 10^{-5}, 1.3 \times 10^{-5}\} = 1.3 \times 10^{-5}$, which will be verified with the simulation result in Section 2.4.5.

2.4.4.2.3 Amplifier with ASE Noise

Following a procedure similar to that of zero detuning (Section 2.4.4.1), we obtain the required gain for lasing, and the power of the signal after n round trips as

$$G = G_I r = \frac{\sqrt{1+4\alpha\beta}}{\Gamma}e^{\left(t_d^2/4\beta\right)}r \tag{2.53}$$

$$P_n = \frac{\sigma_N^2}{\sqrt{8\pi\beta}}\frac{(1+4\alpha\beta)^n}{\sqrt{S_n Q_n}}e^{\frac{t_d^2}{2\beta}\left(n-\frac{R_n}{Q_n}\right)}r^{2n} \tag{2.54}$$

where G_I in (2.54) is the intrinsic gain.

The noise is given as

$$P_n^{noise} = \frac{1}{2\pi}\int_{-\infty}^{+\infty}\left|\rho_n^{noise}(\omega)\right|^2 d\omega$$

$$= P_n\left[1 + 2\sqrt{\alpha\beta a_n} - 4\sqrt{\frac{2\alpha\beta a_n}{1+4\alpha\beta a_n}}e^{-4\alpha(t_d a_n - t_s)^2/(1+4\alpha\beta a_n)}\right] \tag{2.55}$$

Therefore, the SNR of the laser is

$$
\text{SNR} = \frac{\sum_{n=1}^{\infty}\left(P_n - P_n^{noise}\right)}{\sum_{n=1}^{\infty}\left(P_n^{noise}\right)}
$$

$$
= \frac{\sum_{n=1}^{\infty} P_n\left(4\sqrt{\dfrac{2\alpha\beta a_n}{1+4\alpha\beta a_n}}\,e^{-4\alpha(t_d a_n - t_s)^2/(1+4\alpha\beta a_n)} - 2\sqrt{\alpha\beta a_n}\right)}{\sum_{n=1}^{\infty} P_n\left(1+2\sqrt{\alpha\beta a_n} - 4\sqrt{\dfrac{2\alpha\beta a_n}{1+4\alpha\beta a_n}}\,e^{-4\alpha(t_d a_n - t_s)^2/(1+4\alpha\beta a_n)}\right)} \qquad (2.56)
$$

The evolution of the signal also consists of two phases: shaping phase and fading phase. In the shaping phase, filtering and modulating have a strong effect and shape the noise to the steady-state Gaussian pulse. This phase is characterized by analyzing the following three series: $a_n = S_n/Q_n$, $b_n = (1+4\alpha\beta)^n/\text{sqrt}(S_n Q_n)$, and $c_n = \exp((n-R_n/Q_n)t_d^2/2\beta)$. The a_n series controls the evolution of the pulse width and pulse position, the b_n and c_n series control the evolution of the power. When a_n, b_n, and c_n obtain their steady states, the shaping phase completes and the signal gets the shape of the steady-state pulse as determined in the noiseless amplifier case. In the fading phase, the $r_n = r^{2n}$ series plays a dominant role and the signal power decreases exponentially with r^2. The value of r is determined from solving the equation

$$
P = \sum_{n=1}^{\infty} P_n = \frac{G_I^2 r^2}{G_I^2 r^2 - 1} P_{sat}\ln\frac{G_{ss}}{G_I^2 r^2} \qquad (2.57)
$$

Figure 2.20 plots the series of a_n, b_n, c_n, r_n, P_n for the MLL with a frequency detuning of $\delta f_m = 10^{-5}$ ($\Delta f_m = 100\,\text{kHz}$). During the first round trips (shaping phase), the noise is strongly shaped to the steady-state pulse due to the filtering, modulation, and delaying effects. This reflects on the rapid evolution and convergence of a_n, b_n, and c_n. They reach 99% of their steady-state values after 47, 19, and 63 round trips, respectively. Compared to the exact tuning case, the shaping phase takes more round trips since it involves an additional series evolution c_n. In contrast to the exact tuning case, the value of the gain-reduced factor r in detuning is rather small, especially when the detuning is large. This causes the signal to rapidly decrease in the fading phase. It can be seen from Figure 2.20b that the signal power falls to nearly zero after just 20,000 round trips. As the noise is stronger in the shaping phase than in the fading phase, the long shaping phase and short fading phase in detuning reduces the laser's SNR. Using the previously given parameter values of the laser, the SNR is only 28.4 dB, which is largely reduced from the value of 53.7 dB when zero detuning is used.

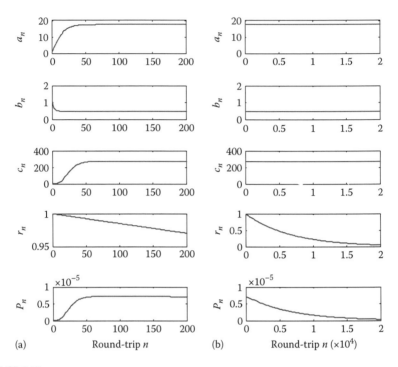

FIGURE 2.20
Evolution of the pulse width and pulse position depends on the controlling series a_n; power controlling series b_n, c_n; gain reduced factor series r_n; and power series P_n in a detuned actively MLL with frequency detune of $\delta f_m = 10^{-5}$: (a) first 200 round-trip evolutions; (b) long-term evolution; the additional power controlling series c_n when detuning requires the pulse to make more round trips to reach the steady state; small gain reduced factor r causes the power series P_n to decrease faster; longer shaping phase and shorter decaying phase reduce the SNR of the laser when detuned.

The dependence of SNR on the frequency detuning is shown in Figure 2.21. As the detuned frequency increases, the SNR decreases rapidly. SNR drops to 18.6 dB with just a normalized detune of 1.2×10^{-5}. In practice, the laser's SNR must satisfy a minimum value to be useful in applications. A minimum value of 25 dB is required for telecommunication applications. Using this requirement, we can determine the locking range of the laser in the presence of the ASE noise. Figure 2.22 shows the locking range of an MLL with different filter bandwidths and NFs. The position limited locking range, which is specified in (2.52), is also plotted for comparison. It is shown that with the presence of the ASE noise and a requirement for the minimum laser's SNR, the locking range of the laser is limited to a narrower range compared to that limited by the maximum shifted position.

For different requirement of the SNR of the laser, we can use the SNR contour plotted in Figure 2.23 to determine the locking range. For a given SNR quality, the filter bandwidth and detuning frequency should be kept

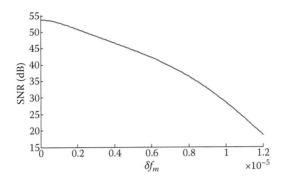

FIGURE 2.21
Degrading of the MLL's SNR with an increase of the frequency detuning.

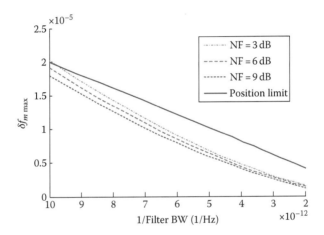

FIGURE 2.22
Maximum locking range of the MLL with an SNR threshold of 25 dB; the locking range with the ASE noise considered is narrower than that in the noiseless situation.

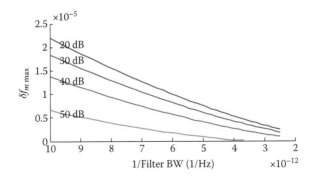

FIGURE 2.23
SNR contours of the actively MLL.

in the region under the specific SNR contour line. As the filter bandwidth increases, not only the maximum frequency detuning is decreased, but also the spacing between the contour line is decreased. Hence, the laser quality is more sensitive to the detuning; a small detuning amount can degrade the laser's SNR from high quality to very low quality.

2.4.5 Simulation

The simulation model consists of a gain saturated amplifying medium with white Gaussian noise, a BPF, and a modulator (see Figure 2.18). The accumulation process of the pulse in the simulation is similar to that in the real MLL, which is a self-started process from the ASE noise. The initial signal is a set with random complex numbers, which represent the ASE white noise. The signal is then amplified in a saturated gain medium with a white Gaussian noise. Filtering is implemented in the frequency domain using fast Fourier transform (FFT) algorithm. The signal is first converted to the frequency domain and then multiplied with the filter transfer function. The signal is then converted back to the time domain using the inverse fast Fourier transform (IFFT) algorithm and passed through the modulator. The Mach–Zehnder modulator transmission function expressed in (2.11) is used in the simulation, except where it is specified in the text.

After passing through the modulator, the signal completes one round trip and is fed back to the amplifier for the next round. This process continues until equilibrium is obtained.

First of all, the laser is simulated with a noiseless amplifier in which the ASE noise source is turned off. The theoretical Gaussian modulation transmission function is used for easy comparison with the analytical results. Figure 2.24a and b show the simulation of the buildup process of the pulse inside the cavity when the modulation frequency is detuned.

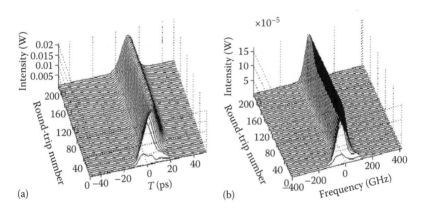

(a)   (b)

FIGURE 2.24
Pulse evolution in an actively MLL with a normalized detuned frequency of 5×10^{-6} ($\Delta f_m = 50\,\text{kHz}$): (a) time domain response, (b) frequency domain response.

Initially, the pulse position is shifted away from the transmission peak for every round trip due to detuning. After about 100 round trips, the pulse position inclines to a constant, which is the steady-state value t_s, as predicted by the theory described in Section 2.4.4. At this position, the retiming effect of the modulation completely compensates for the delay due to detuning and hence the pulse position is fixed for every round trip afterward (Figure 2.25).

The simulation also confirms that the pulse width is independent of detuning and has the same value as that due to exact tuning. The variation of the pulse position and the gain are shown in Figure 2.26a and b, respectively.

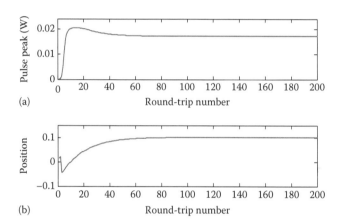

(a)

(b)

FIGURE 2.25
Evolution of (a) the pulse peak value and (b) pulse position during the pulse buildup process in the detuned MLL.

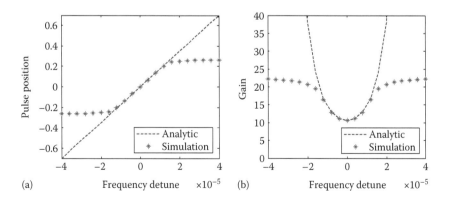

(a)

(b)

FIGURE 2.26
(a) Normalized pulse position and (b) gain of the MLL when detuned; the pulse position linearly increases with the frequency detuning within the locking range $\pm \delta f_{m\,max} = \pm 1.3 \times 10^{-5}$.

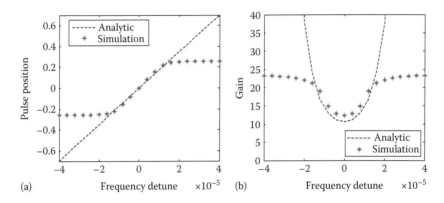

FIGURE 2.27
(a) Normalized pulse position and (b) gain of the MLL with an MZIM when detuned; the results are nearly the same as that of the Gaussian modulation function.

Within the calculated locking range, $\pm\delta f_{m\,max}=\pm1.3\times10^{-5}$, the simulation results are in good agreement with the analytical results. As predicted in theory, the pulse position linearly increases while the required gain exponentially increases with detuning. When the detuned frequency exceeds $\delta f_{m\,max}$, the theoretical calculation based on the Gaussian approximation is no longer valid. The pulse position does not increase linearly with detuning but inclines to a constant of about $T_m/4$.

Simulation with the MZIM transmission function is shown in Figure 2.27. The results are nearly the same as that when an approximated Gaussian transmission function is used. The small difference in the required gain is due to the difference between the two transmission functions.

Figure 2.28 shows the pulse's buildup process, its peak power, and position evolution for over 10,000 round trips of the laser with the ASE noise when the frequency detune is zero. Compared to the noiseless case, the steady-state peak power and position have slight fluctuations due to the noise generated by the amplifier. The standard deviations of the power and position are 8.8×10^{-5} and 4.3×10^{-7}, respectively. Figures 2.29 through 2.32 show the evolution of the pulse with different detuned frequencies. As the detune increases, the fluctuation of the pulse increases. This is also observed in Figure 2.30, which shows the pulse position when frequency detuning changes from -2×10^{-5} to 2×10^{-5}. The simulation is run for over 10,000 round trips for every frequency detuning value. The pulse positions after a few hundred transient round trips are recorded and used to calculate the means and standard deviations. In Figure 2.32, the dotted-line curve is the means of the pulse position in steady state while the curve with bars is the standard deviations. It can be seen from the simulation results that the pulse quality has been critically decreased before the detuned frequency reaches some certain level, the limit for the noiseless case. This

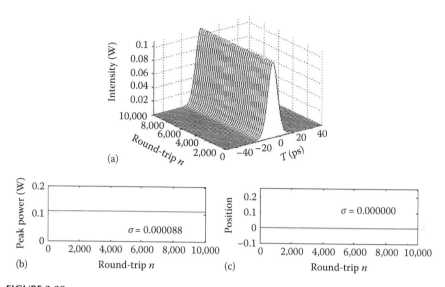

(a)

(b)

$\sigma = 0.000088$

(c)

$\sigma = 0.000000$

FIGURE 2.28
Simulation of zero-detuned MLL under the influence of the ASE noise: (a) pulse buildup process, (b) pulse peak power, and (c) normalized pulse position.

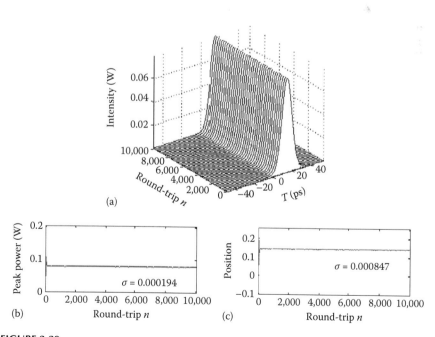

(a)

(b)

$\sigma = 0.000194$

(c)

$\sigma = 0.000847$

FIGURE 2.29
Simulation of the MLL under the influence of the ASE noise with the normalized frequency detune $\delta f_m = 8 \times 10^{-6}$: (a) pulse buildup process, (b) pulse peak power, and (c) normalized pulse position.

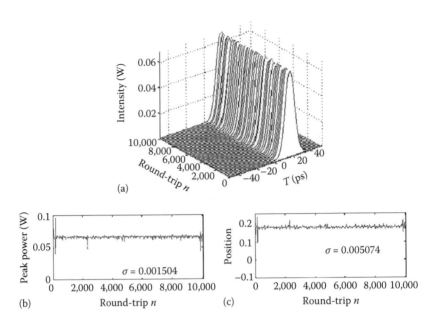

FIGURE 2.30
Simulation of the MLL under the influence of the ASE noise when the normalized frequency detune is $\delta f_m = 10 \times 10^{-6}$: (a) pulse buildup process, (b) pulse peak power, and (c) normalized pulse position.

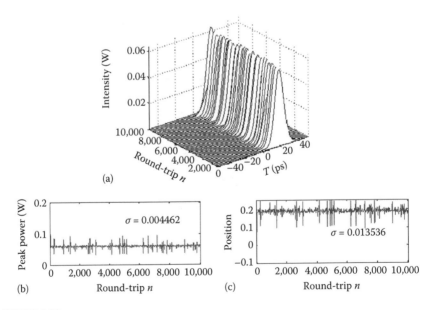

FIGURE 2.31
Simulation of the MLL under the influence of the ASE noise when the normalized frequency detune is $\delta f_m = 11 \times 10^{-6}$: (a) pulse buildup process, (b) pulse peak power, and (c) normalized pulse position.

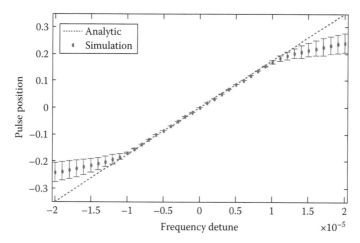

FIGURE 2.32
Normalized pulse position of the MLL simulated with the ASE noise included; the error bar shows the standard variation of the simulated position.

confirms that the locking range when considering the ASE noise is narrower than that derived from the pulse position limiting.

2.5 Conclusions

In this chapter, the principles of fiber laser and mode locking have been presented. A detailed analysis of the actively MLL using the mathematical series has also been described to trace the evolution of the signal in the laser cavity. Using this approach, we can obtain not only the steady state but also the transient evolution processes of the mode-locked pulse. The noiseless amplifier laser model has been analyzed and the results obtained are consistent with those obtained using classical analysis. Steady-state pulse parameters in either the exact frequency tuning or frequency detuning have been derived. It has been shown that the pulse width when the frequency is detuned is exactly the same as that when the frequency is tuned. The pulse position is shifted linearly with the frequency detune and this limits the locking range of the laser.

The advantage of our proposed series method is its capability to allow the analysis of the laser under the influence of the ASE noise that is normally ignored in classical analyses. Using the series method, we have calculated the SNR of the laser and studied the dependence of SNR on the cavity design parameters such as the cavity loss and amplifier noise. The decrease of SNR when the laser is detuned has also been studied. It is found that the presence of ASE noise narrows down the locking range compared to the case in which an ideal noiseless amplifier can be used.

References

1. W. Sibbett and J. Taylor, Passive mode locking in the near infrared, *IEEE J. Quantum Electron.*, 20, 108–110, 1984.
2. D. Kunimatsu, S. Arahira, Y. Kato, and Y. Ogawa, Passively mode-locked laser diodes with bandgap-wavelength detuned saturable absorbers, *IEEE Photon. Technol. Lett.*, 11, 1363–1365, 1999.
3. H.A. Haus and E.P. Ippen, Self-starting of passively mode-locked lasers, *Opt. Lett.*, 16, 1331–1333, 1991.
4. H.A. Haus, J.G. Fujimoto, and E.P. Ippen, Structures for additive pulse mode-locking, *J. Opt. Soc. Am. B: Opt. Phys.*, 8, 2068–2076, 1991.
5. H.A. Haus, J.G. Fujimoto, and E.P. Ippen, Analytic theory of additive pulse and Kerr lens mode-locking, *IEEE J. Quantum Electron.*, 28, 2086–2096, 1992.
6. I.N. Duling, Subpicosecond all-fiber erbium laser, *Electron. Lett.*, 27, 544–545, 1991.
7. K. Tamura, H.A. Haus, and E.P. Ippen, Self-starting additive pulse mode-locked erbium fiber ring laser, *Electron. Lett.*, 28, 2226–2228, 1992.
8. F.X. Kartner, U. Morgner, R. Ell, E.P. Ippen, J.G. Fujimoto, V. Scheuer, G. Angelow, and T. Tschudi, Few-cycle-pulse generation and its applications, in *Lasers and Electro-Optics, 2001. CLEO/Pacific Rim 2001. The 4th Pacific Rim Conference*, Makuhari, Japan, 2001.
9. U. Morgner, F.X. Kartner, S.H. Cho, Y. Chen, H.A. Haus, J.G. Fujimoto, E.P. Ippen, V. Scheuer, G. Angelow, and T. Tschudi, Sub-two-cycle pulses from a Kerr-lens mode-locked Ti: Sapphire laser, *Opt. Lett.*, 24, 411–413, 1999.
10. L.E. Hargrove, R.L. Fork, and M.A. Pollack, Locking of He-Ne laser modes induced by synchronous intracavity modulation, *Appl. Phys. Lett.*, 5, 4–5, 1964.
11. O. Guy, V. Kubecek, and A. Barthelemy, Mode-locked diode-pumped Nd: YAP laser, *Opt. Commun.*, 130, 41–43, 1996.
12. H.J. Eichler, J. Albertz, F. Below, A. Kummrow, T. Leitert, A.A. Kaminskii, and L. Jakab, Acoustooptic mode-locking of 3-Mu-M Er lasers, *Appl. Opt.*, 31, 4909–4911, 1992.
13. U. Keller, K.D. Li, B.T. Khuriyakub, D.M. Bloom, K.J. Weingarten, and D.C. Gerstenberger, High-frequency acoustooptic mode locker for picosecond pulse generation, *Opt. Lett.*, 15, 45–47, 1990.
14. J.F. Pinto, C.P. Yakymyshyn, and C.R. Pollock, Acoustooptic mode-locked soliton laser, *Opt. Lett.*, 13, 383–385, 1988.
15. C.C. Davis, *Laser and Electro-optics: Fundamentals and Engineering*, Cambridge University Press, Cambridge, U.K., 1996.
16. K. Iga, *Encyclopedic Handbook of Integrated Optics*, CRC Press, New York, 2006.
17. D. Malacara, *Handbook of Optical Engineering*, Marcel Dekker, New York, 2004.
18. L.A. Kulakova and E.Z. Yakhkind, Acousto-optic modulator for IR diode laser radiation, *J. Opt. A: Pure Appl. Opt.*, 3, S9–S11, 2001.
19. J. He and K.T. Chan, All-optical actively modelocked fibre ring laser based on cross-gain modulation in SOA, *Electron. Lett.*, 38, 1504–1505, 2002.
20. K. Vlachos, C. Bintjas, N. Pleros, and H. Avramopoulos, Ultrafast semiconductor-based fiber laser sources, *IEEE J. Sel. Top. Quantum Electron.*, 10, 147–154, 2004.

21. M.J.F. Digonnet, *Rare-Earth-Doped Fiber Lasers and Amplifiers*, Marcel Dekker Inc., New York, 2001.
22. B. Pedersen, A. Bjarklev, J.H. Povlsen, K. Dybdal, and C.C. Larsen, The design of erbium-doped fiber amplifiers, *J. Lightwave Technol.*, 9(9), 1105–1112, 1991.
23. New Focus, *Application Note 1—Insights into High-Speed Detectors and High-Frequency Techniques*, New Focus, San Jose, CA, 2002.
24. R.L. Jungerman, G. Lee, O. Buccafusca, Y. Kaneko, N. Itagaki, R. Shioda, A. Harada, Y. Nihei, and G. Sucha, 1-THz bandwidth C- and L-band optical sampling with a bit rate Agile timebase, *IEEE Photon. Technol. Lett.*, 14(8), 1148–1150, 2002.
25. G. Zhu, H. Chen, and N. Dutta, Time domain analysis of a rational harmonic mode-locked ring fiber laser, *J. Appl. Phys.*, 90(5), 2143–2147, 2001.
26. R.L. Jungerman, C. Johnsen, D.J. McQuate, K. Salomaa, M.P. Zurakowski, R.C. Bray, G. Conrad, D. Cropper, and P. Hernday, High-speed optical modulator for application in instrumentation, *J. Lightwave Technol.*, 8(9), 1363–1370, 1990.
27. G. Zhu, Q. Wang, H. Chen, H. Dong, and N.K. Dutta, High-quality optical pulse train generation at 80 Gb/s using a modified regenerative-type mode-locked fiber laser, *IEEE J. Quantum Electron.*, 40, 721–725, 2004.
28. G. Lin, J. Wu, and Y. Chang, 40 GHz rational harmonic mode-locking of erbium-doped fiber laser with optical pulse injection, in *Optical Fiber Communications Conference, 2003 (OFC 2003)*, Atlanta, GA, 2003.
29. D.L.A. Seixas and M.C.R. Carvalho, 50 GHz fiber ring laser using rational harmonic mode-locking, in *Microwave and Optoelectronics Conference, 2001. IMOC 2001. Proceedings of the 2001 SBMO/IEEE MTT-S International*, Belem, Brazil, 2001.
30. E. Yoshida and M. Nakazawa, 80~200 GHz erbium doped fibre laser using a rational harmonic mode-locking technique, *Electron. Lett.*, 32, 1370–1372, 1996.
31. P.H. Wang, L. Zhan, Z.C. Gu, Q.H. Ye, and Y.X. Xia, Generation of the 11th order rational harmonic mode-locked pulses with an arbitrary numerator in fiber-ring lasers, *Opt. Commun.*, 238, 345–349, 2004.
32. O. Pottiez, P. Megret, and M. Blondel, Environmentally induced noises in an actively mode-locked erbium fibre laser operating in the second-order rational harmonic mode locking regime, *Opt. Commun.*, 213, 103–119, 2002.
33. X. Shan, S.G. Edirisinghe, and A.S. Siddiqui, A stabilized, rationally mode-locked 10 GHz erbium fiber laser, *Fiber Integr. Opt.*, 20, 341–346, 2001.
34. K.K. Gupta, N. Onodera, K.S. Abedin, and M. Hyodo, Pulse repetition frequency multiplication via intracavity optical filtering in AM mode-locked fiber ring lasers, *IEEE Photon. Technol. Lett.*, 14, 284–286, 2002.
35. Z. Ahmed and N. Onodera, High repetition rate optical pulse generation by frequency multiplication in actively modelocked fibre ring lasers, *Electron. Lett.*, 32, 455–457, 1996.
36. K.K. Gupta, N. Onodera, and M. Hyodo, Equal amplitude optical pulse generation at higher-order repetition frequency in fibre ring lasers using intra-cavity Fabry-Perot etalon and rational harmonic mode-locking, in *Microwave Photonics, 2001 (MWP '01). 2001 International Topical Meeting*, Long Beach, CA, 2002.
37. K.K. Gupta, N. Onodera, and M. Hyodo, Technique to generate equal amplitude, higher-order optical pulses in rational harmonically modelocked fibre ring lasers, *Electron. Lett.*, 37, 948–950, 2001.

38. K.K. Gupta, N. Onodera, M. Hyodo, M. Watanabe, and J. Ravikumar, Evaluation of amplitude-stabilized optical pulse trains from rational harmonically mode-locked fiber ring lasers, *J. Lightwave Technol.*, 22, 1935–1945, 2004.
39. Z.X. Wang, T. Wang, L. Huo, C.Y. Lou, and Y.Z. Gao, Pulse-amplitude-equalization methods for the repetition rate multiplication using F-P filter, *Int. J. Infrared Milli. Waves*, 24, 399–407, 2003.
40. X. Shan, D. Cleland, and A. Ellis, Stabilising Er fibre soliton laser with pulse phase locking, *Electron. Lett.*, 28, 182–184, 1992.
41. M. Nakazawa, E. Yoshida, and K. Tamura, 10 GHz, 2 ps regeneratively and harmonically FM mode-locked erbium fibre ring laser, *Electron. Lett.*, 32, 1285–1287, 1996.
42. M. Nakazawa, E. Yoshida, and K. Tamura, Ideal phase-locked-loop (PLL) operation of a 10 GHz erbium-doped fibre laser using regenerative modelocking as an optical voltage controlled oscillator, *Electron. Lett.*, 33, 1318–1320, 1997.
43. K.S. Abedin, M. Hyodo, and N. Onodera, Active stabilization of a higher-order mode-locked fiber laser operating at a pulse-repetition rate of 154 GHz, *Opt. Lett.*, 26, 151–153, 2001.
44. G. Zhu and N.K. Dutta, Regeneratively stabilized 4th order rational harmonic mode-locked erbium doped fiber laser operating at 40 Gb/s, *Opt. Express*, 13, 3371–3375, 2005.
45. M. Nakazawa, E. Yoshida, and Y. Kimura, Ultrastable harmonically and regeneratively modelocked polarisation-maintaining erbium fibre ring laser, *Electron. Lett.*, 30, 1603–1605, 1994.
46. B. Bakhshi, P.A. Andrekson, and X.P. Zhang, A polarization-maintaining and dispersion-managed 10-GHz mode-locked erbium fiber ring laser providing both sech(2)- and Gaussian-shaped pulses, *Opt. Fiber Technol.*, 4, 293–303, 1998.
47. H. Takara, S. Kawanishi, M. Saruwatari, and K. Noguchi, Generation of highly stable 20 GHz transform-limited optical pulses from actively mode-locked Er[3+]-doped fibre lasers with an all-polarisation maintaining ring cavity, *Electron. Lett.*, 28, 2095–2096, 1992.
48. Y.M. Jhon, Y.T. Byun, and D.H. Woo, Pulse-amplitude equalization using a polarization-maintaining laser resonator, *Opt. Lett.*, 31, 2678–2680, 2006.
49. R. Hayashi, S. Yamashita, and T. Saida, 16-wavelength 10-GHz actively mode-locked fiber laser with demultiplexed outputs anchored on the ITU-T grid, *IEEE Photon. Technol. Lett.*, 15, 1692–1694, 2003.
50. M. Nakazawa and E. Yoshida, A 40-GHz 850-fs regeneratively FM mode-locked polarization-maintaining erbium fiber ring laser, *IEEE Photon. Technol. Lett.*, 12, 1613–1615, 2000.
51. T. Pfeiffer and G. Veith, 40 GHz pulse generation using a widely tunable all-polarisation preserving erbium fibre ring laser, *Electron. Lett.*, 29, 1849–1850, 1993.
52. M. Yoshida, T. Hirayama, M. Yakabe, M. Nakazawa, Y. Koga, and K. Hagimoto, An ultrastable PLL mode-locked fiber laser with a hydrogen maser clock, in *Lasers and Electro-Optics, 2005. CLEO/Pacific Rim 2005. Pacific Rim Conference*, Tokyo, Japan, 2005.
53. N. Onodera, Supermode beat suppression in harmonically mode-locked erbium-doped fibre ring lasers with composite cavity structure, *Electron. Lett.*, 33, 962–963, 1997.

54. M. Nakazawa, Ultrafast optical pulses and solitons for advanced communications, in *Lasers and Electro-Optics, 2003. CLEO/Pacific Rim 2003. The 5th Pacific Rim Conference*, Taipei, Taiwan, 2003.
55. L. Schares, R. Paschotta, L. Occhi, and G. Guekos, 40-GHz mode-locked fiber-ring laser using a Mach-Zehnder interferometer with integrated SOAs, *J. Lightwave Technol.*, 22, 859–873, 2004.
56. D. Kuizenga and A. Siegman, FM and AM mode locking of the homogeneous laser—Part I: Theory, *IEEE J. Quantum Electron.*, 6, 694–708, 1970.
57. H. Haus, A theory of forced mode locking, *IEEE J. Quantum Electron.*, 11, 323–330, 1975.
58. J. Fontana, Theory of spontaneous mode locking in lasers using a circuit model, *IEEE J. Quantum Electron.*, 8, 699–703, 1972.
59. Y. Li, C. Lou, and Y. Gao, Theoretical study on a detuned AM mode-locked laser, in *Lasers and Electro-Optics Society 1999 12th Annual Meeting. LEOS '99, IEEE*, San Francisco, CA, 1999.
60. G. Zhu and N.K. Dutta, Dispersion effects on the detuning properties of actively harmonic mode-locked fiber lasers, *Opt. Express*, 13, 2688, 2005.
61. H.A. Haus, Mode-locking of lasers, *IEEE J. Sel. Top. Quantum Electron.*, 6, 1173–1185, 2000.
62. P. Becker, A. Olsson, and J. Simpson, *Erbium-Doped Fiber Amplifiers*, Elsevier Academic Press, San Diego, CA, 1999.
63. E. Desurvire, *Erbium-Doped Fiber Amplifiers Principles and Applications*, John Wiley & Sons, New York, 1994.
64. G.R. Walker, Gain and noise characterisation of erbium doped fibre amplifiers, *Electron. Lett.*, 27, 744–745, 1991.
65. M.A. Mahdi, S. Selvakennedy, P. Poopalan, and H. Ahmad, Saturation parameters of erbium doped fibre amplifiers, in *Semiconductor Electronics, 1998. Proceedings. ICSE '98. 1998 IEEE International Conference*, Isobe, Japan, 1998.
66. G.P. Agrawal, *Applications of Nonlinear Fiber Optics*, Academic Press, San Diego, CA, 2001.
67. T.R. Clark, T.F. Carruthers, I.N. Duling III, and P.J. Matthews, Sub-10 femtosecond timing jitter of a 10-GHz harmonically mode-locked fiber laser, in *Optical Fiber Communication Conference, 1999, and the International Conference on Integrated Optics and Optical Fiber Communication. OFC/IOOC '99. Technical Digest*, San Diego, CA, 1999.
68. M.E. Grein, H.A. Haus, Y. Chen, and E.P. Ippen, Quantum-limited timing jitter in actively modelocked lasers, *IEEE J. Quantum Electron.*, 40, 1458–1470, 2004.
69. J.S. Wey, J. Goldhar, and G.L. Burdge, Active harmonic modelocking of an erbium fiber laser with intracavity Fabry-Perot filters, *J. Lightwave Technol.*, 15, 1171–1180, 1997.

3

Active Mode-Locked Fiber Ring Lasers: Implementation

A fiber amplifier can be converted into a laser by placing it inside a cavity designed to provide positive optical feedback. Such lasers are called fiber lasers. A fiber laser is an important element in all optical fiber communication systems, sensors, and photonic switching systems. There are a number of potential operating characteristics that make the fiber lasers particularly attractive. The lasers have been shown to have quantum-limited noise, narrow pulse width, both active and passive MLFLs, higher repetition rate (active MLFL), transform-limited pulses, etc. Comparing with laser diodes, fiber lasers have many advantages, such as low threshold value, high average optical power, low timing jitter, low pulse dropout ratio, wide wavelength tunable range, high efficiency, and compatibility with fiber devices [1,2]. Despite the advantages of the fiber lasers, there are also some difficulties, such as environmental sensitivity, rare components, unpredictable pulse spacing, bistable operation, and extra spectral components. As a result, a large number of designs have been examined, attempting to optimize the laser operation to suit desired applications.

Mode locking, a well-known technique as described in Chapter 2, has been used to generate short optical pulses from a laser. It first appeared in the work of [3] on ruby lasers and [4] on He–Ne lasers. With the invention of Mollenauer's soliton laser [5], many MLL configurations started to boom in this area, out of which are the fiber simple ring lasers [6,7], sigma lasers [8,9], Sagnac loops [10], figure-eight lasers [11], etc., either with active or passive mode-locking techniques.

3.1 Building Blocks of Active Mode-Locked Fiber Ring Laser

With the basic laser knowledge from Chapter 2, we shall now move on to the actual design parameters and working mechanism of an active MLFRL, i.e., a pulsed fiber ring laser. In this section, we will focus on the main building blocks that make up the MLFRL, which includes the laser cavity design, active or gain medium, filter, and modulator mechanism.

3.1.1 Laser Cavity Design

The most common type of laser cavity is known as the Fabry–Perot cavity, which is made by placing the gain medium in between two high-reflecting mirrors. In the case of fiber lasers, mirrors are often butt-coupled to the fiber ends to avoid diffraction losses. Ring cavities are often used to realize unidirectional operation of a laser, with the help of an optical isolator. In the case of fiber lasers, an additional advantage is that a ring cavity can be made without using mirrors, resulting in an all-fiber cavity.

The total cavity length is an important design parameter, which determines the repetition frequency of the laser. There could be many different segments of fiber, such as EDF, DSF, SMF, etc., in a fiber ring cavity, which might have different refractive indices. Hence, the actual longitudinal mode separation is given by

$$f_c = \frac{c}{\sum_j n_j L_j} \tag{3.1}$$

where n_j and L_j are the refractive index and length of the jth segment of the fiber. However, for simplicity's sake, the variation of the refractive index between the fiber segments is assumed to be small. Therefore, the simple relation $f_c = c/nL$ can be used, with L being the total cavity length.

By selecting a proper choice of the refractive index profile, the cavity can be designed to support only one electromagnetic mode, i.e., single-mode operation, with two possible polarizations. Field patterns of a greater transverse variation are not guided but lost to radiation. However, there will be two polarized states of propagating lightwaves in the ring [12], hence two simultaneously propagating rings, which could interfere or hop between one another. This is due to the difference in the group velocity of two different lightwave polarizations. In principle, there is no exchange of power between the two polarization components. If the power of the light source is delivered into one polarization only, the power received remains in that polarization. In practice, however, slight random imperfection or uncontrollable strains in the fiber could result in random power transfer between the two polarizations [13].

If this polarization effect is not treated properly, random pulse splitting will occur, degrading the system performance and challenging the system stability. Therefore, for a better laser performance, polarization-maintaining fiber (PMF) and all-polarization-maintaining optical components [14] could be used to maintain a constant polarization state of the lightwave traveling in the laser cavity, but with higher loss. Alternatively, a polarization controller could be used in the ring to resolve this problem.

The minimum pulse width occurs when the cavity group velocity dispersion (GVD) is zero in a homogeneous MLL [15]; however, in an inhomogeneous MLL, the generation of a single soliton-like pulse is limited within a certain range of negative GVD [16]. In the small negative GVD region, the gain bandwidth limits the spectral broadening, whereas in the large negative GVD region, the nonsoliton spectral components destabilize the soliton pulse–shaping process, and the laser generates only the incoherent pulses. Hence, small GVD is preferred for good laser performance. Generally, the resonant dispersion from the gain medium has a negligible effect on the mode-locking process for a homogeneously broadened laser [17].

3.1.2 Active Medium and Pump Source

In order to compensate for the losses within the laser cavity, an amplifying element must be included, and EDFA is commonly used in this regard. The main transition of interest in EDFA is the high gain $^4I_{13/2} \rightarrow {}^4I_{15/2}$ transition centered around 1550 nm, because the $^4I_{13/2}$ is the only metastable state for common oxide glasses at room temperature and gain is only available for these materials at 1500 nm $^4I_{13/2} \rightarrow {}^4I_{15/2}$ emission band. This is shown in Figure 3.1. The peak absorption and emission wavelengths occur around 1530 nm; therefore, the operating wavelength must be chosen in the band so that the emission cross section is much greater than the absorption cross section, which is around 1550 nm, the third optical communication window.

Under current technology, an EDFA pumped with a laser diode is possible with lasers operating at around 810, 980, and/or 1480 nm. However, the pump wavelength of 810 nm suffers from the strong ESA, which causes an undesirable waste of pump photons. As a result, the 980 and 1480 nm are more

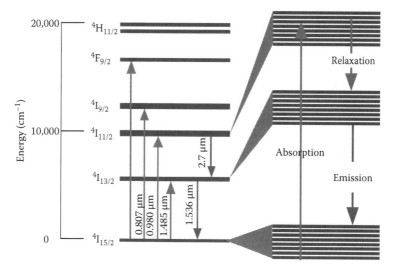

FIGURE 3.1
Erbium energy transition diagram.

commonly used as the pumping wavelengths. 980 nm (InGaAs/GaAs-based lasers) pumping wavelength provides higher gain efficiency and SNR for small-signal amplifiers [18]. It also gives better noise figures and quantum conversion efficiencies for power amplifiers. Quantum conversion efficiency is defined as number of photons added to the signal for every pump photon coupled into the fiber. The power conversion efficiency for power amplifier is higher for 1480 nm (InGaAsP/InP-based lasers) pumping because the energy per pump photon is lower, based on $E_{ph} = hc/\lambda$, where E_{ph} is the photon energy, h is Planck's constant, c is the speed of light in vacuum, and λ is the wavelength.

Most of the pump power is consumed within the first portion from the ends, leaving the middle part of the fiber present with signal and ASE. At this part of the EDF, the ASE acts as a pump source on the signal. This is possible only when the signal wavelength at 1550 nm is on the long wavelength side of the ASE peak at 1530 nm. Thus, the bidirectional pumping scheme may be viewed as three serially connected amplifiers, with the middle amplifier being pumped by forward and backward propagating ASEs generated in the surrounding amplifiers. The major part of the gain improvement is indirectly achieved through the strong ASE attenuation in the middle part of the EDF [18]. This causes fewer pump photons to be used to amplify the spontaneous emission, thereby increasing the pump efficiency.

3.1.3 Filter Design

By introducing a filter in the MLFRL, the noise generated in the system could be reduced, especially the ASE generated by EDFA. It also provides stabilization against the energy fluctuation of the pulses propagating through the cavity. When the pulse energy increases beyond the design average, it will shorten the pulse width and broaden the spectrum due to strong nonlinear effects. Pulses with a broader spectrum experience excess loss and thus energy increases are reduced by filtering. The reverse is true for the decrease of energy. This energy equalization by filtering acts against the gain variation over the erbium bandwidth. However, filtering is associated with the noise penalty. The pulses generated by the laser require higher gain to compensate for the insertion loss of the filter, and the noise at the center frequency is not affected by the filter and sees the excess gain. This noise eventually affects the pulse generated.

Besides the noise reduction capability, the filter is also used for mode selection. The tunability of the filter facilitates the selectability of the operating center wavelength of the laser. The wavelength tuning range is chosen in such a way that it falls within the erbium amplification spectral region, which is in the conventional band (C-band) of the communication systems. With a 3 dB bandwidth of 1 nm ($\approx$125 GHz @ 1550 nm), there will be 12 longitudinal modes locked in the laser if a 10 GHz repetition pulse train is desired. However, this number is reduced to 3 for a 40 GHz pulse train. Based on the mode-locking equations shown in Chapter 2, the more longitudinal modes involved in the mode locking, N, the narrower the width and the higher the intensity of the pulse generated. As a result, this filter may not be suitable for 40 GHz operation. By changing to a wider bandwidth filter, a narrower pulse train could be obtained, however with an increase in the system noise level. This is the trade-off between the pulse width and the SNR of the system. Alternatively, the 40 GHz operation can be achieved by using the linear repetition rate multiplication, such as direct multiplication by fractional temporal Talbot effect, which will be evaluated in Chapter 9.

3.1.4 Modulator Design

Modulator is the essential component in the MLFRL. Amplitude modulator is selected due to its ability to generate a more stable pulse as compared to phase modulator, where the pulses have the tendency to switch randomly between two maximum phases of the modulation signal. Structurally, there are two main types of modulators, namely, bulk and integrated optic. Bulk modulators are made out of discrete pieces of a nonlinear optical crystal. They feature very low insertion losses and high power handling capability. Integrated-optic modulators use waveguide technology to lower the drive voltage. They are fiber pigtailed and compact. Due to their small size and compatibility with single-mode fiber, they have been chosen in our design of an active MLFL [19].

Both X-cut and Z-cut LiNbO$_3$ (integrated-optic modulator) are commonly used to fabricate modulators by in-diffusion of titanium to form waveguides. The other process for patterning single-mode optical waveguide on LiNbO$_3$ is annealed proton exchange (APE). In z-cut designs, the coplanar traveling wave electrode is placed over the optical waveguide. It is possible to reverse the polarity of the modulating electric field by offsetting the electrode, forming a phase reversal electrode structure. Velocity mismatch between the microwave and optical fields in the traveling wave device can be overcome over a large modulation frequency range by suitably choosing a pattern of phase reversals. In x-cut devices, phase reversal designs are not possible since the waveguide must be placed in the gap between the electrodes to utilize the strong r_{33} electro-optic coefficient. To achieve fast operation, the device must be made short to overcome the velocity mismatch.

The high-frequency response is limited mainly by the microwave-optical velocity mismatch, since the microwave loss in the short active length of the device is not large. The response at high frequencies can be improved by using a SiO$_2$ buffer layer between the electrode and the waveguide. The low dielectric constant of the buffer layer increases the microwave velocity nearer to the optical velocity, which reduces the velocity mismatch. Also, the buffer layer reduces the optical insertion loss by shielding the optical mode from the conductive electrodes [20].

At a lower frequency (<40 GHz), the x-cut structure is preferred due to its better modulation response. It is because the x-cut structure has a better overlap of the applied electric and the optical field, and the absence of phase reversals. However, when the operating frequency approaches 40 GHz, the z-cut device is more efficient than the x-cut one.

The optical transfer function of a LiNbO$_3$ electro-optic modulator can be expressed by Zhu et al. [21]

$$T(t) = 1 + \cos\left(\pi \frac{V_b + V_m \sin(\omega_m t)}{V_\pi}\right) \tag{3.2}$$

where

V_b is the bias voltage
V_m is the modulation amplitude
V_π is the half wavelength voltage that creates a π phase shift

For the most linear operation, the device must be biased half-on, i.e., in quadrature. Variation in the bias point with changing environmental conditions is a serious problem in LiNbO$_3$ devices.

Three main types of bias drift are generally reported: bias variation due to temperature changes via the pyroelectric effect, long-term drift after applying an external electrical signal to set the bias point, and optical damage to the device due to the photorefractive effect causing bias point changes [22]. The pyroelectric effect is the change of electric polarization in a crystal due to the change of temperature, it is much more severe in z-cut than in x-cut LiNbO$_3$. Over-coating z-cut devices with a conductive layer to bleed off surface charges significantly reduces the pyroelectric bias drift. However, the drift in x-cut devices is very small; hence, no temperature control is required. Electrically biasing a modulator produces a slow drift to the final bias value. This drift depends on the bias voltage applied and the time over which the bias is applied and the temperature. Finally, the photorefractive effect caused by the high optical power may contribute to the bias point drift too.

3.2 AM and FM Mode-Locked Erbium-Doped Fiber Ring Laser

3.2.1 AM Mode-Locked Fiber Lasers

Having known the main building blocks of an active MLFRL, we should now proceed to the actual setup of the laser. In an active MLFRL, the gain medium, modulator, filter, optical isolator, polarization controller, optical fiber couplers, and other associated optics are interconnected to form a closed loop optical path, and the details are given in the following section.

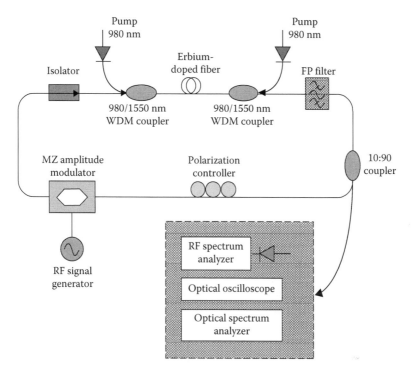

FIGURE 3.2
An active mode-locked erbium-doped fiber ring laser (similar to Figure 2.9).

By combining all the above design considerations, a dual-pump EDF ring laser, as shown in Figure 3.2, has been constructed. The laser design is based on a fiber ring cavity where a ~25 m EDF with Er+ ion concentration of 4.7×10^{24} ions/m^3 is pumped by two diode lasers at 980 nm: SDLO-27-8000-300 and CosetK1116 with maximum forward pump power of 280 mW and backward pump power of 120 mW. Bidirectional pumping will lead to higher gain efficiency than co-pumping and counter-pumping schemes. An optimum EDF length is chosen for maximal gain to ensure good performance of the amplifier, and its calculation is given in Appendix A.

The responsivities of the pump diode lasers are 0.7 and 0.6, respectively. The dark currents measured at -5 V is about 0.005 µA. The pump lights are coupled into the cavity by the 980/1550 nm WDM couplers; with insertion losses for the 980 and 1550 nm signals of about 0.48 and 0.35 dB, respectively. A polarization-independent optical isolator ensures the unidirectional lasing. The isolator provides a peak isolation of 47 dB and has an insertion loss of 0.3 dB. Besides that, the isolator can also be used to minimize the backward traveling ASE. The birefringence of the fiber is compensated by a PC. A tunable FP filter with 3 dB bandwidth of 1 nm and wavelength tuning range from 1530 to 1560 nm is inserted into the cavity to select the center wavelength of the generated signal as well as to reduce the noise in the system. In addition, it is used for the selection of longitudinal modes in the mode-locking process.

Pulse operation is achieved by introducing a JDS Uniphase 10 Gbps lithium niobate, Ti:LiNbO$_3$ Mach–Zehnder amplitude modulator into the cavity with half-wave voltage, V_π of 5.8 V. The modulator is DC biased near the quadrature point and not more than the V_π such that it operates on the linear region of its characteristic curve and driven by the sinusoidal signal derived from an Anritsu 68347 C Synthesized Signal Generator. The modulating depth should be <1 to avoid signal distortion. The modulator has an insertion loss of ≤7 dB. The output coupling of the laser is optimized using a 10/90 coupler. Ninety percent of the optical power is coupled back into the cavity ring loop, while the remaining portion is taken out as the output of the laser and is analyzed using a New Focus 1014B 40 GHz photodetector, Ando AQ6317B Optical Spectrum Analyzer, Textronix CSA 8000 80E01 50 GHz Communications Signal Analyzer, or Agilent E4407B RF Spectrum Analyzer.

By reducing the output coupling, it reduces the cavity loss and thus the round-trip gain, thereby lowering the ASE power. The environmental perturbations such as temperature change, mechanical vibration, pressure variation, etc., will introduce some random phase fluctuations that broaden the line width. The laser is most susceptible to these perturbations when the output coupling and, hence, the loss are high, so that by reducing the intra-cavity loss, the effect of these perturbations is minimized.

All connections between the components are made using connectors for easy access to different points of the system and fast component replacement. The loss introduced by each connector is assumed to be less than 0.5 dB. The splicing loss in the system is assumed small and can be neglected. Nevertheless, these "easy access connections" will give rise to certain amount of back reflections, which will disturb the system performance. With the help of an isolator, the effect of back reflection can be significantly reduced. The dispersion value of SMF used in our fiber laser construction is about 17 ps/nm/km (-21.7 ps^2/km) around 1550 nm, which is in the anomalous dispersion region, and hence suitable for soliton pulse generation.

3.2.2 FM or PM Mode-Locked Fiber Lasers

Optical ultrashort pulse sources are the most attractive block in the photonic signal processing systems. FM mode locking can be selected because the generated pulse is shorter than that in the AM mode locking scheme. A schematic of the experimental setup of the phase-modulated or FM MLFL is shown in Figure 3.3a. Two optical amplifiers (EDFAs) pumped at 980 nm can be used in the fiber loop to ensure the saturation of the circulating optical power in the loop under mode locking. The average power inside the cavity is controlled by changing the pump power or the driving current of the pump lasers. A phase modulator with 8 GHz bandwidth integrated in the

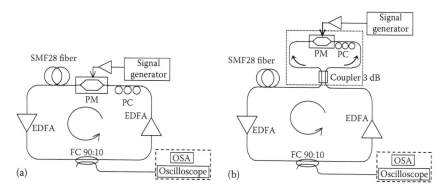

FIGURE 3.3
Experimental setup of (a) the FM MLFRL. PM, phase modulator; PC, polarization controller; OSA, optical spectrum analyzer. (b) MLFRL using a phase-modulated Sagnac loop. PM, phase modulator.

optically amplified fiber ring assumes the mode locking role and controls the phase-matching conditions of the generation of optical pulse sequence. In the current setup, we use the signal generator HP 8476A to drive the phase modulator by a sinusoidal electrical signal of about 1 GHz frequency after it is amplified 18 dB by a radio frequency amplifier (RFA). A PC at its input is used to control the polarization of the lightwaves that is very important to optimize the polarization and its effects on the mode-locking condition. A 50 m Corning SMF-8 fiber length can be inserted after the phase modulator to change the average dispersion of the cavity and to ensure that the average dispersion in the loop is anomalous. By tuning the modulation frequency at different harmonics order, the fundamental frequency of the fiber loop is determined to be 1.7827 MHz, thus the total loop length is equivalent to 114 m of single-mode optical fiber. The outputs of the MLL is obtained from the 90:10 fiber coupler and are monitored by an optical spectrum analyzer (OSA) HP70952B and a high-speed sampling oscilloscope Agilent DCA-J 86100C, which has an optical bandwidth of 65 GHz. We note that a wideband optical filter can be used in the ring in order to determine the center of the passband of the periodic pulse sequence. The bandwidth of this optical filter must be wide enough to accommodate the overall spectrum of the narrow pulses. The insertion of an optical filter will add a loss of energy in the loop, which requires a higher saturated power amplifier.

From above setup on the FM MLFL, we conducted an experimental investigation of FM mode locking. Under conventional conditions with a modulation frequency of 998.4 MHz, the laser was mode locked and the output repetition rate was around 1 GHz. When the amplified RF driving power was increased higher than 19 dBm, distortion of the driving signal waveform was observed and the higher-order harmonic components were enhanced in the electrical spectrum. At these high driving powers, rational harmonic mode locking is easily obtained with the multiplied repetition rate from two to four times when we detune the modulation frequency by proper amounts. By optimizing the mode-locking condition, the pulse width of output pulses of sub 10 ps can be attained in our setup.

By optimizing the polarization states together with optical power in the cavity, an interesting phenomenon is observed in our experimental system that is the multi-soliton bound states. When increasing average optical power levels in the cavity, the multi-soliton bound states such as the dual-, triple-, and quadruple-soliton bound states are observed one after the other. We believe that this is the first time that the multi-soliton bound states have been experimentally observed in the active MLFL. The significant advantage of this fiber laser is the easy generation of a multi-soliton bound sequence at the modulation frequency. Bound states are shown in Figure 3.4. Under the presence of active PM in the cavity, the formation mechanism of multi-soliton bound states in our setup is quite distinct from that in passively MLFLs. The stabilization of multi-soliton bound states through the balance of the repulsive and attractive forces between adjacent pulses is followed by splitting into multi pulses when the power in the cavity is raised above a specific mode-locking threshold. The influence of active PM on the bound states has also been initially investigated, which shows the important role of PM to the features of bound states such as time separation and splitting power levels. This preliminary research shows that many research issues of the active FM MLFL still remain to be dealt with.

The MLFL can be constructed using a phase-modulated Sagnac loop, as shown in Figure 3.3b. The phase-modulated Sagnac loop is characterized before inserting it into the fiber ring. In this configuration, the common mode-locked state is more stable than that of the FM MLFL due to the features of the loop mentioned above together with the filtering effect in the Sagnac loop. However, under high RF driving powers, the mode-locked pulses

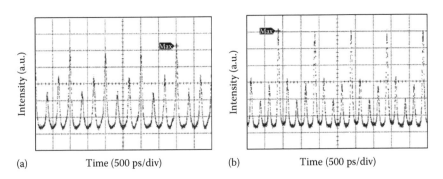

(a)   Time (500 ps/div)   (b)   Time (500 ps/div)

FIGURE 3.4
The unusual pulse sequences observed from the MLFL using the Sagnac loop. (a) Shows the temporal formation at the output of the MLFL which is about to reach the stable mode-locked sequence. Note the uneven distribution of the pulse sequence in the time domain. (b) Also displays the pulse sequence in which there are more smaller pulses between two peak pulses compared to (a); the pulse sequence is very near the final locking state.

showed the unusual behaviors when detuning the modulation frequency, as displayed in Figure 3.4. The waveforms in the time domain looked like the chaos, which could be due to the complicated transmittance characteristic of the Sagnac loop; however, the investigations on this characteristic can be considered as future work. The modeling of this PM MLFL is presented in Chapter 4. Further details on the generation of bound solitons by FM MLFLs are described in Chapter 10.

3.3 Regenerative Active Mode-Locked Erbium-Doped Fiber Ring Laser

One problem with the MLFRL is the thermal drift of the cavity length. With $\partial n/\partial T = 1.1 \times 10^{-5}/°C$ for silica fiber, where n is the refractive index and T is the temperature; for a typical fiber length of 30 m, the optical path length fluctuates by 0.33 mm/°C. This translates to a frequency fluctuation of about 110 kHz/°C for a 10 GHz modulation frequency. Accordingly, slight thermal variations are able to cause the pulses to lose synchronism with the modulator. Other than the thermal fluctuations, fiber lasers are sensitive to small environmental perturbations, such as polarization change, acoustic and mechanical vibrations.

Hence, the lasers do not exhibit long-term stability without stabilization techniques. One of the most obvious stabilization schemes is the dynamic adjustment of the cavity length. In 1993, Shan and Spirit [23] demonstrated a stabilization technique using a PZT. A portion of the fiber in the cavity was wound around a piezoelectric drum. By driving the drum with the proper error signal, the cavity length was adjusted to keep the cavity frequency synchronous with the fixed modulator drive frequency. The control of the cavity technique in MHz range would be difficult due to the need for either a high-voltage amplifier to drive the PZT or a PZT with a high expansion coefficient. Another technique utilizing the NPR effect to suppress the pulse amplitude fluctuations, has been proposed by Doerr et al. in 1994 [24]. Alternatively, regenerative feedback where the modulation frequency is derived directly from the pulse train can be used to stabilize the modulation. Any change in the cavity length automatically adjusts the modulation frequency to maintain the pulse-modulator synchronism. The use of the regenerative feedback technique to stabilize the fiber laser was first demonstrated by Nakazawa et al. [25].

Both amplitude and phase noise of an active harmonically MLFRL and a regenerative mode-locked fiber ring laser (RMLFRL) have been well studied by Gupta et al. [7] using the spectral domain technique. It has been found that the regenerative configuration exhibits superior performance compared to the active structure, both in terms of reduced amplitude noise and phase noise.

3.3.1 Experimental Setup

The experimental setup of the proposed RMLFRL, shown in Figure 3.5, is constructed. Figure 3.6a and b illustrates the hardware arrangement of four different types of MLFLs with repetition rate from 10 to 40 GHz. Figure 3.6c and d shows a typical temporal sequence and its corresponding spectrum as observed at the output of the fiber laser. Compared with Figure 3.2, the newly added components are the clock recovery unit, RFA and modulator driver. The basic operation of the RMLFRL is similar to that of MLFRL, shown in Figure 3.2. The difference between these two schemes is the source of the modulation signal. In MLFRL, the modulating signal is directly taken from the signal generator; however, the modulating signal of RMLFRL is the feedback signal generated by the clock recovery unit. This will improve the laser stability as described before. The clock recovery unit is used to generate an RF sinusoidal wave at about 9.953 GHz (OC-192 standard) in order to drive the modulator prior to the modulator driver and RFA, which are used to amplify the RF signal in the feedback loop. Higher frequency clock recovery circuits (OC-768 standard) can be used for 40 GHz operation. However, cost is one of the main considerations for this high-frequency setup.

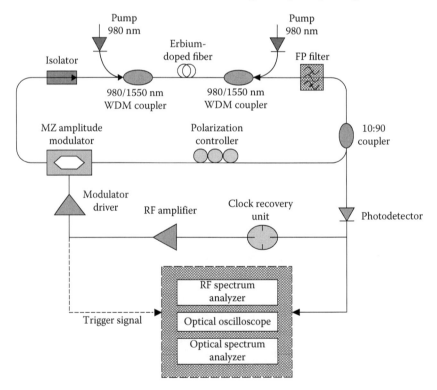

FIGURE 3.5
Regenerative active MLFRL.

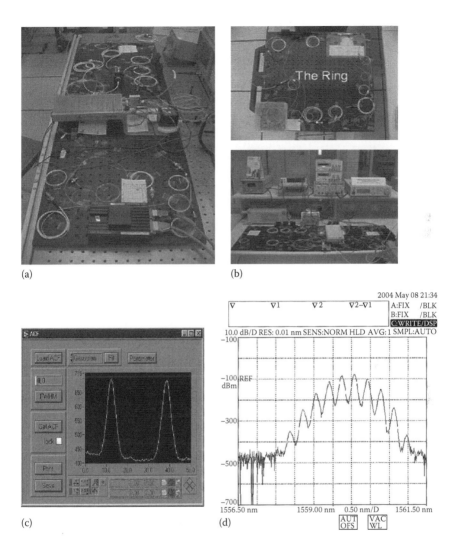

FIGURE 3.6
Experimental arrangement of MLFLs (a) 40 GHz, (b) 10 GHz and experimental test bed, (c) pulse sequence at 40 GHz repetition rate, and (d) spectrum of generated pulse sequence at 40 GHz repetition rate.

By using this optical-RF feedback to control the modulation of the intensity modulator, the optical noise at the output of the laser must be significantly greater than that of the electronic noise for the startup of the mode-locking and lasing. In other words, the loop gain of the optical-electronic feedback loop must be greater than unity. Thus, it is necessary for the EDFA to operate in the saturation mode and the total average optical power circulating in the loop must be sufficiently adequate for detection at the photodetector and electronic preamplifier. Under this condition, the optical quantum shot noise dominates over the electronic shot noise.

3.3.2 Results and Discussion

The experimental procedures and precautions for the MLFRL and RMLFRL are given in Appendix A. A systematic experimental procedure must be followed in order to ensure the accuracy of the results and to avoid unnecessary faults or problems in the later stage of the experiments. Extra care must be given to the handling of the equipment and components to prevent possible damages, due to their high operating frequency and sensitivity.

3.3.2.1 Noise Analysis

There are two major sources of noise generated at the input of the photodetector, namely, the optical quantum shot noise generated by the detection of the optical pulse trains and the electronic random thermal noise generated by the small-signal electronic amplifier following the detector. The optical quantum shot noise spectral density (A^2/Hz) is defined as $S_S = 2qRP_{av}$ where R is the responsivity of the photodetector, q is the electronic charge, and P_{av} is the average optical power of the input signal. Suppose that a 1 mW (or 0 dBm) average optical power is generated at the output of the MLL and with a detector responsivity of 0.8, then the optical quantum shot noise spectral density will be 2.56×10^{-22} A^2/Hz.

Usually, the electronic amplifier will have a 50 Ω equivalent input resistance, R_{eq}, referred to the input of the optical preamplifier as evaluated at the operating repetition frequency, this gives a thermal noise spectral density of $S_R = 4kT/R_{eq}$ where k is the Boltzmann's constant, which equals to 3.312×10^{-22} A^2/Hz at 300 K.

Depending on the electronic bandwidth, B_e of the electronic preamplifier and optical bandwidth, B_o of the photodetector, the total equivalent optical quantum noise and electronic thermal noise (square of noise current) are given as follows:

$$i_{NT}^2 = S_R B_e$$

$$i_{NS}^2 = S_S B_o$$

(3.3)

Under the worst-case scenario, when a wideband amplifier of a 3 dB electrical and optical bandwidth of 10 GHz, the equivalent electronic noise at the input of the electronic amplifier is 3.312×10^{-11} A^2, i.e., an equivalent noise current of 5.755 μA and the equivalent optical noise is 2.56×10^{-11} A^2, i.e., an equivalent noise current of 5.060 μA are present at the input of the clock recovery unit. This corresponds to an SNR of 40.4 dB. If a narrowband amplifier of 50 MHz 3 dB bandwidth centered at 10 GHz is employed, the equivalent electronic and optical noises current will be 0.129 and 0.113 µA, respectively, and the corresponding SNR will be 73.3 dB, which is an impressive value.

The average optical power generated by the MLL must be high in order to obtain a high SNR at the input of the photodetector. In order to counteract with the noises and losses in the ring laser, an EDFA with high output power and saturation power should be employed.

3.3.2.2 Temporal and Spectral Analysis

With a +17 dBm ($V_{rms} = 1.58$ V, $V_{peak} = 2.24$ V) modulating signal of about 9.954365 GHz applied to the MZM that is biased near the quadrature point (around 2.9 V) on its transfer characteristic curve, the observed pulse train and spectrum of MLFRL are shown in Figure 3.7. As for RMLFRL, the frequency and power level of the modulating signal are set by the clock recovery

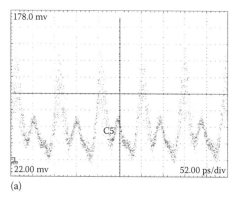

(a)

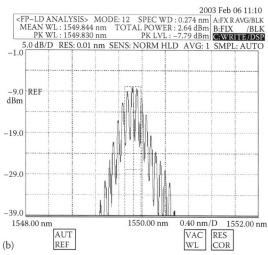

(b)

FIGURE 3.7
(a) Optical pulse train and (b) spectrum of MLFRL.

unit with clock frequency of about 9.953 GHz ± 10 MHz and modulator driver with saturation power of +23 dBm. The corresponding output waveforms are shown in Figure 3.8. The polarization controller prior to the modulator was adjusted to a desired position that gave maximum output intensity. The EDFA was pumped at the deep saturation region to ensure full inversion of the erbium ions. The shortest and most stable optical pulses were obtained by the careful adjustment of the RF drive frequency for the MLFRL and the polarization of the circulating light. As discussed before, RMLFRL provides better stability as compared to MLFRL. Higher peak power is obtained in the RMLFRL configuration, which reduces the number of pulses being dropped out, and has better system stability.

The time-bandwidth products for MLFRL and RMLFRL are 0.38 and 0.50, respectively; close to the transform-limited value (0.315 for a sech^2 pulse and

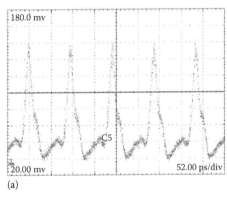

(a)

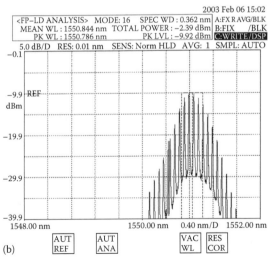

(b)

FIGURE 3.8
(a) Optical pulse train and (b) spectrum of RMLFRL.

0.441 for a Gaussian pulse), however, with a bit of frequency chirping. This nonlinear chirping in the initial mode-locked pulses will degrade the pulse quality of the multiplied pulses if a repetition rate multiplication technique (i.e., fractional temporal Talbot effect) is employed to increase the system repetition rate.

Figure 3.7a and b shows the optical pulse train and its corresponding spectrum of the MLFRL. While Figure 3.8a and b shows the optical pulse train and its corresponding spectrum of an RMLFRL.

A slight wavelength or frequency shift was observed in the RMLFRL. This is because the RF modulating frequency may vary within the bandwidth of the clock recovery unit; and hence, the lasing mode may be locked to another mode of the ring cavity.

A signal must be generated and used for the purpose of triggering the oscilloscope for the observations of the optical pulse train. For Tektronix CSA 8000 80E01 digital sampling oscilloscope, the minimum trigger signal must be >600 mV. This is critical for the RMLFRL, because the RF signal detected and phase locked via the clock recovery circuitry have to be split to generate this triggering signal. As a result, the RF signal generated by the clock recovery unit that will be amplified by the RF amplifier must be at least 3 dB more than the minimum power required by the oscilloscope.

3.3.2.3 Measurement Accuracy

To maintain the fidelity of the measurements, every component in the system needs to have a bandwidth greater than the 3 dB bandwidth of the signal, or, equivalently, an impulse response faster than the fastest part of the signal. For a transform-limited secant or Gaussian pulse, the equipment's 3 dB bandwidth should be greater than 0.315/FWHM (for secant pulse) or 0.441/FWHM (for Gaussian pulse), where FWHM is the full width at half maximum. From the measurements taken (as shown in Figures 3.7 and 3.8), the signal bandwidth of both configurations are ~34 and ~45 GHz, respectively, which are within the 3 dB bandwidth of the photodetector (45 GHz) and the oscilloscope (50 GHz). As for the signals' bandwidths that are greater than the equipment's 3 dB bandwidth, the measurement made by the equipment was incorrect because the pulse width shown on the scope depends on the convolution of many bandwidths, including those of the signal, PD and the oscilloscope. However, the pulse width of the signal can still be estimated using the following relationship:

$$\tau_{measured} = \sqrt{\tau_{optical}^2 + \tau_{photodetector}^2 + \tau_{jitter}^2 + \tau_{electrical}^2} \qquad (3.4)$$

where

$\tau_{measured}$ is the measured signal pulse width on the oscilloscope
$\tau_{optical}$ is the optical pulse width

$\tau_{\text{photodetector}}$ is the response time (or rise/fall time) of the photodetector

τ_{jitter} is the timing jitter, including pulse-to-pulse timing fluctuation of the laser, synthesizer jitter, etc.

$\tau_{\text{electrical}}$ is the response time (or rise/fall time) of the electrical equipment

- Rise time (10%–90%) of the photodetector, $\tau_{\text{photodetector}} = 9\,\text{ps}$
- Rise time of the Tektronix CSA 8000, 80E01 module, $\tau_{\text{electrical}} = 7\,\text{ps}$
- Time jitter of CSA, $\tau_{\text{jitter}} = 2.5\,\text{ps}$ when locked to a 10 MHz reference frequency

Other important factors to keep in mind are the bandwidths of the cables, connectors, and pulse-to-pulse jitter of the laser. Because the sampled oscilloscope trace is made up of data taken from many different pulses, timing jitter can broaden the measured signal. The pulse widths measured for nonfeedback and feedback loop are 11.12 and 11.10 ps, respectively.

3.3.2.4 EDF Cooperative Up-Conversion

The output characteristics of EDFA used in the systems are shown in Figure 3.9. A saturation power of about 18 dBm and a maximum gain of about 30 dB and a noise figure of about 7 dB were obtained through measurements. The power conversion efficiency of EDF is about 50%.

Both nonfeedback and feedback configurations are operating in the deep saturation mode to ensure a sufficiently high power level to sustain the nonlinear effects within the ring. However, it may have some side effects due to this action. The amplifier efficiency may degrade due to $Er^{3+}-Er^{3+}$ ion–ion interaction, through cooperative up-conversion. When two excited ions interact, one can transfer its energy to the other, leaving itself in the ground state and the other in the higher $^4I_{9/2}$ state. In oxide glasses, the $^4I_{9/2}$ level quickly relaxes through multi-phonon emission back to the $^4I_{13/2}$ state, the net result of the process is to convert one excitation into heat. Since cooperative up-conversion requires two interacting ions in the excited state, it will not be evident at low pumping levels. Nevertheless, at high pump powers it will appear as accelerated and nonexponential decay. This up-conversion process leads to loss of inversion without the emission of a stimulated photon.

Polarization hole burning is another important effect that is related to the saturation properties of EDFA. Normally, the amplifier gain is polarization insensitive. However, if one polarization saturates the amplifier, a slight excess gain is left over in the other polarization. Noise can grow in this polarization and affect the system performance.

3.3.2.5 Pulse Dropout

When the energy in the cavity is very low, the nonlinear effects can be neglected. This results in the noisy pulses, exhibiting the amplitude and

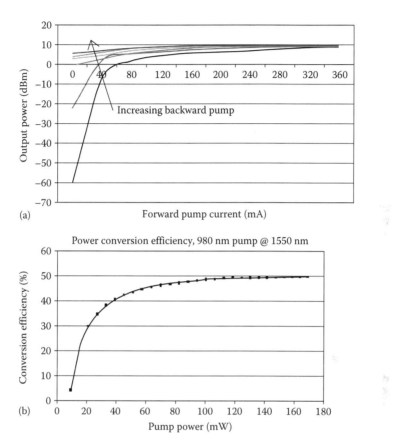

FIGURE 3.9
(a) Output characteristics of a 980 nm dual-pump 25 m EDF with input power = −20 dBm @ 1550 nm and (b) power conversion efficiency.

pulse shape fluctuations. This noisy behavior, which leads to supermode competition in the frequency domain, is due to the lack of stabilization from the nonlinear effects, so that the pulse amplitude can change without affecting its bandwidth. When the pump power is increased to a level where the nonlinear effects become important, it will eventually decrease the pulse duration. The direct interaction between pulses in the active MLL is weak due to the presence of the modulator; however, the pulses all interact indirectly via the amplifier. Because of the slow response time of the EDFA (millisecond life time of erbium), the pulses all affect the amplifier saturation, leading to highly complex dynamics. The complex interaction of all the pulses with the gain medium and the mode locker leads to the drop out of some pulses and others to stabilize with similar shapes and amplitudes. The dropout occurs since some pulses can decrease their loss in the mode locker by increasing their energy and decreasing their

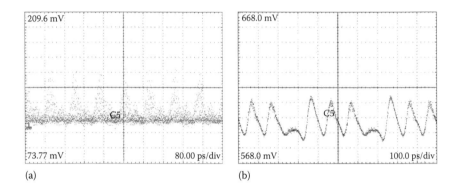

FIGURE 3.10
(a) Noisy pulses due to insufficient pump power; (b) pulse dropout in an MLL (i.e., two pulses disappear from this plot).

duration due to the soliton shortening effect. The increase of the energy of those pulses decreases the amplifier gain due to the amplifier saturation effects and causes the net loss in a round trip to other pulses that will eventually drop, while the remaining pulses will possess nearly identical shapes and amplitudes. Figure 3.10 shows the average output pulse train with the pulse dropout effect. As the pump power increases, the fraction of pulses that drop out decreases. However, it is expected that the number of filled slots also depends on the previous state of the laser. The amplitude of the pulses will be stabilized when all the slots are filled, and this is achieved by a higher pump power. When the repetition rate increases, the minimum average power that is needed for stable operation will increase exponentially.

The output pulse train of the harmonically MLFRL will experience some pulse dropouts if the operating power is below the minimum required power for the stable operation [26]:

$$P_{min} = \sec h^{-1}\left(\frac{1}{\sqrt{2}}\right)\frac{ED\omega_m^{3/2}}{\pi\gamma_{nl}}\left[\frac{8M\omega_g^2\langle x^2\rangle}{G\langle\omega^2\rangle}\right]^{1/4} \tag{3.5}$$

where
    E is the energy enhancement factor
    D is the average dispersion
    M is the modulation depth
    ω_m is the modulation frequency
    ω_g is the amplifier bandwidth
    γ_{nl} is the nonlinear coefficient
    G is the average gain coefficient

$$\left\langle x^2 \right\rangle = \frac{\int_{-\infty}^{\infty} |f(x)|^2 \, x^2 dx}{\int_{-\infty}^{\infty} |f(x)|^2 \, dx}, \quad \left\langle \omega^2 \right\rangle = \frac{\int_{-\infty}^{\infty} |df(x)/dx|^2 \, dx}{\int_{-\infty}^{\infty} |f(x)|^2 \, dx}, \tag{3.6}$$

and $f(x)$ is the pulse shape function.

For hyperbolic secant pulses,

$$\frac{\left\langle \omega^2 \right\rangle}{\left\langle x^2 \right\rangle} = \frac{4 \left[2 \sec h^{-1} (2^{-1/2}) \right]^4}{\pi^2}. \tag{3.7}$$

For Gaussian pulses,

$$\frac{\left\langle \omega^2 \right\rangle}{\left\langle x^2 \right\rangle} = [4 \ln 2]^2 \tag{3.8}$$

Another possible reason for this pulse dropout is the supermode competition [25] between the modes locked within the laser. With a total cavity length of about 30 m, which comprises ~25 m of EDF and ~5 m of SMF; the fundamental cavity frequency is about 6.8966 MHz. Note that n is assumed to take a typical value of 1.45 for a silica-based fiber. For a repetition rate of 10 GHz, the locking will occur around the 1450th harmonic mode. Approximately, there are 1450 pulses simultaneously propagating inside the laser cavity; some of them may drop from the pulse train and cause errors in the system. In RMLFRL, the output pulses will be used to drive the amplitude modulator of the laser. The clock information will be successfully extracted by the clock recovery unit even with the pulse dropout.

3.4 Ultrahigh Repetition-Rate Ultra-Stable Fiber Mode-Locked Lasers

This section gives a detailed account of the design, construction, and characterization of a MLFRL. The MLL structure employs in-line optical fiber amplifiers, a guided-wave optical MZIM and associate optics to form a ring resonator structure, generating optical pulse trains of several gigahertz repetition rate with pulse duration in the order of picoseconds. Long-term stability of amplitude and phase noises has been achieved that indicates that the optical pulse source can produce an error-free Pseudo Random Binary

Sequence (PRBS) pattern in a self-locking mode for more than 20 h. An MLL operating at 10 GHz repetition rate has been designed, constructed, tested, and packaged. The laser generates an optical pulse train of 4.5 ps pulse width when the modulator is biased at the phase quadrature quiescent region. Preliminary experiment of a 40 GHz repetition rate MLL has also been demonstrated. Although it is still unstable in long term, without an O/E feedback loop, optical pulse trains have been observed.

3.4.1 Regenerative Mode-Locking Techniques and Conditions for Generation of Transform-Limited Pulses from a Mode-Locked Laser

3.4.1.1 Schematic Structure of MLRL

Figures 3.11 and 3.12 show the composition of an MLRL without and with feedback loop used in this study, respectively. It principally consists of a nonfeedback ring, an optical close loop with an optical gain medium, an optical modulator (intensity or phase type), an optical fiber coupler, and associated optics. An O/E feedback loop detecting the repetition rate signal and generating RF sinusoidal waves to electro-optically drive the intensity modulator is necessary for the regenerative configuration, as shown in Figure 3.12.

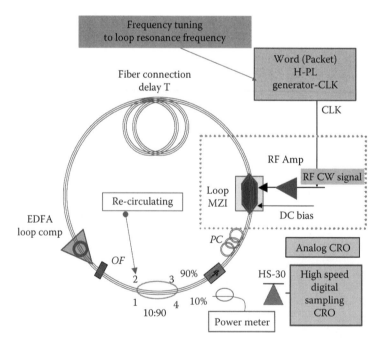

FIGURE 3.11
Schematic arrangement of a mode-locked ring laser without the active feedback control.

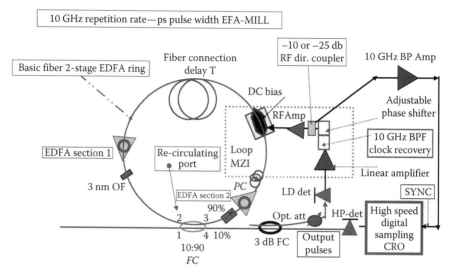

FIGURE 3.12
Schematic arrangement of a mode-locked ring laser with an O/E-RF electronic active feedback loop.

3.4.1.2 Mode-Locking Conditions

The basic conditions for MLRL to operate in the pulse oscillation mode are described below:

For nonfeedback optical mode locking

Condition 1: The total optical loop gain must be greater than unity when the modulator is in the ON-state, i.e., when the optical waves transmitting through the MZIM is propagating in phase.

Condition 2: The optical lightwaves must be depleted when the optical modulator is in the OFF-state, i.e., when the lightwaves of the two branches of the MZIM are out of phase with each other (e.g., 180°) or in a destructive interference mode.

Condition 3: The frequency repetition rate at a locking state must be a multiple integer of the fundamental resonant frequency of the ring.

For optical-RF feedback mode locking—Regenerative mode locking.

Condition 4: To control the modulation of the intensity modulator under an O/E-RF feedback, the optical noise at the output of the laser must be significantly greater than that of the electronic noise for the startup of the mode locking and lasing. In other words, the loop gain of the optical-electronic feedback loop must be greater than unity.

Thus, it is necessary that the EDF amplifiers are operated in the saturation mode and the total average optical power of the lightwaves circulating in the loop must be sufficiently adequate for detection by the photodetector and the electronic preamplifier. Under this condition, the optical quantum shot noise dominates the electronic shot noise.

3.4.1.3 Factors Influencing the Design and Performance of Mode Locking and Generation of Optical Pulse Trains

The locking frequency is a multiple integer of the fundamental harmonic frequency of the ring, which is defined as the inverse of the traveling time around the loop and is given by

$$f_{RF} = \frac{Nc}{n_{eff}L} \tag{3.9}$$

where

f_{RF} is the RF frequency required for locking and the required generation rate
N is an integer and indicates mode number order
c is the velocity of light in vacuum
n_{eff} is effective index of the guided propagating mode
L is the loop length including those of the optical amplifiers

Under the requirement of the OC-192 standard bit rate, the locking frequency must be in the region of 9.95 Giga-pulses per second. That is, the laser must be locked to a very high order of the fundamental loop frequency that is in the region of 1–10 MHz depending on the total ring length. For an optical ring of length of about 30 m and a pulse repetition rate of 10 GHz, the locking occurs at approximately the 1400th harmonic mode.

It is also noted that the effective refractive index n_{eff} can be varied in different sections of the optical components forming the laser ring. Furthermore the two polarized states of propagating lightwaves in the ring, if the fibers are not of a polarization maintaining type, would form two simultaneously propagating rings, and they could interfere or hop between these dual polarized rings.

The pulse width, denoted as $\Delta\tau$, of the generated optical pulse trains can be found to be given by Siegman [1]

$$\Delta\tau = 0.45 \left(\frac{\alpha_t G_t}{\Delta_m} \right)^{1/4} \frac{1}{(f_{RF}\Delta v)^{1/2}} \tag{3.10}$$

where

$\alpha_t G_t$ is the round trip gain coefficient, which is the product of all the loss and gain coefficients of all optical components including their corresponding fluctuation factor
Δ_m is the modulation index
Δv is the overall optical bandwidth (in units of Hz) of the laser

Hence the modulation index and the bandwidth of the optical filter influence the generated pulse width of the pulse train. However the optical characteristics of the optical filters and optical gain must be flattened over the optical bandwidth of the transform limit for which a transform-limited pulse must satisfy, for a sech² pulse intensity profile, the relationship

$$\Delta\tau\Delta v = 0.315 \tag{3.11}$$

Similarly, for a Gaussian pulse shape, the constant becomes 0.441.

The fluctuation of the gain or loss coefficients over the optical flattened region can also influence the width and mode-locking condition of the generated optical pulse.

In the regenerative mode-locking case as illustrated in Figure 3.12, where the optical output intensity is split and O/E detected, we must consider the sensitivity of and noises generated by the photodetector (PD). Two major sources of noises are generated by the PD: first, the optical quantum shot noises generated by the detection of the optical pulse trains and the random thermal electronic noises of the small-signal electronic amplifier following the detector. Usually, the electronic amplifier would have a 50 Ω equivalent input resistance R referred to the input of the optical preamplifier as evaluated at the operating repetition frequency, this gives a thermal noise spectral density of

$$S_R = \frac{4kT}{R} A^2/Hz \tag{3.12}$$

with k the Boltzmann's constant. This equals to 3.312×10^{-22} A²/Hz at 300 K. Depending on the electronic bandwidth B_e of the electronic preamplifier, i.e., wideband or narrow-band type, the total equivalent electronic noise (square of noise "current") is given by, $i_{NT}^2 = S_R B_e$. Under the worst case when a wideband amplifier of a 3 dB electrical bandwidth of 10 GHz is used, the equivalent electronic noise at the input of the electronic amplifier is 3.312×10^{-11} A², i.e., an equivalent noise current of 5.755 µA is present at the input of the "clock" recovery circuit. If a narrow band-pass amplifier of 50 MHz 3 dB bandwidth centered at 10 GHz is employed, this equivalent electronic noise current is 0.181 µA.

Now considering the total quantum shot noise generated at the input of the "clock" recovery circuit, suppose that a 1.0 mW (or 0 dBm) average optical power is generated at the output of the MLRL, then a quantum shot noise* of approximately 2.56×10^{-22} A²/Hz (that means an equivalent electronic noise current of 16 nA) is present at the input of the clock recovery circuit. This quantum shot noise current is substantially smaller than that of the electronic noise.

---

* A quantum shot noise can be determined using the relationship of the quantum noise spectral density of $2qRP_{av}$ with P_{av} the average optical power, q the electronic charge, and R the responsivity of the detector.

In order for the detected signal at the optical receiver incorporated in the "clock" recovery circuit to generate a high SNR, the optical average power of the generated pulse trains must be high, at least at a ratio of 10. We estimate that this optical power must be at least 0 dBm at the PD in order for the MLRL to lock efficiently to generate a stable pulse train.

Given that a 10% fiber coupler is used at the optical output and an estimated optical loss of about 12 dB due to coupling, connector loss and attenuation of all optical components employed in the ring, the total optical power generated by the amplifiers must be about 30 dBm. To achieve this, we employ two EDFAs of 16.5 dBm output power each positioned before and after the optical coupler; one is used to compensate for the optical losses and the other for generating sufficient optical gain and power to dominate the electronic noise in the regenerative loop.

3.4.2 Experimental Setup and Results

The experimental setups for MLL and RMLL are as shown, again, in Figures 3.11 and 3.12. The associated equipment used for the monitoring of the mode locking and the measurement of the lasers are also included. However, we note the following: (1) in order to lock the lasing mode of the MLL to a certain repetition rate or a multiple of the harmonic of the fundamental ring frequency, a synthesizer is required to generate the required sinusoidal waves for modulating the optical intensity modulator and tuning it to the harmonic of the cavity fundamental frequency; and (2) a signal must be created for the purpose of triggering the digital oscilloscope to observe the locking of the detected optical pulse train. For the HP-54118A, the amplitude of this signal must be >200 mV. This is also critical for the RMLL setup as the RF signal detected and phase locked via the clock recovery circuitry must be split to generate this triggering signal.

Typical experimental procedures are as follows: (1) after the connection of all optical components with the ring path broken, ideally at the output of the fiber coupler, a CW optical source can be used to inject optical waves at a specific wavelength to monitor the optical loss of the ring; (2) close the optical ring and monitor the average optical power at the output of the 90:10 fiber coupler and hence the optical power available at the PD was estimated at about −3 dBm after a 50:50 fiber coupler; (3) determine whether an optical amplifier is required for detecting the optical pulse train or whether this optical power is sufficient for O/E RF feedback condition as stated above; (4) set the biasing condition and hence the bias voltage of the optical modulator; and (5) tune the synthesizer or the electrical phase to synchronize the generation and locking of the optical pulse train.

The following results can be obtained: (1) Optical pulse train generated at the output of the MLL or RMLL. Experimental setup is shown in Figures 3.13 and 3.14. (2) Synthesized modulating sinusoidal waveforms can be monitored, as shown in Figures 3.15 through 3.17. Figure 3.15 illustrates the

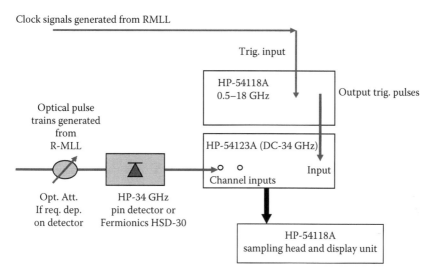

FIGURE 3.13
Experimental setup for monitoring the locking of the photonic pulse train.

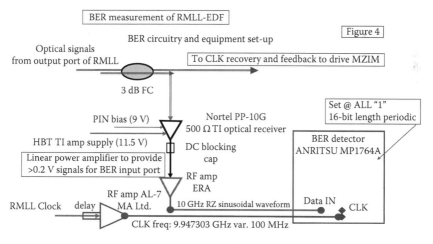

FIGURE 3.14
Experimental setup for monitoring the BER of the photonic pulse train.

mode locking of an MLL operating at around 2 GHz repetition rate with the modulator driven from a pattern generator while Figures 3.16 and 3.17 show the generated sinusoidal waveforms when the MLRL was operating at the self-mode-locking state. (3) The interference of other supermodes of the MLL without RF feedback for self-locking is indicated in Figure 3.15. (4) Observed optical spectrum (not available in electronic form). (5) Electrical spectrum of the generated pulse trains was observed, showing a −70 dB supermode suppression under the locked state of the RMLRL.

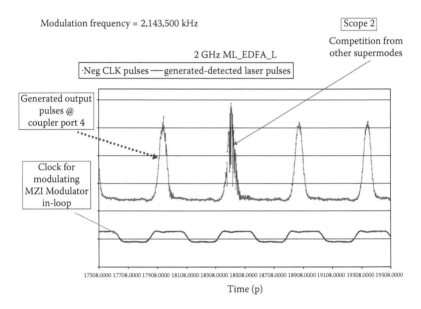

FIGURE 3.15
Detected pulse train at the MLRL output tested at 2 GHz repetition frequency. Note that the unit p in the time axis represents picosecond.

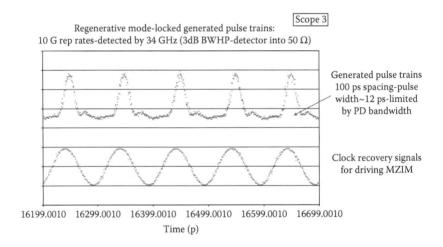

FIGURE 3.16
Output pulse trains of the regenerative MLRL and the RF signals as recovered for modulating the MZIM for self-locking. Note that the unit p in the time axis represents picosecond.

(6) Figures 3.16 and 3.18 show that the RMLRL can be operated under the cases when the modulator is biased either at the positive or at the negative going slope of the optical transfer characteristics of the Mach–Zehnder modulator. (7) Optical pulse width was measured using an optical auto-correlator (slow or fast scan mode). Typical pulse width obtained with the

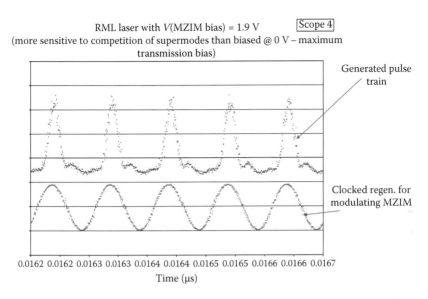

FIGURE 3.17
Detected output pulse trains of the regenerative MLRL and the recovered clock signal when the MZIM is biased at a negative going slope of the operating characteristics of the modulator.

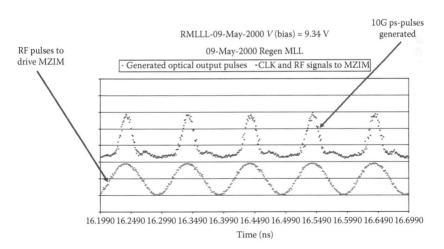

FIGURE 3.18
Output pulse trains and clock-recovered signals of the 10 GHz RMLRL when the modulator is biased at the positive going slope of the modulator's operating transfer characteristics curve.

slow scan auto-correlator is shown in Figure 3.19. Minimum pulse duration obtained was 4.5 ps with a time-bandwidth product of about 3.8, showing that the generated pulse is near transform limited. (8) BER measurement was used to monitor the stability of the RMLRL. The BER error detector was then programmed to detect all "1" at the decision level at a

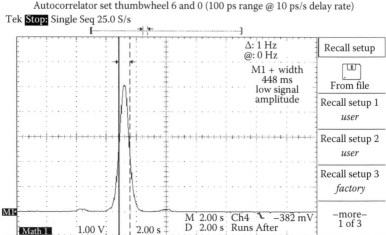

FIGURE 3.19
Autocorrelation trace of output pulse trains of the 9.95 GHz regenerative MLRL.

tuned amplitude level and phase delay. The clock source used is that produced by the laser itself. This setup is shown in Figure 3.14. The opto-electrically detected waveform of the output pulse train for testing the BER is shown in Figure 3.20, after 20 h operation, where the recorded waveform is obtained under the infinite persistence mode of the digital oscilloscope. (9) A drift of clock frequency of about 20 kHz over in 1 h in open laboratory environment is observed. This is acceptable for a 10 GHz repetition rate. (10) The "clock" recovered waveforms were also monitored at the initial locked state and after the long-term test, as shown in Figures 3.16 and 3.17, respectively. Figure 3.17 was obtained under the infinite persistence mode of the digital oscilloscope.

We note the following factors that are related to the above measurements (Figures 3.15 through 3.17): (1) all the above measurements have been conducted with two distributed optical amplifiers (GTi EDF optical amplifiers) driven at 180 mA and a specified output optical power of 16.5 dBm and (2) optical pulse trains are detected with 34 GHz 3 dB bandwidth HP pin detector directly coupled to the digital oscilloscope without using any optical preamplifier.

3.4.3 Remarks

An MLL operating under an open loop condition and with O/E RF feedback providing regenerative mode locking can be constructed. The O/E feedback

BER measurement
BER = 0 × 10−15 measured for over 20 h-
RF (CLK) frequency varied from 9.954 to 9.952 GHz gradually over measurement period
BER measurement-clock signal-infinite persistence

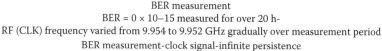

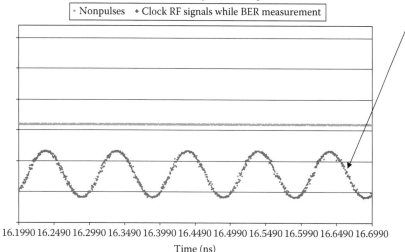

16.1990 16.2490 16.2990 16.3490 16.3990 16.4490 16.4990 16.5490 16.5990 16.6490 16.6990

Time (ns)

FIGURE 3.20

BER measurement—O/E detected signals from the generated output pulse trains for BER test set measurement. The waveform was obtained after a continuous duration of 20 h.

can certainly provide a self-locking mechanism under the condition that the polarization characteristics of the ring laser are manageable. This is done by ensuring that all fiber paths are under constant operating condition. The RMLRL can self-lock even under the DC drifting effect of the modulator bias voltage (over 20 h).* The generated pulse trains of 4.5 ps duration can be, without any difficulty, compressed further to less than 3 ps for 160 Gbps optical communication systems.

The RMLRL can be an important source for all-optical switching of an optical packet switching system.

We recommend the following for the RMLRL: (1) eliminating the polarization drift through the use of Faraday mirror or all-polarization maintaining (PM) optical components, for example, polarized Er-doped fiber amplifiers, PM fibers at the input and output ports of the intensity modulator; (2) stabilizing the ring cavity length with an appropriate packaging and via piezoelectric or thermal control to improve the long-term frequency drift; (3) controlled and automatic tuning of the DC bias voltage of the intensity modulator; (4) developing an electronic RF "clock"-recovery circuit for the RMLRL to operate at 40 GHz repetition rate together with an appropriate

---

* Typically, the DC bias voltage of an LiNbO$_3$ intensity modulator is drifted by 1.5 V after 15 h of continuous operation.

polarization control strategy; (5) studying the dependence of the optical power circulating in the ring laser by varying the output average optical power of the optical amplifiers under different pump power conditions; and (6) incorporating a phase modulator, in lieu of the intensity modulator, to reduce the complexity of the polarization dependence of the optical waves propagating in the ring cavity, thus minimizing the bias drift problem of the intensity modulator.

Current system operations for long-haul optical communications systems require higher bit rate, 40 Gbps or even higher up to 100 and 160 Gbps. The construction and testing of a RMLRL at 40 GHz repetition rate, the RMLL demands much more care on the electrical path of the feedback signals in order to ensure the matching of the sinusoidal waves applied to the modulator.

3.5 Conclusions

Mode-locked fiber ring lasers operating under an open loop condition and with RF feedback providing regenerative mode locking have been demonstrated. The O/E RF feedback can certainly provide a self-locking mechanism under the condition that the polarization characteristic of the ring laser is controllable. This could be implemented by ensuring that all the fiber paths are under the constant operating condition. The RMLFRL can self-lock itself even under the DC drifting effect of the modulator bias voltage (over several hours).

In order to successfully generate stable ultrafast and ultrashort pulses, the modes in the MLFL must retain their phase relationship all the time so that the mode mocking is stable. This requires that all modes experience the same traveling time around the cavity. In other words, the optical path length of the cavity is independent of the signal frequency. In general, this will not be the case due to the presence of frequency-dependent dispersion in the cavity. That is, the pulse will experience GVD as it travels through the cavity. Therefore, the dispersion value of the cavity should be kept at minimum.

References

1. A.E. Siegman, *Lasers*, University Science Books, Mill Valley, CA, 1986.
2. K.S. Abedin and F. Kubota, Wavelength tunable high-repetition-rate picosecond and femtosecond pulse sources based on highly nonlinear photonic crystal, *IEEE J. Sel. Top. Quantum Electron.*, 10(5), 1203–1210, 2004.

3. K. Gürs and R. Müller, Breitband-modulation durch Steuerung der emission eines optischen masers (Auskopple-modulation), *Phys. Lett.*, 5, 179–181, 1963.

4. H. Statz and C.L. Tang, Zeeman effect and nonlinear interactions between oscillating modes, in *Quantum Electronics III*, P. Grivet and N. Bloembergen (eds.), Columbia University Press, New York, 1964, pp. 469–498.

5. L.F. Mollenauer, R.H. Stolen, and J.P. Gordon, Experimental observation of picosecond pulse narrowing and solitons in optical fibers, *Phys. Rev. Lett.*, 45(13), 1095–1098, 1980.

6. Z. Ahmed and N. Onodera, High repetition rate optical pulse generation by frequency multiplication in actively mode-locked fiber ring lasers, *Electron. Lett.*, 32(5), 455–457, 1996.

7. K.K. Gupta, D. Novak, and H. Liu, Noise characterization of a regeneratively mode-locked fiber ring laser, *IEEE J. Quantum Electron.*, 36(1), 70–78, 2000.

8. T.F. Carruthers and I.N. Duling III, 10-GHz, 1.3-ps erbium fiber laser employing soliton pulse shortening, *Opt. Lett.*, 21(23), 1927–1929, 1996.

9. M. Horowitz, C.R. Menyuk, T.F. Carruthers, and I.N. Duling, Theoretical and experimental study of harmonically modelocked fiber lasers for optical communication systems, *J. Lightwave Technol.*, 18(11), 1565–1574, 2000.

10. G. Das and J.W.Y. Lit, Wavelength switching of a fiber laser with a Sagnac loop reflector, *IEEE Photon. Technol. Lett.*, 16(1), 60–62, 2004.

11. N.H. Seong and D.Y. Kim, A new figure-eight fiber laser based on a dispersion-imbalanced nonlinear optical loop mirror lumped dispersive elements, *IEEE Photon. Technol. Lett.*, 14(4), 459–461, 2002.

12. Y. Chen and H.A. Haus, Solitons and polarization mode in dispersion, *Opt. Lett.*, 25(5), 290–292, 2000.

13. B.E.A. Saleh and M.C. Teich, *Fundamentals of Photonics*, John Wiley & Sons Inc., New York, 1991, Chapter 8.

14. M. Nakazawa and E. Yoshida, A 40 GHz 850 fs regeneratively FM mode-locked polarization maintaining erbium fiber ring, *IEEE Photon. Technol. Lett.*, 12(12), 1613–1615, 2000.

15. D.J. Kuizenga and A.E. Siegman, FM and AM mode locking of the homogeneous laser—Part I: Theory, *IEEE J. Quantum Electron.*, 6(11), 694–708, 1970.

16. W. Lu, L. Yan, and C.R. Menyuk, Dispersion effects in an actively mode-locked inhomogeneously broadened laser, *IEEE J. Quantum Electron.*, 38(10), 1317–1324, 2002.

17. J. Zhou, G. Taft, C.P. Huang, M.M. Murnane, and H.C. Kapteyn, Pulse evolution in broad bandwidth Ti: Sapphire laser, *Opt. Lett.*, 19(15), 1149–1151, 1994.

18. M.J.F. Digonnet, *Rare-Earth-Doped Fiber Lasers and Amplifiers*, Marcel Dekker Inc., New York, 2001.

19. New Focus, *Application Note 1—Insights into High-Speed Detectors and High-Frequency Techniques*, New Focus, San Jose, CA, 2002.

20. R.L. Jungerman, C. Johnsen, D.J. McQuate, K. Salomaa, M.P. Zurakowski, R.C. Bray, G. Conrad, D. Cropper, and P. Hernday, High-speed optical modulator for application in instrumentation, *J. Lightwave Technol.*, 8(9), 1363–1370, 1990.

21. G. Zhu, H. Chen, and N. Dutta, Time domain analysis of a rational harmonic mode-locked ring fiber laser, *J. Appl. Phys.*, 90(5), 2143–2147, 2001.

22. R.L. Jungerman, G. Lee, O. Buccafusca, Y. Kaneko, N. Itagaki, R. Shioda, A. Harada, Y. Nihei, and G. Sucha, 1-THz bandwidth C- and L-band optical sampling with a bit rate agile timebase, *IEEE Photon. Technol. Lett.*, 14(8), 1148–1150, 2002.

23. X. Shan and D.M. Spirit, Novel method to suppress noise in harmonically modelocked erbium fiber lasers, *Electron. Lett.*, 29(11), 979–981, 1993.
24. C.R. Doerr, H.A. Haus, E.P. Ippen, M. Shirasaki, and K. Tamura, Addictive pulse limiting, *Opt. Lett.*, 19(1), 31–33, 1994.
25. M. Nakazawa, E. Yoshoda, and Y. Kimura, Ultrastable harmonically and regeneratively mode-locked polarization-maintaining erbium fiber ring laser, *Electron. Lett.*, 30(19), 1603–1605, 1994.
26. M. Horowitz and C.R. Menyuk, Analysis of pulse dropout in harmonically mode-locked fiber lasers by use of the Lyapunov method, *Opt. Lett.*, 25(1), 40–42, 2000.

4

NLSE Numerical Simulation of Active Mode-Locked Lasers: Time Domain Analysis

4.1 Introduction

As described in Chapters 2 and 3, the designs based on the fundamental principles of mode-locked lasers have been demonstrated experimentally in three types of mode-locked lasers, namely, harmonic, rational harmonic, and regenerative, for the generation of short repetitive pulse trains. However, the simulation of the generation of mode-locked pulses is essential to confirm with those pulse trains observed in experiments. Furthermore, a simulation package would allow a detailed design and analysis of the dynamic behavior of such lasers. Thus, this chapter presents the development of the simulation software package and simulated results.

The pulse envelopes propagating in the ring laser follow the nonlinear Schrödinger equation (NLSE). Furthermore, as the pulses pass through an optical filter, the optical modulator, as well as the erbium-doped fiber amplifier (EDFA), the representation of the lightwave propagation and transmission through these photonic elements is necessary.

This chapter presents a novel approach of simulation by representing the propagation of the pulse envelopes all in the temporal domain. The linear dispersion and nonlinear self-phase modulation (SPM) can be approximated accurately using the finite-difference method. Indeed, simulated results have been compared and confirmed with those obtained experimentally.

Section 4.2 outlines the temporal domain representation of the laser model, the optical filter, the EDFA, the fiber, and the optical modulator. Section 4.3 describes the mathematical models using the finite-difference approach for all these elements. Simulation results are then presented first for the propagation of light pulses through an optical fiber to demonstrate the validity of our modeling technique. The models of other photonic elements are then integrated with the fiber propagation model to form a generic simulation package for the generation of mode-locked pulses. Using this simulation package, the effects of the modulation frequency, modulation depth, filter bandwidth, and pump power on the generated pulses are studied and confirmed with experimental observations. Furthermore, the developed simulation models are demonstrated with the generation and mode locking of pulses in a rational harmonic mode-locked laser.

4.2 The Laser Model

The laser configuration used in the experiment is modeled, as shown in Figure 4.1. All optical and photonic components are indicated and the signal-flow paths are noted by straight lines with end arrows. The model consists of an EDFA, which is the gain medium for the laser, a length of optical fiber of a dispersion shifted type (DSF), an optical bandpass filter (BPF), and an optical intensity modulator. The optical waves are amplified in the EDFA and then propagated through the DSF. The lightwaves intensity is then modulated by the optical modulator. The BPF ensures that a certain optical frequency band is allowed to pass through to determine the central wavelength and the pulse width of the generated signals. The output of the modulator is fed back into the EDFA to create a loop. The signals are then circulating through the optical loop. The detailed modeling of each photonic component of the cavity is described in the following sections.

4.2.1 Modeling the Optical Fiber

The DSF fiber can be modeled by the NLSE [1]:

$$\frac{\partial A}{\partial z}+\frac{j}{2}\beta_2\frac{\partial^2 A}{\partial T^2}+\frac{\alpha}{2}A = j\gamma \,|\,A\,|^2 A \qquad (4.1)$$

where
$j = \sqrt{-1}$
A is the slowly varying envelope of the optical field
z is the axial distance
T is the delayed time ($T = t - z/v_g$), v_g is group velocity
α is the linear attenuation factor of the fiber and accounts for the loss

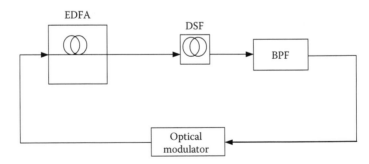

FIGURE 4.1
General schematic model of the fiber laser.

γ is the nonlinear coefficient of the fiber, which accounts for the SPM effect
β_2 is the second-order derivative of the propagation constant β and can be calculated from the fiber dispersion parameter as

$$\beta_2 = \frac{\lambda^2 D}{2\pi c} \tag{4.2}$$

where
λ is the operating wavelength
D is the chromatic dispersion of the fiber (or the GVD)
c is the speed of light in vacuum

4.2.2 Modeling the EDFA

The EDFA can be modeled by an EDF with ions pumped into an excited state for providing gain when the signal travels through it. The propagation equation for the signal traveling through the EDF is

$$\frac{\partial A}{\partial z} + \frac{j}{2}\left(\beta_2 + jg_0 T_2^2\right)\frac{\partial^2 A}{\partial T^2} + \frac{1}{2}\left(g_0 - \alpha\right)A = j\gamma \mid A \mid^2 A \tag{4.3}$$

This equation follows a similar form of Equation 4.1 except that the gain factors g_0 and T_2 are included. T_2 accounts for the decrease of the gain coefficient at the wavelength located far from the gain peak and is usually defined as the inverse of the 3 dB bandwidth of the gain spectrum. The saturation of the EDFA is usually modeled by presenting the gain g_0 as a function of the signal average power

$$g_0 = \frac{g_{ss}}{1 + P_{av}/P_{sat}} \tag{4.4}$$

where
g_{ss} is small-signal gain
P_{av} is the signal average power
P_{sat} is the saturation power level

4.2.3 Modeling the Optical Modulation

The optical modulator can be modeled by the transmission function

$$T = \alpha_m \cos^2\left(\frac{\pi\left(v_m - V_{sh}\right)}{2V_\pi}\right) \tag{4.5}$$

where
V_π is the voltage applied on the modulator that causes a π phase shift in one
    arm of the integrated optical interferometer
V_{sh} accounts for the DC drift of the modulator
α_m is the insertion loss
$v_m(t)$ is the modulating voltage signal and can be given by

$$v_m(t) = V_m \cos(\omega_m t) + V_b \qquad (4.6)$$

where
V_m is the amplitude of the modulating signal
ω_m is the modulating frequency
V_b is the bias voltage

Substituting (4.6) into (4.5) and noting that V_{sh} can be assumed to be zero
without any effect on the final result, the transmission function of the modulator can be written as

$$T = \alpha_m \cos^2\left(\frac{\pi}{4}\Delta_m \cos(\omega_m t) + \frac{\pi}{2}\frac{V_b}{V_\pi}\right) \qquad (4.7)$$

where $\Delta_m = 2V_m/V_\pi$ is the modulation depth. When the modulator is biased at
the quadrature point $V_b = V_\pi/2$, Equation 4.7 becomes

$$T = \alpha_m \cos^2\left[\frac{\pi}{4}\left(\Delta_m \cos(\omega_m t)\right) + 1\right] \qquad (4.8)$$

4.2.4 Modeling the Optical Filter

The optical filter BPF can be described by the transfer function following a
Gaussian profile as

$$H(f) = \alpha_F \exp\left(-\frac{1}{2}\left(\frac{f}{B_0}\right)^2\right) \qquad (4.9)$$

where
α_F is the insertion loss
B_0 is half of the $(1/e)$ bandwidth of the filter

4.3 The Propagation Model

4.3.1 Generation and Propagation

The most popular method used for solving the NLSE describing the propagation of the pulse in an optical fiber is the well-known split-step Fourier (SSF) method [2]. The equation can be split into two parts: the linear and nonlinear parts. The linear part is solved in the frequency domain by using the Fourier transform method. The results are then converted back to the time domain by taking the inverse Fourier transform. The nonlinear part is solved in the time domain.

The SSF is claimed to be a fast method for studying the propagation of the pulse in the fiber and has been applied to study the formation of a single pulse in an active mode-locked laser [3–6]. However, since SSF requires converting the signal from the time domain into the frequency domain by using fast Fourier transform (FFT) algorithm, the signal must be windowed and the number of samples in the window must be limited. Although each FFT operation is relatively fast, a large number of FFT operations on a large-size array make the computation impossible.

To simulate every pulse that exists in a harmonic mode-locked laser without windowing, a fully time domain method is used. The propagation equations of the pulses in the optical fiber are solved using the finite element method.

First, the propagation equation of the pulse in the EDF is solved to obtain the signal at the output of the EDF. The time and space are discretized by using the following equations:

$$z = k\Delta z \tag{4.10}$$

$$T = m\Delta T \tag{4.11}$$

where
Δz and ΔT are the steps in the space domain and time domain, respectively
k and m are integers

The derivatives of A in Equation 4.3 are approximated as follows:

$$\frac{\partial A(z,T)}{\partial z} = \frac{A_{k+1,m} - A_{k,m}}{\Delta z} \tag{4.12}$$

$$\frac{\partial A(z,T)}{\partial T} = \frac{A_{k,m+1} - A_{k,m}}{\Delta T} \tag{4.13}$$

$$\frac{\partial A(z,T)}{\partial T^2} = \frac{A_{k,m+1} - 2A_{k,m} + A_{k,m-1}}{\Delta T^2} \tag{4.14}$$

where $A_{k,m} = A(k\Delta z, m\Delta T)$. Substituting (4.12) through (4.14) into Equation 4.3 gives

$$\frac{A_{k+1,m} - A_{k,m}}{\Delta z} + \frac{j}{2}\left(\beta_2 + jg_0 T_2^2\right)\frac{A_{k,m+1} - 2A_{k,m} + A_{k,m-1}}{\Delta T^2}$$

$$+ \frac{1}{2}\left(g_0 - \alpha\right)A_{k,m} = j\gamma \mid A_{k,m} \mid^2 A_{k,m} \tag{4.15}$$

Therefore the field at $(z + \Delta z)$ can be calculated from the field values at the preceding position z as follows:

$$A_{k,m+1} = A_{k,m} + \Delta z \left[\begin{array}{c} j\gamma \mid A_{k,m} \mid^2 A_{k,m} - \frac{j}{2}\left(\beta_2 + jg_0 T_2^2\right)\frac{A_{k,m+1} - 2A_{k,m} + A_{k,m-1}}{\Delta T^2} \\ \\ -\frac{1}{2}\left(g_0 - \alpha\right)A_{k,m} \end{array} \right]$$

$$\tag{4.16}$$

The process is repeated till the end of the fiber is reached. The output optical field then travels through the DSF. The same method is used to solve Equation 4.1 for getting the optical field at the end of the DSF.

The optical field output from the DSF is then filtered using the filter's transfer function in Section 4.2.4. The filter is implemented in the time domain as well. The filter output is calculated as follows:

$$A_F(t) = A_{DSF}(t) * h(t) \tag{4.17}$$

where
 * denotes the convolution operation
 $A_F(t)$ is the optical field at the output of the filter
 $A_{DSF}(t)$ is the optical field at the output of the DSF
 $h(t)$ is the impulse response of the filter's transfer function $H(f)$ in (4.9)
   and can be determined by performing the inverse Fourier transform
   on $H(f)$

The output of the filter $A_F(t)$ is then convolved with the modulator's transmission function described in Equation 4.7. The output of the modulator is looped back to the input of the EDFA and the calculation is repeated till the steady state is reached.

4.3.2 Results and Discussions

4.3.2.1 Propagation of Optical Pulses in the Fiber

First, the fiber model is used to study the propagation of the pulses in an optical fiber. A *sech* profile pulse is launched into the fiber and the changes of the pulse along the traveling distance z are recorded. In Sections 4.3.2.1.1 and 4.3.2.1.2, the effects of dispersion and nonlinearity on the pulses are discussed.

4.3.2.1.1 Dispersion Effect

The effect of GVD on the pulse propagation along an optical fiber is studied by setting other terms in Equation 4.1 to zero except the GVD term. The fiber attenuation is assumed to be zero (i.e., $\alpha = 0$ in Equation 4.1), so that the GVD effect on the pulse propagation can be isolated and studied. The pulse envelope $A(z, T)$ satisfies the following equation:

$$\frac{\partial A}{\partial z} = -\frac{j}{2}\beta_2 \frac{\partial^2 A}{\partial T^2} \qquad (4.18)$$

Figure 4.2 shows the simulation results for a hyperbolic-secant pulse with a full width at half maximum (FWHM) of 20 ps propagating in a standard single-mode fiber (SMF) under the GVD effect. Taking the value of the SMF's dispersion $D = 17$ ps/nm/km at 1550 nm, β_2 can be calculated as follows: $\beta_2 = -21.7$ ps²/km.

This can be seen in Figure 4.2a, in which the pulses are broadened in the temporal domain under the GVD effect. The pulse width increases along the fiber and this broadening causes the pulse's peak power drop by 65% at $z = 25$ km. However there is no change in the pulse spectrum as seen in Figure 4.2b because the GVD effect is a linear effect that does not change the spectrum of the pulse.

The pulse-broadening effect at different traveling distances is illustrated in Figure 4.3. As the traveling distance increases, the pulse width gets wider and the pulse peak power decreases. The pulse width is nearly double that of the initial pulse at $z = 25$ km, and the peak power reduces to 35% of the initial value. The propagation of the same pulse in a nonzero dispersion shifted fiber (NZ-DSF) is illustrated in Figures 4.4 and 4.5. $\beta_2 = -3.82$ ps²/km is calculated from the dispersion value of $D = 3$ ps/nm/km at 1550 nm for NZ-DSF.

Pulse broadening is less severe in NZ-DSF than in SMF because the dispersion ($D = 3$ ps/nm/km) of NZ-DSF is smaller than the dispersion ($D = 17$ ps/nm/km) of SMF. The pulse width (FWHM) just increases slightly after traveling a distance of 25 km at 21 ps. This can be understood by comparing the traveling distance of the pulse to the dispersion length when traveling in SMF and in NZ-DSF. The dispersion length is defined as

$$L_D = \frac{T_0^2}{|\beta_2|} \qquad (4.19)$$

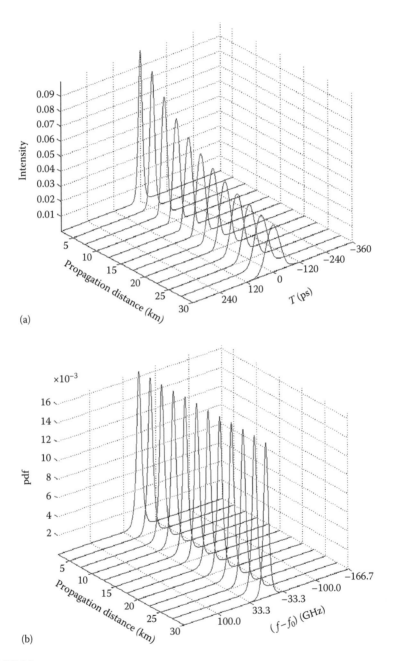

(a)

(b)

FIGURE 4.2
Hyperbolic-secant pulse propagation in an SMF under the GVD effect in the temporal domain (a) and in the frequency domain (b); $\lambda = 1550\,\text{nm}$, $\beta_2 = -21.7\,\text{ps}^2/\text{km}$. In plot (b), pdf on the vertical axis means probability density function.

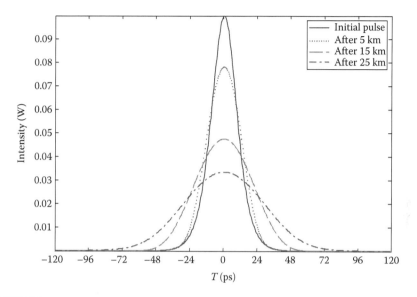

FIGURE 4.3
Pulse broadening after traveling through the SMF at $z = 5$, 15, and 25 km.

where T_0 is the $1/e$ pulse width related to the FWHM T_{FWHM} as $T_{FWHM} = 1.665T_0$.
For the pulse width of $T_{FWHM} = 20$ ps, the dispersion lengths for SMF and NZ-DSF are 6.66 and 37.7 km, respectively.

4.3.2.1.2 Self-Phase Modulation Effect

The SPM effect is studied by setting other terms in Equation 4.1 to zero except the SPM term. The fiber attenuation is also assumed to be zero (i.e., $\alpha = 0$ in Equation 4.1), so that the SPM effect on the pulse propagation can be isolated and investigated. The pulse propagation is described by following equation:

$$\frac{\partial A}{\partial z} = j\gamma |A|^2 A \qquad (4.20)$$

Figure 4.6 shows the propagation of a hyperbolic-secant pulse with a pulse width of $T_{FWHM} = 20$ ps and a peak power of $P_0 = 0.1$ W. The typical value of γ is 1.53/W/km. There is no change to the pulse shape in the temporal domain due to the absence of the GVD effect, and the pulse shape and pulse width do not vary along the traveling distance. However, the pulse spectrum is spread out severely and distorted due to the nonlinear phase shift (or nonlinear chirping) induced by the SPM.

The spreading of the pulse spectrum is illustrated clearly in Figure 4.7, in which the pulse spectrum is plotted at different traveling distances. The pulse spectrum gets wider as the pulse propagates along the fiber. The two

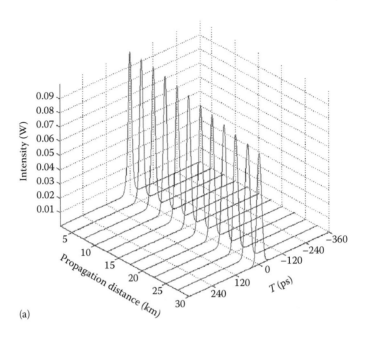

(a)

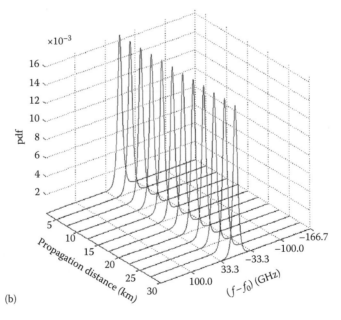

(b)

FIGURE 4.4
Hyperbolic-secant pulse propagation in an NS-DSF under the GVD effect in the temporal domain (a) and in the frequency domain (b); $\lambda = 1550\,\text{nm}$, $\beta_2 = -3.82\,\text{ps}^2/\text{km}$.

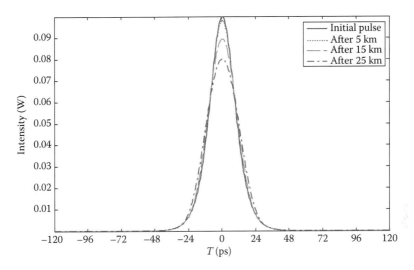

FIGURE 4.5
Pulse broadening effect in the NZ-DSF at $z = 5$, 15, and 25 km.

peaks appear at $z = 25$ km. There are two peaks in the pulse spectrum at $\xi = 4.5$, where ξ is the normalized propagation length and is defined as

$$\xi = \frac{z}{L_{NL}} \quad (4.21)$$

where
  z is the propagation length
  L_{NL} is the nonlinear length

$$L_{NL} = \frac{1}{\gamma P_0} \quad (4.22)$$

Using $\gamma = 1.53/\text{W/km}$ and $P_0 = 0.1$ W, the nonlinear length is $L_{NL} = 6.54$ km. The normalized propagation length for $z = 25$ km is $\xi = 3.8$.

The pulse spectrum broadening is also studied at different values of γ and P_0. The results are plotted in Figures 4.8 and 4.9. The pulse spectrum broadening at $z = 25$ km decreases as γ decreases or P_0 decreases due to the presence of a smaller SPM effect.

4.3.2.1.3 Combined Effect of GVD and SPM Effects: A Soliton Pulse

The NLSE (or the soliton equation) for the pulse propagation under GVD and SPM effects is given as

$$\frac{\partial A}{\partial z} = -\frac{j}{2} \beta_2 \frac{\partial^2 A}{\partial T^2} - j\gamma |A|^2 A \quad (4.23)$$

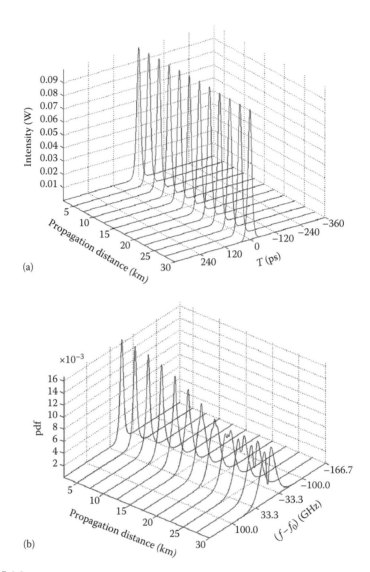

FIGURE 4.6
Hyperbolic-secant pulse propagation in an SMF under the SPM effect in the temporal domain (a) and in the frequency domain (b); $\lambda = 1550\,\text{nm}$, $\gamma = 1.53/\text{W/km}$.

where the fiber attenuation is also assumed to be zero (i.e., $\alpha = 0$ in Equation 4.1), so that the GVD and SPM effects on the pulse propagation can be studied.

The optical soliton theory shows that when a hyperbolic-secant pulse (with a pulse width T_0 and peak power P_0) is launched into the fiber and satisfies the condition $T_0^2/|\beta_2| = 1/\gamma P_0$, the GVD and SPM effects can be designed to balance each other by choosing the appropriate values of the dispersion D and the nonlinear coefficient γ of the fiber, and hence the pulse will

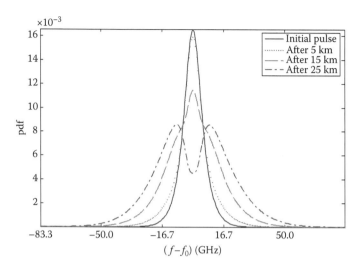

FIGURE 4.7
Pulse spectrum broadening in the SMF under the SPM effect at $z = 5$, 15, and 25 km.

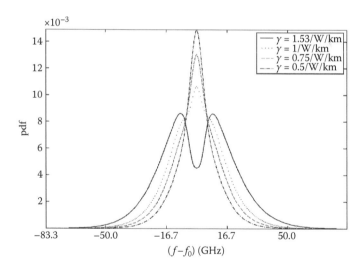

FIGURE 4.8
Hyperbolic-secant pulse spectrum after propagating a distance of $z = 25$ km in the SMF for different values of γ.

propagate without any change in its pulse shape and spectrum (an optical soliton). The GVD effect broadens the pulse in the time domain while the SPM compresses the pulse in the time domain (or broadens the pulse spectrum in the frequency domain, as shown in Figures 4.6 through 4.9). Thus, when the GVD effect is well balanced with the SPM effect, the pulse will not

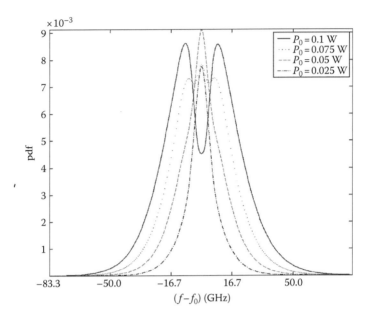

FIGURE 4.9
Hyperbolic-secant pulse spectrum after propagating a distance of $z = 25$ km in the SMF for different values of the pulse's initial peak power P_0.

change its shape along the fiber length. The SPM effect acts as a "dispersion compensator." Another reason for this unchanging pulse shape is that the hyperbolic-secant pulse (a soliton pulse) is a solution of the NLSE. However, it is worth noting that, in practice, it is extremely difficult to generate an optical modulated signal with the exact hyperbolic-secant pulse and thus, a practical pulse is, strictly speaking, not a solution of the soliton equation and will therefore change its pulse shape slightly as it propagates through the fiber.

Figure 4.10 shows the simulation result of the propagation of a hyperbolic-secant pulse in a SMF. The pulse width is $T_{FWHM} = 20$ ps ($T_0 = 12$ ps) and the peak power is $P_0 = 0.1$ W. Therefore, the soliton condition is satisfied and hence the pulse shape as well as the pulse spectrum remain unchanged when the pulse propagates along the fiber.

4.4 Harmonic Mode-Locked Laser

4.4.1 Mode-Locked Pulse Evolution

The laser model is described in Figure 4.1. Unlike other simulations on the mode-locked fiber laser reported so far [1,2], which are solely based on the SSF method for solving the NLSE equation and only one pulse in the ring is

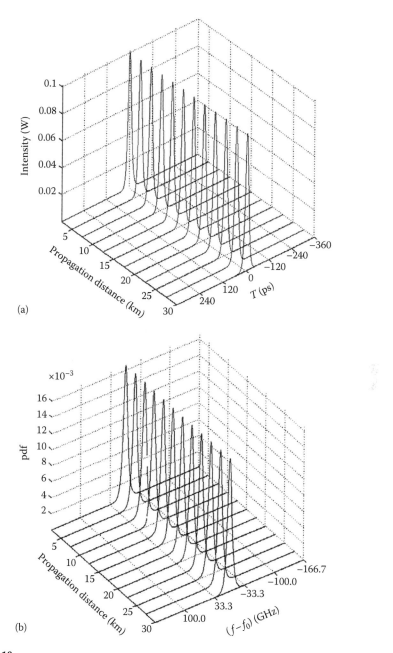

(a)

(b)

FIGURE 4.10
Hyperbolic-secant pulse propagation in an SMF under the GVD and SPM effects in the temporal domain (a) and in the frequency domain (b); $\lambda = 1550$ nm, $\gamma = 1.53$/W/km, $\beta_2 = -21.7$ ps^2/km, $P_0 = 0.1$ W, $T_{FWHM} = 20$ ps.

simulated, the simulation method reported here is modeled fully in the time domain and hence there is no need to convert the pulses into the frequency domain. This makes the model simulate not just one pulse but all pulses traveling in the ring, which actually happens in a harmonic mode-locked ring laser.

The laser is excited from an initial flat amplitude, which is regarded as a continuous wave. A hyperbolic-secant pulse shape of less than ps width is launched into the laser. After about 3000 round trips, steady-state output harmonic mode-locked pulses are reached, as shown in Figure 4.11. It can be seen that there are four pulses present in one time period of the ring as the modulation frequency is set to be four times the ring frequency. The laser is locked to the fourth harmonic frequency. The ring period is the time required for the pulses to propagate a round trip and can be calculated from the lengths of the fibers (0.1 m of EDF and 2 m of DSF) with the assumption that there is no delay induced by other components such as the modulator and filter. Without loss of generality, the lengths of the fibers are chosen to be short to reduce the simulation time. The insertion losses of the modulator and filter are set to zero to minimize the total loss of the cavity. The other settings for the ring are amplification factor $g_{ss}=3$, saturation power $P_{sat}=10\,\text{mW}$, filter bandwidth $BW_f=400\,\text{GHz}$ (0.32 nm), modulation depth $\Delta_m=0.5$, modulation frequency $f_m=0.39\,\text{GHz}$ (the fourth harmonic of the ring frequency), the dispersion of the EDF and DSF at 1550 nm are 17 and 3 ps/nm/km, respectively, and the nonlinear coefficients of both fibers are 1.53/W/km. Figure 4.12a and b give a closer look at one of the four mode-locked pulses (Figure 4.11) and its spectrum, respectively. From Figure 4.12a,

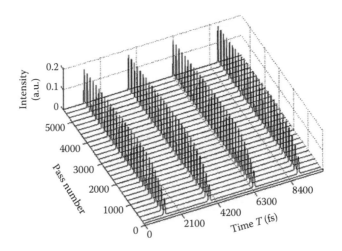

FIGURE 4.11
Pulse evolution in a harmonic active mode-locked fiber laser, four pulses traveling in the ring for the fourth harmonic order. The pass number indicates the number of passes through the ring laser.

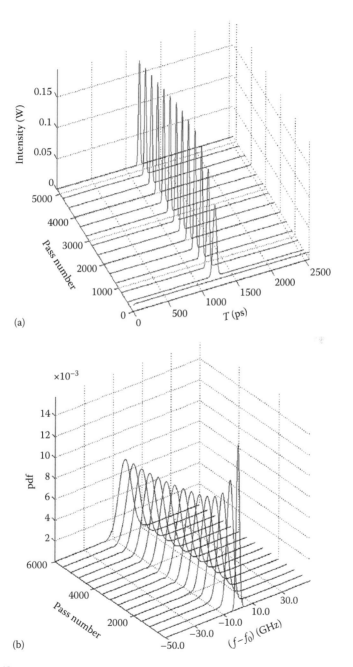

FIGURE 4.12
Pulse evolution in an active harmonic mode-locked fiber laser: (a) temporal domain and (b) frequency domain.

it can be seen that the mode-locked pulse maintains its shape after several thousands of round trips due the well balanced interplay between the GVD and SPM effects, as discussed in Section 4.3. This explanation can also be seen in Figure 4.12b where the pulse spectrum is not distorted because the SPM effect is cancelled out by the GVD effect.

4.4.2 Effect of Modulation Frequency

The effect of modulation frequency on the mode-locked pulses is studied by running the simulation with different values of the modulation frequency. Other settings of the laser are the same as those in Section 4.4.1.

Steady-state pulses and their spectra with modulation frequencies of 0.39, 0.78, 1.16, and 2.32 GHz (corresponding to the 4th, the 8th, the 12th, and the 24th harmonic of the ring/fundamental frequency $f_F = 97.6$ MHz) can be obtained. The peak power of the pulse reduces as the modulation frequency increases. This is consistent with the energy conservation. As the modulation increases, more pulses travel in the ring sharing the same energy provided by the pump signal in one ring period. The energy provided by the pump in one ring period is constant, as the pump power is kept unchanged. When the number of pulses traveling simultaneously in the ring increases, the total energy is divided among those pulses and hence each pulse envelope accumulates less energy.

This fact can also be explained from Figure 4.13a and b, which shows the pulse width of the steady-state pulse train and its bandwidth versus the modulation frequency. As the modulation frequency increases by four times, the pulse width decreases by half and the bandwidth is double. The simulation results show that the bandwidth increases and the pulse width decreases proportionally to the square root of the modulation frequency. Therefore higher modulation frequency not only produces pulses with higher repetition rates but also makes the pulses shorter. This is consistent with the experiment results shown in Chapter 3. However, the modulation frequency

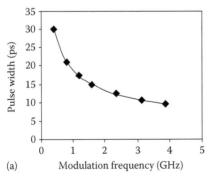
(a)

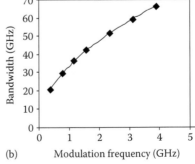
(b)

FIGURE 4.13
Pulse width (a) and bandwidth (b) as a function of the modulation frequency.

cannot increase infinitely but is limited by the bandwidth of the modulator. A higher repetition rate can be obtained using the detuning technique, which will be simulated and discussed in detail in Chapter 5.

4.4.3 Effect of Modulation Depth

In this section, the effect of modulation depth is examined by studying the laser output under different settings of the modulation depth Δ_m. The pulse evolutions with $\Delta_m = 0.5$, $\Delta_m = 0.2$, and $\Delta_m = 0.1$ are shown in Figures 4.14 through 4.16, respectively. The modulation frequency is $f_m = 1.56\,GHz$. Other settings are the same as those shown in Section 4.4.1.

The simulation results show that with a decrease of the modulation depth, the damped oscillation process during pulse evolution lasts longer. This can be easily explained by comparing these simulated results with experimental results. Deeper modulation depth shapes the pulse effectively and the mode-locking process is enhanced. However, there is only a slight difference in both the pulse width and its bandwidth of the steady-state pulses, as shown in Figures 4.17 and 4.18a and b. When the modulation depth is reduced to 0.001 (or 0.1%), no mode-locked pulse is generated. This is the modulation depth threshold for mode locking. This value is quite different to the threshold obtained from the experiment described in Chapter 3, where the threshold value of the modulation depth is found to be 0.08. The difference may be due to the noise (e.g., EDFA noise) and environmental conditions in the experimental setup, which are not included in our simulation model.

4.4.4 Effect of the Optical Filter Bandwidth

The filter is inserted into the laser cavity to select the central wavelength and help to shape the pulse. As the pulse passes through the modulator during every round trip, it is compressed in the temporal domain and spread out in the frequency domain due to the sidebands induced by the modulation. However, the pulse cannot be compressed endlessly. The filter bandwidth limits the pulse spectrum width and thus balances the temporal domain pulse compression. In this section, the mode-locked pulses are studied with the variation of the filter bandwidth. The cavity settings are the same as before except that the filter bandwidth is varied. The simulation results are shown in Figures 4.19 and 4.20.

Figure 4.19 shows the steady-state pulses for various values of the filter bandwidths. The cavity with a wider filter bandwidth generates pulses with narrower pulse width and wider bandwidth, and vice versa. As the filter bandwidth increases, the pulse width becomes shorter with a higher peak power. On the other hand, the pulse bandwidth increases with wider filter bandwidth.

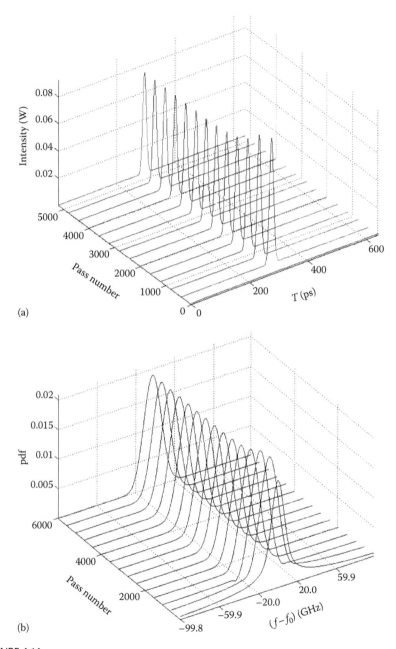

FIGURE 4.14
Pulse evolution in the temporal domain (a) and the frequency domain (b) under modulation depth of 0.5.

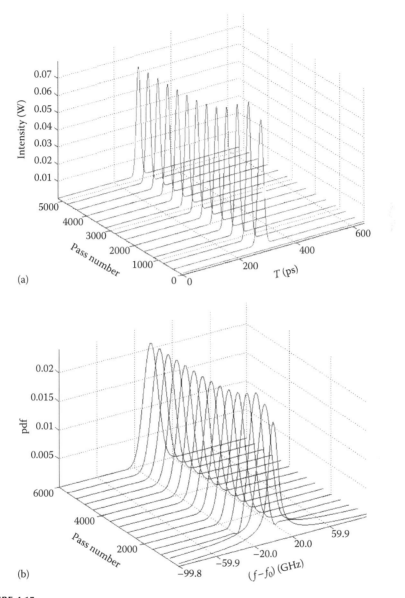

FIGURE 4.15
Pulse evolution in the temporal domain (a) and the frequency domain (b) under modulation depth of 0.2.

Pulse width and bandwidth are plotted against the filter bandwidth in Figure 4.20. The pulse width decreases exponentially with the increase of filter bandwidth. Therefore, a wider filter bandwidth is preferred as it generates shorter pulse. However in practice, mode locking is easier to obtain with a narrow filter bandwidth. It is difficult to lock the laser when

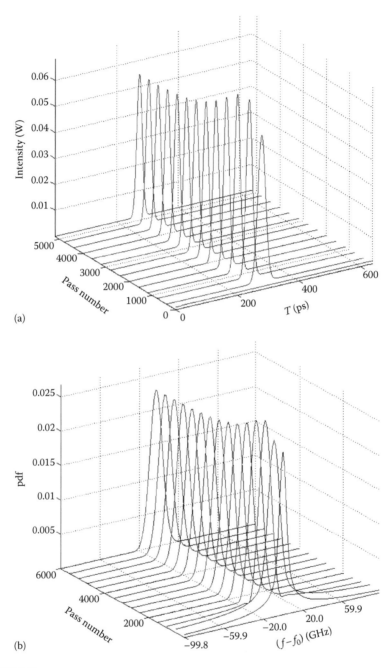

(a)

(b)

FIGURE 4.16
Pulse evolution in the temporal domain (a) and the frequency domain (b) under modulation depth of 0.1.

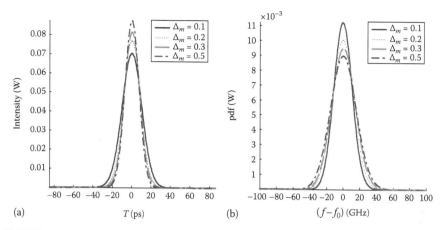

FIGURE 4.17
Mode-locked pulses with different modulation depths: (a) temporal domain and (b) frequency domain.

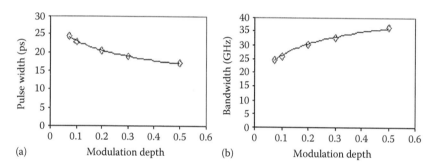

FIGURE 4.18
Pulse width (a) and bandwidth (b) as a function of the modulation depth.

the filter bandwidth is very wide because neither the center frequency selection nor supermode filtering can be obtained with a wide bandwidth filter.

4.4.5 Effect of Pump Power

The affect of the pump power on the mode-locked laser is also studied by varying the values of P_{sat} in the EDFA model and keeping other settings the same as before.

Figure 4.21 shows the mode-locked pulses for various values of P_{sat}. When P_{sat} varies, only the peak power of the pulses is affected while the pulse width and bandwidth remain nearly unchanged. This is consistent with the result obtained in the experiment given in Chapter 3. The pulse width and

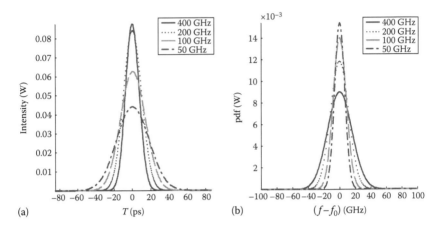

FIGURE 4.19
Mode-locked pulses with different filter bandwidths: (a) temporal domain and (b) frequency domain.

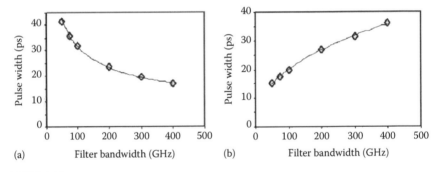

FIGURE 4.20
Pulse width (a) and bandwidth (b) as a function of the filter bandwidth.

bandwidth do not vary with the change of the pump power in the short-length ring setup. The shortening of the pulse due to the increase of the pump power is just observed when the length of the ring is 118 m.

4.4.6 Rational Harmonic Mode-Locked Laser

As discussed above, the repetition rate of the active mode-locked laser is limited by the bandwidth of the modulator. To increase the repetition rate, the rational harmonic mode-locking (RHML) technique is applied. In the RHML laser, the modulation frequency is not an integer number of the harmonic of the fundamental frequency f_R but is detuned by an amount of f_R/N, that is,

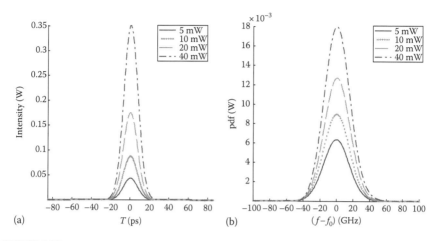

FIGURE 4.21
Mode-locked pulses with different values of P_{sat}: (a) temporal domain and (b) frequency domain.

$$f_m = \frac{mf_R + f_R}{N} \tag{4.24}$$

where m and N are integer numbers and N is defined as the multiplying factor. The output pulse's repetition rate is now no longer f_m but Nf_m. In other words, the repetition rate has been multiplied by a factor of N.

The doubling repetition rate in RHML by detuning the modulation frequency has been experimentally demonstrated in Chapter 3.

The cavity settings are the same as before, except that the modulation depth is $\Delta_m = 0.1$, and the gain factor is $g_{ss} = 6$. This higher gain factor is required because in RHML the pulses do not pass through the modulator at the minimum loss position but at different loss positions for every round trip. Therefore the pulses in the RHML experience more loss than those of the HML and hence require higher gain to compensate for the loss. When the modulation frequency is set to 0.39 GHz, which corresponds to the fourth harmonic of the fundamental frequency $f_F = 97.6$ MHz, harmonic mode locking is obtained with four pulses traveling in the ring, as shown in Figure 4.22.

When the modulation frequency is detuned by $f_F/2$ to become 0.439 GHz, doubling of repetition rate is obtained, as shown in Figure 4.23. It can be seen that there are totally nine pulses traveling in the ring during one ring period T_R. It is interesting that the amplitude-unequal pulses are also observed. This again confirms the experimental result that the RHML can be used to increase the repetition rate but the output pulses suffer from amplitude fluctuation from pulse to pulse unless pulse amplitude equalization is applied.

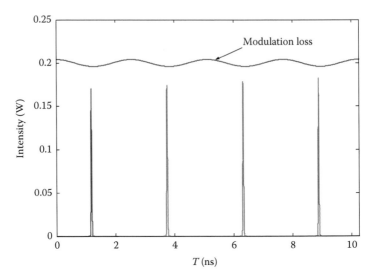

FIGURE 4.22
Harmonic mode locking with four pulses in one ring period T_R when the modulation frequency is set to four times the fundamental frequency.

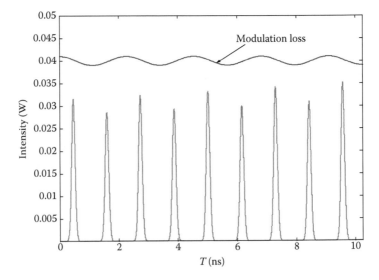

FIGURE 4.23
Rational harmonic mode locking with nine pulses in one ring period T_R when the modulation frequency is set to 4.5 times the fundamental frequency.

4.5 FM or PM Mode-Locked Fiber Lasers

In an effort to understand and investigate the behavior of mode-locked fiber lasers as well as future proposed lasers, it is helpful to develop a model for such laser systems that allows us to get an insight of laser dynamics. The physical properties of propagation medium that must be included in the model consist of (1) dispersion, (2) nonlinearity, (3) polarization, (4) finite bandwidth, saturable gain (which describes amplification in a fiber amplifier such as EDFA), (5) phase modulation from an electro-optic phase modulator, and (6) amplitude modulation from an intensity modulator using, for example, a phase-modulated Sagnac loop.

The model we expect includes all active mode-locking mechanisms as well as the effects on pulse shaping and propagation. The model must also be flexible to cater for the change in the laser configuration, but simple in computation. Hence a recirculation model of the fiber ring is used to simulate mode locking in the fiber cavity. This model consists of the basic components of an actively mode-locked fiber laser, as shown in Figure 4.24, in other words, the cavity of the actively mode-locked fiber laser is modeled as a sequence of different elements such as optical fibers (including optical fiber and doped fibers with gain medium), optical filter, and mode locker (phase modulator or intensity modulator using, e.g., a phase-modulated Sagnac loop).

The pulse propagation in the optical fibers is governed by the NLSE, as described in Chapter 2 and also in Equation 4.1. To include all mode-locking mechanisms in the numerical model, two coupled partial differential equations or a vector NLSE is solved by using the well-known split-step technique. However the mode-locking mechanism relating to polarizations such as nonlinear polarization rotation is negligible in the active mode-locked fiber laser, so we first start with a scalar model without including the polarization effect. The scalar equation to describe pulse propagation in optical fiber, fiber amplifiers, or fiber lasers is given as

$$\frac{\partial u}{\partial z} + \frac{j}{2}\left(\beta_2 + jg_{sat}T_2^2\right)\frac{\partial^2 u}{\partial T^2} - \frac{\beta_3}{6}\frac{\partial^3 u}{\partial T^3} = j\gamma|u|^2 u + \frac{1}{2}\left(g_{sat} - \alpha\right)u \qquad (4.25)$$

where
  u represents the slowly varying envelope of the optical field
  β_2 and β_3 account for the second- and third-order fiber dispersion, respectively
  γ is the nonlinear coefficient

The fiber attenuation is denoted by α and the gain by g_{ss}. The finite bandwidth is given by $\Delta\omega = 2/T_2$, where T_2 is the dipole relaxation time. The gain including the gain saturation effect is as follows:

$$g_{sat}(z) = \frac{g_0}{1 + \dfrac{P_{ave}(z)}{P_{sat}}} \qquad (4.26)$$

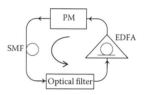

FIGURE 4.24

A circulating model for simulating the FM mode-locked fiber ring laser. SMF, single mode fiber; EDFA, erbium-doped fiber amplifier; PM, phase modulator.

where g_0 is the small-signal gain that is approximately constant, and the location-dependent average power is given by

$$P_{ave}(z) = \frac{1}{T_m} \int_{-T_m/2}^{T_m/2} |u(t,z)|^2 dt \qquad (4.27)$$

where $T_m = 1/f_m$, with f_m being the modulation frequency. The model of common single-mode fibers is simply obtained by setting g_0 to zero in Equation 4.26.

When the active erbium-doped fiber length is short or the dispersion and the nonlinearity in this fiber are negligible, the amplification of the signal including the saturation of the EDFA can be modeled as a black box with the amplified output as

$$u_{out} = \sqrt{G} u_{in} \qquad (4.28)$$

and

$$G = G_0 \exp\left(-\frac{G-1}{G}\frac{P_{out}}{P_s}\right) \qquad (4.29)$$

where

G is the amplification factor
G_0 is the unsaturated amplifier gain
P_{out} and P_{sat} are the output power and saturation power, respectively

The phase modulator is treated as if it modifies an input field u_{in} according to

$$u_{out} = u_{in} e^{j\Omega(t)} \qquad (4.30)$$

where $\Omega(t) = m\cos[\omega_m(t + T_s) + \varphi_0]$, m is the phase modulation index and $\omega_m = 2\pi f_m$ is angular modulation frequency, which is assumed to be a harmonic of the fundamental frequency of the fiber loop, T_s is the time delay between the modulator and the pulse reference frame that is accounted for the detuning effect. In the case of rational harmonic mode locking, the $\Omega(t)$ can be replaced by a summation of a series of cosine functions: $\Omega(t) = \sum_{n=1}^{\infty} A_n \cos[n\omega_m t + \varphi_n]$, where A_n and φ_n are the modulation index and phase delay bias for each frequency component, respectively.

We also have verified our program by comparing its results with analytically predicted ones in some simplified cases [7]. A typical result obtained from the model can be seen in Figure 4.25. The model also allows us to simulate the fiber laser not only in a single-pulse time window (Figure 4.25a) but also in a multi-pulse one (Figure 4.25b). The detuning effect can be easily investigated by using this model.

The model has been used in this simulation program to investigate the bound states in the FM mode-locked fiber laser. The results obtained from our model verified the formation of multi-soliton bound states in the FM

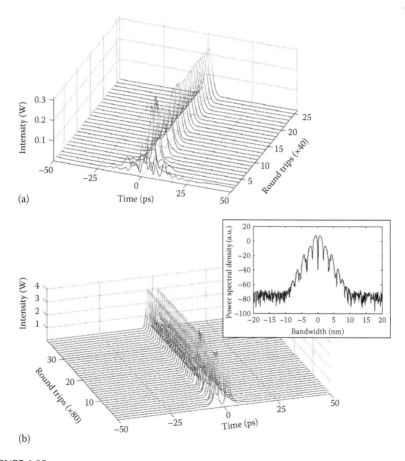

(a)

(b)

FIGURE 4.25
Numerical simulation of the building of a mode-locked pulse (a) and a bound soliton pair (b) from the noise inside a FM mode-locked fiber laser (Inset: spectrum of the bound state).

mode-locked fiber laser. Furthermore, the evolution of multi-bound solitons under various conditions of the fiber cavity has been also investigated.

Besides, this program has been employed to investigate the temporal imaging effects in the fiber systems. The duality-based applications such as repetition-rate multiplication and adaptive equalization have been simulated. By using active phase modulation, we have proposed a simple method to diminish the laser sensitivity to the phase mismatch of the Talbot condition and improve the quality of multiplied pulse sequence. More details of simulation and experimental results on bound solitons generated by FM mode-locked fiber lasers are described in Chapter 10.

4.6 Concluding Remarks

A novel temporal domain simulation has been developed to generate and propagate photonic pulses in mode-locked fiber lasers and presented in this chapter. The NLSE has been solved fully in both time and spatial domains using the finite-difference method without the need of the conversion of optical signals into the frequency domain, unlike that employed in the SSF method commonly used by several published works [8]. This development and demonstration by the results we have obtained are very significant and allow the simulation of mode-locked lasers, especially for the generation of multiple pulses (e.g., multi-bound solitons), in such fiber laser structures. To the best of our knowledge, this simulation has been developed for the first time.

The main findings that can be drawn from this chapter are summarized as follows. First, the propagation of an optical pulse inside a fiber has been simulated to ensure the validity of our numerical approach. The propagation of lightwave pulses has been tested and compared when the fiber is under linear and nonlinear operating conditions that correspond to the linear dispersion effect and/or under the SPM effect. It is noted that only the SPM effect is included in this simulator. Other nonlinear effects such as Raman scattering and four-wave mixing effects can be easily incorporated without any difficulty. It is found that the GVD and SPM effects play important roles in the propagation of the pulse. Under the GVD effect, different chromatic components of the pulse travel with different velocities, which cause the pulse to spread out in the temporal domain while its spectrum remains unchanged. On the other hand, the nonlinear phase shift induced by the SPM effect widens the pulse spectrum while keeping the pulse shape in the temporal domain unchanged. When both these effects are combined together on a hyperbolic-secant pulse, their effects cancel out each other and the result is that the pulse shape and spectrum remains unchanged while the pulse propagates inside the fiber.

Second, the pulse formation in a harmonic mode-locked fiber ring laser can be simulated via the excitation of lightwaves via the NLSE. The formation of N mode-locked pulses traveling in the ring occurs when the laser is locked to the Nth harmonic, which has been demonstrated in our simulation technique. The effects of the modulation frequency on the output pulse characteristics such as pulse width and bandwidth have also been studied. The pulse width decreases linearly while the bandwidth increases linearly with the square root of the modulation frequency. The effect of the modulation depth on the output pulses has also been examined. It is found that as the modulation depth increases, the pulse width decreases slightly and the bandwidth increases a little. The filter bandwidth has been found to have a great effect on the pulse width and bandwidth. The pulse width decreases exponentially with an increase in the filter bandwidth. Hence, a wider filter bandwidth is preferred for shorter pulse generation. However, in practice, our fiber laser can easily be locked using a narrower filter bandwidth than with a wider one.

The output pulse characteristic has also been examined with the variation of the pump power. As the pump power increases, the output pulse power increases but the pulse width and bandwidth remain the same. This is consistent with the experiment result that the pump power does not affect the pulse width and bandwidth as the fiber ring length is short compared to the nonlinear length of the fiber.

Finally, rational harmonic mode locking has been numerically demonstrated. By detuning the modulation frequency to half the fundamental frequency, doubling of the repetition rate has been realized. The simulation results of rational harmonic mode locking have been demonstrated with the multiplication of the repetition rate of the mode-locked pulse trains. However, there is amplitude fluctuation in the output pulses and hence amplitude equalization should be employed in the RHML laser.

In summary, this chapter has demonstrated the simulation scheme that can generate mode locking and propagate lightwave pulse trains. The technique presented here can be applied not only to mode-locked lasers but also in other fiber structures such as optical fiber filters [9]. Furthermore, it has demonstrated the numerical implementation of three fundamental structures of mode-locked lasers including the harmonic, regenerative, and rational harmonic types. Furthermore, we have also observed a new phenomenon in which an ultrahigh peak pulse train has been generated.

A simulation package has been described to numerically study the generation of pulse trains by the mode-locking technique. Our approach is based on the finite difference in the temporal and spatial domains. To the best of our knowledge, this is the first time that mode locking has been modeled using the propagation of lightwave via the NLSE, the optical transmission characteristics of photonic elements incorporated in the ring resonator, and the complete temporal domain approach. Modeling results are confirmed with those obtained experimentally. Theory on an active harmonic mode-locked fiber laser has been presented. An active HML-FL has been implemented.

The stability of the laser can be improved by employing a fiber ring length in the control feedback loop. The fiber is wounded on a PZT drum whose diameter can be controlled by applying an electrical voltage on the control electrodes. The output of the laser is monitored and compared with a reference signal. The difference is then used to control the PZT. Therefore, any change of the fiber length due to the variation of environment is fed back to the system and it automatically adjusts the PZT to make the fiber length unchanged.

In addition, the amplitude fluctuation of the output pulse train in the RHML can be equalized by employing an intracavity Fabry–Perot filter (FFP) whose free spectral range (FSR) is equal to the repetition rate of the pulse train. With the use of FFP, all supermodes and the mode induced by the modulation frequency can be filtered out, resulting in a highly stable amplitude-equalized mode-locked pulse train.

Finally, the experimental confirmation of simulated pulses and spectra can be conducted, as described in Chapter 3. With advances in desktop computers with parallel processors, the parallel computing technique can be applied to increase the simulation speed so that the characteristics of the mode-locked fiber lasers can be intensively studied within a small temporal window.

References

1. G.P. Agrawal, *Applications of Nonlinear Fiber Optics*, Academic Press, San Diego, CA, 2001.
2. G.P. Agrawal, *Nonlinear Fiber Optics*, San Francisco, CA, 2001.
3. B. Bakhshi and P.A. Andrekson, 40 GHz actively modelocked polarization maintaining erbium fiber ring laser, *Electron. Lett.*, 36(5), 411–412, 2000.
4. M. Nakazawa, E. Yoshida, and Y. Kimura, Ultrastable harmonically and regeneratively modelocked polarisation-maintaining erbium fibre ring laser, *Electron. Lett.*, 30, 1603–1605, 1994.
5. (a) D.-W. Huang, W.-F. Liu, and C.C. Yang, Actively Q-switched all-fiber laser with a fiber Bragg grating of variable reflectivity, in *Optical Fiber Communication Conference, 2000*, Baltimore, MD, 2000; (b) J. Liu, D. Shen, S.-C. Tam, and Y.-L. Lam, Modeling pulse shape of Q-switched lasers, *IEEE J. Quantum Electron.*, 37, 888–896, 2001.
6. (a) P.R. Morkel, K.P. Jedrzejewski, E.R. Taylor, and D.N. Payne, Short-pulse, high-power Q-switched fiber laser, *IEEE Photon. Technol. Lett.*, 4, 545–547, 1992; (b) T.R. Schibli, U. Morgner, and F.X. Kartner, Control of Q-switched mode locking by active feedback, in *Lasers and Electro-Optics, 2000 (CLEO 2000) Conference*, San Francisco, CA, 2000.

7. D.J. Kuizenga and A.E. Siegman, FM and AM mode locking of the homogenous laser—Part I: Theory, *IEEE J. Quantum Electron.*, 6, 694–708, 1970.
8. G.P. Argrawal, *Nonlinear Fiber Optics*, Academic Press, San Diego, CA, 2002.
9. M.K. Christi and J.H. Zhao, Optical filter design, in *Optical Filter Design and Analysis: A Signal Processing Approach*, John Wiley & Sons, New York, 1999.

5

Dispersion and Nonlinearity Effects in Active Mode-Locked Fiber Lasers

Mode-locked fiber lasers are normally operating in an environment in which the guided medium is a single-mode optical waveguide incorporating optical amplifiers for compensating the medium's losses. The energy levels of the optical fields circulating in the fiber laser normally reach the nonlinear threshold level. Thus, the optical fields would experience both linear dispersion and nonlinear distortion effects, and the nonlinear phase noises. These effects result in the phase-to-amplitude conversion of the amplifier noises under nonlinear operating conditions, and hence phase perturbation on optically modulated signals. This chapter presents the dispersion and nonlinearity effects in active MLFLs whose analysis has also been described in Chapter 2.

5.1 Introduction

Fiber lasers have several interesting characteristics that distinguish them from other solid-state or gas lasers. The low loss of the fiber (0.2 dB/km attenuation at 1550 nm) makes it possible to form a low-loss cavity and thus low threshold laser. Since the intensity in the fiber is very high due to its small core area, a nonlinear cavity can be easily obtained. Moreover, the cavity dispersion can also be easily controlled by using fibers with different GVDs. With the introduction of nonlinearity and dispersion into the cavity, lasers with interesting quality and performance can be obtained by optimizing the laser parameters.

In fact, the effects of dispersion and nonlinearity on an optical pulse when it propagates through a fiber have been studied extensively using nonlinear Schrödinger equation (NLSE) [1–8]. By solving the NLSE using the inverse scattering method [9], the authors obtained the solution in form of a soliton whose shape is kept unchanged if the dispersion and nonlinear effects are balanced. Although a soliton in an optical fiber was predicted in 1973 [10], it was not experimentally observed until 1980 [11] due to the lack of a short optical pulse source [12]. Numerically solving the NLSE using the split step Fourier [13] or finite-difference methods shows that the pulse can be compressed by the nonlinear effect or be broadened by the dispersion effect.

Since pulses are formed and travel in actively MLFL, dispersion and nonlinearity of the fiber cavity would play an important role in this mode-locking process. By introducing the master equation for mode locking and numerical simulation, Haus and Silberberg showed that the SPM could shorten the mode-locked pulse by a factor of 2 [14]. Later, Kartner et al. added that the pulse could be further shortened when the anomalous dispersion is introduced into the cavity [15]. The pulse shortening is the result of a soliton pulse forming in the cavity and has been reported in many papers [15–24]. However, the results were obtained by assuming that the mode-locking conditions are satisfied and pulse propagation in the cavity can be approximately described by the master equation. The performance of the laser and the characteristic of the pulse when the laser is detuned were not discussed.

In this chapter, we present a detailed study on the linear dispersion and nonlinear effects on the performance of actively MLFLs. The characteristics of the mode-locked pulses are discussed not only when the laser is exactly tuned to but also when the laser is detuned from the modulation frequency. We show that there is a trade-off between the pulse shortening and the stability of the laser. Cavity dispersion and nonlinearity should be optimized to obtain the shortest pulse while the laser is still stable within a certain amount of detuning.

The structure of this chapter is as follows. Propagation of optical pulses in a fiber is given in Section 5.2 where the effects of dispersion and nonlinearity on the propagating pulse are discussed. Dispersion and nonlinear effects on properties of the mode-locked pulse and locking range of the laser are discussed in Sections 5.3 and 5.4, respectively. Soliton formation under the combined effects of dispersion and nonlinearity is examined in Section 5.5. The numerical results are confirmed with experimental results presented in Section 5.6. And finally, conclusions are given in Section 5.7.

5.2 Propagation of Optical Pulses in a Fiber

Figure 5.1 shows the laser model that consists of an optical amplifier, a BPF, a fiber, and an intensity modulator. The optical wave is amplified in the optical amplifier and then filtered by the BPF. After filtering, the lightwave propagates through the fiber, where it experiences the dispersion and nonlinear

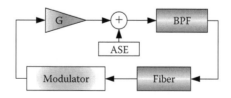

FIGURE 5.1
General schematic model of actively MLFLs.

effects, and is finally modulated by the optical modulator. The output of the modulator is fed back into the amplifier to create a loop.

The modeling equations for the optical propagation through the optical fiber sections, the optical modulator, and the optical filter are described in Chapter 4. The following subsections investigate the linear- and nonlinear-phase evolutionary effects of the fiber on the performance of the generated optical pulses.

5.2.1 Dispersion Effect

When propagating through free space, the lightwave travels at a constant speed c. However, when traveling through a material rather than free space, the light speed is slowed down due to the interaction between the light's electromagnetic field and the material's electron cloud. The degree to which the light is slowed down is given by the material refractive index n, which relates to the speed of the light in the material by

$$v = \frac{c}{n} \tag{5.1}$$

Since the interaction of the electromagnetic field with the material depends on the frequency of the light, the refractive index also depends on the frequency of light. This property is referred to as chromatic dispersion. Chromatic dispersion is also called material dispersion and is given by

$$D_M(\lambda) = \frac{\lambda}{n} \frac{d^2 n}{d\lambda^2} \tag{5.2}$$

in which the refractive index n is a function of wavelength λ and

$$n^2(\lambda) = 1 + \sum_{j=1}^{m} \frac{G_j \lambda^2}{\lambda - \lambda_j^2} \tag{5.3}$$

where G_j are the Sell Meier constants.

A plot of the refractive index over the wavelength region for optical communication for a pure silica fiber is shown in Figure 5.2.

Chromatic dispersion plays a critical role in the propagation of optical pulses. If the optical pulse contained only one frequency component, it would travel through the material without any distortion. In reality, an optical pulse actually includes a range of different frequencies. The shorter the pulse the wider the range of frequency it covers. Different frequency components will travel at different speeds; some travel faster, some lag behind. This results in the broadening of the pulse.

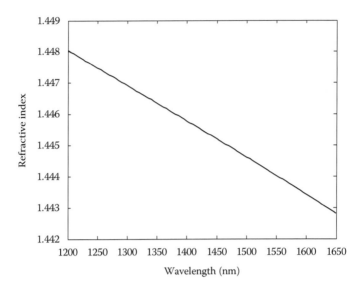

FIGURE 5.2
Variation in the silica refractive index as a function of optical wavelength.

Besides chromatic dispersion, when traveling in an optical fiber, the light also experiences an additional dispersion caused by the structure of the optical fiber/waveguide. This dispersion is called waveguide dispersion and is given by

$$D_W(\lambda) = -\frac{n_2(\lambda)\Delta}{c\lambda} V \frac{d^2(Vb)}{dV^2} \tag{5.4}$$

where
$n_2(\lambda)$ is refractive index of the core
Δ is the relative refractive index difference
b is the normalized propagation constant
V is the normalized frequency

The total dispersion of the fiber is the combination of chromatic dispersion and waveguide dispersion:

$$D(\lambda) = D_M(\lambda) + D_W(\lambda) \tag{5.5}$$

A plot of the fiber dispersions over the wavelength region for optical communication is shown in Figure 5.3.

The propagation of an optical pulse in a fiber is characterized by a mode propagation constant β. The mode propagation constant β can be expanded in a Taylor series about the center frequency ω_0

$$\beta(\omega) = \frac{n(\omega)\omega}{c} = \beta_0 + \beta_1(\omega - \omega_0) + \frac{1}{2}\beta_2(\omega - \omega_0)^2 + \cdots \tag{5.6}$$

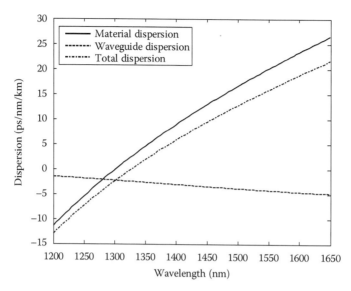

FIGURE 5.3
Fiber dispersion as a function of optical wavelength.

where

$$\beta_m = \left.\frac{d^m \beta}{d\omega^m}\right|_{\omega=\omega_0} \quad (m = 0, 1, 2, \ldots) \tag{5.7}$$

The pulse envelope moves at the group velocity ($v_g = 1/\beta_1$) while the parameter β_2 is responsible for pulse broadening. β_2 is generally referred to as the GVD parameter and relates to the chromatic dispersion parameter D by

$$\beta_2 = -\frac{\lambda^2 D}{2\pi c} \tag{5.8}$$

The fiber is said to exhibit normal dispersion for $\beta_2 > 0$, and to exhibit anomalous dispersion for $\beta_2 < 0$. In the normal dispersion regime, high frequency (blue-shifted) components of an optical pulse travel slower than the low frequency (red-shifted) components. The opposite occurs in the anomalous dispersion regime.

Dispersion can be a major problem in fiber optic communication systems due to its effect on pulse broadening. Broadened pulse has lower peak intensity than the initial pulse's intensity, making it more difficult to detect. Worse still, pulse broadening can cause two adjacent pulses to overlap, leading to errors at the receiver. However, dispersion is not always a harmful effect. When combined with the SPM effect, it can lead to the formation of an optical soliton, the wave that keeps its shape unchanged during its propagation through the fiber.

5.2.2 Nonlinear Effect

The fiber refractive index not only varies with wavelength but also varies with the intensity of the light traveling inside the fiber. The interaction between the light and the material electron cloud increases nonlinearly with the light intensity, so the high-intensity light travels slower than the low-intensity one. The result is that the refractive index of the fiber increases with the intensity of light. This is known as the fiber nonlinearity or the Kerr effect.

The fiber refractive index can be written as a function of wavelength and light intensity as

$$n(\lambda, |E|^2) = n(\lambda) + n_2 |E|^2 \tag{5.9}$$

where
$n(\lambda)$ is the linear part given by Equation 5.3
$|E|^2$ is the optical intensity inside the fiber
n_2 is the nonlinear-index coefficient

There are several effects of the fiber nonlinearity on the optical pulse but the most concern to the pulse propagation theory is the SPM effect. Under the SPM effect, the optical pulse experiences a phase shift induced by the intensity-dependent refractive index while the pulse shape remains unchanged. The nonlinear phase shift induced to the optical pulse traveling a distance L is given by

$$\phi_{NL}(L, T) = |U(0, T)|^2 \left(\frac{L}{L_{NL}} \right) \tag{5.10}$$

where
$U(0,T)$ is the normalized pulse envelope at $z=0$
L_{NL} is the nonlinear length defined as

$$L_{NL} = (\gamma P_0)^{-1} \tag{5.11}$$

where
P_0 is the peak power of the incident pulse
γ is the nonlinear coefficient
γ can be calculated from the nonlinear-index coefficient n_2 as

$$\gamma = \frac{n_2 \omega_0}{c A_{eff}} \tag{5.12}$$

where A_{eff} is the effective area of the fiber.

The temporally varying phase ϕ_{NL} implies that the instantaneous optical frequency differs from its central frequency ω_0. The frequency difference $\delta\omega$ is given by

$$\delta\omega(T) = -\frac{\partial\phi_{NL}}{\partial T} = -\left(\frac{L}{L_{NL}}\right)\frac{\partial|U(0,T)|^2}{\partial T} \tag{5.13}$$

It can be seen from Equation 5.13 that new frequency components are continuously generated as the pulse propagates down the fiber. Those new frequencies broaden the spectrum over its initial value at $z=0$. The time dependence of $\delta\omega$ is referred to as frequency chirping.

5.2.3 Soliton

Consider the case that an optical pulse is launched into a fiber with $\beta_2<0$, anomalous dispersion. SPM leads to lower frequency components at the leading edge of the pulse and higher frequency components at the trailing edge. In the anomalous dispersion regime, lower frequency components travel slower than higher frequency components. Therefore, the anomalous dispersion causes the leading edge of the pulse to slow down while the trailing edge travels faster. Effectively, anomalous dispersion compresses the pulse and undoes the frequency chirp induced by SPM. If the pulse has the shape so that the effects of SPM and GVD cancel out with each other, then the pulse maintains its initial width along the entire length of the fiber. This solitary wave pulse is called a soliton.

The pulse that has the above property is the sech profile pulse, which is a solution of the NLSE and is defined as

$$U(T) = P_0 \operatorname{sech}\left(\frac{T}{T_0}\right) \tag{5.14}$$

where
P_0 is the peak power
T_0 is the pulse width related to the FWHM T_{FWHM} as

$$T_0 = \frac{T_{FWHM}}{1.665} \tag{5.15}$$

The peak power P_0 required to support the fundamental (or first-order) soliton is given by

$$P_0 = \frac{|\beta_2|}{T_0^2} \tag{5.16}$$

In a real system, it is difficult to generate pulses with exact parameters specified above for soliton formation. In other words, the practically generated
pulse is not a solution of the NLSE and thus, strictly speaking, a soliton cannot be formed. However, theoretical and experimental results show that if the
pulse does not have the hyperbolic-secant shape and the exact power level as
specified in Equation 5.16, soliton pulse can still be formed with the condition that the initial parameter values of the soliton must be close enough to
the final parameter values of the soliton [12,25]. The interplay between the
GVD and SPM effects can molt the pulse into a hyperbolic-secant shape with
an appropriate peak power.

5.2.4 Propagation Equation in Optical Fibers

Propagation of optical pulses in an optical fiber, including all linear and nonlinear effects, can be described by the following NLSE:

$$\frac{\partial A}{\partial z} + \beta_1 \frac{\partial A}{\partial t} + \frac{j}{2}\beta_2 \frac{\partial^2 A}{\partial t^2} - \frac{1}{6}\beta_3 \frac{\partial^3 A}{\partial t^3} + \frac{\alpha}{2} A = j\gamma \,|A|^2 A - k_1 \frac{\partial(|A|^2 A)}{\partial t} - k_2 A \frac{\partial |A|^2}{\partial t}$$

$$(5.17)$$

where

$j = \sqrt{-1}$

A is the slowly varying pulse envelope

z is the axial distance

t is the time

β_m is the mth order derivative of the mode-propagation constant β defined
in Equation 5.7

α is the fiber loss

γ is the nonlinear coefficient

k_1 is given by $k_1 = \gamma/\omega_0$, and k_2 is given by $k_2 = j\gamma T_R$

T_R is related to the slope of the Raman gain spectrum and can be approximated to be about $3\,fs$

Making the transformation

$$T = \frac{t - z}{v_g} = t - \beta_1 z \tag{5.18}$$

Equation 5.17 becomes

$$\frac{\partial A}{\partial z} + \frac{j}{2}\beta_2 \frac{\partial^2 A}{\partial T^2} - \frac{1}{6}\beta_3 \frac{\partial^3 A}{\partial T^3} + \frac{\alpha}{2} A = j\gamma \,|A|^2 A - k_1 \frac{\partial(|A|^2 A)}{\partial T} - k_2 A \frac{\partial |A|^2}{\partial T}$$

$$(5.19)$$

On the left-hand side of the equation, the second term is responsible for the GVD, the third term governs the effects of third-order dispersion, and the last term accounts for the loss. On the right-hand side, the first term governs the nonlinear effects such as the SPM and plays an important role in the formation of solitons. The term proportional to k_1 is responsible for the self-steepening and shock formation. The last term is related to Raman scattering and accounts for the self-frequency shift.

For picosecond pulses, the effects of self-steepening and self-frequency shifts are so small that the last two terms on the right-hand side can be neglected. Since the contribution of the third-order dispersion term is also quite small (when the operating or channel wavelengths are far away from the zero-dispersion wavelength of the fiber, which is often the case for WDM systems), the third term on the left-hand side can also be neglected. Using these assumptions, the propagating equation can be simplified to

$$\frac{\partial A}{\partial z} + \frac{j}{2}\beta_2\frac{\partial^2 A}{\partial T^2} + \frac{\alpha}{2}A = j\gamma\,|A|^2\,A \tag{5.20}$$

In the special case where $\alpha=0$, Equation 5.20 becomes the well-known NLSE describing the propagation of moderately short pulses. The fundamental soliton solution for NLSE is given by

$$A(z,T) = \sqrt{P_0}\,\mathrm{sec}\,h\left(\frac{T}{T_0}\right)\exp\left(\frac{jz}{2L_D}\right) \tag{5.21}$$

where
  T_0 is the pulse width
  P_0 is the peak-power-satisfied Equation 5.16
  L_D is the dispersion length defined as

$$L_D = \frac{T_0^2}{|\beta_2|} \tag{5.22}$$

5.3 Dispersion Effects in Actively Mode-Locked Fiber Lasers

5.3.1 Zero Detuning

To study the dispersion effects in actively MLFLs, we set the nonlinearity parameter in the fiber model to zero so that the laser cavity exhibits only the dispersion property. The cavity dispersion in ps/nm is determined as

$$D_{cav} = D_f L \qquad (5.23)$$

where
D_f is the fiber dispersion in ps/nm/km
L is the fiber length

For convenience, we normalize it using

$$D_n = \frac{D_{cav}}{D_{ref}} \qquad (5.24)$$

in which

$$D_{ref} = \frac{2\pi c (\tau_G / 1.703)^2}{\lambda_0^2} \qquad (5.25)$$

where τ_G is the pulsewidth of the Gaussian pulse generated when the fiber effect is ignored. For given values of the modulation frequency f_m, modulation index m, and filter bandwidth Ω, we have

$$\tau_G = 2(2\ln 2)^{3/4} \sqrt[4]{\frac{1}{m\omega_m^2 \Omega^2}} \qquad (5.26)$$

The simulation parameters are as follows: central wavelength $\lambda = 1550\,\text{nm}$, filter bandwidth $BW = 1.2\,\text{nm}$, modulation frequency $f_m = 10\,\text{GHz}$, modulation index $m = 0.5$, small signal gain $G_{ss} = 20\,\text{dB}$, noise figure $NF = 3\,\text{dB}$, and the total cavity loss $\Gamma = 6\,\text{dB}$.

Figure 5.4 shows the buildup process of the pulse in an actively MLFL with a normalized cavity dispersion of 0.2. Similarly to the case of zero-dispersion cavity shown in Figure 5.4b, the pulse rapidly reaches its steady state after about 100 round trips. However, the pulse width for the dispersion cavity is much larger than that of the zero-dispersion cavity. Steady-state pulses with different cavity dispersion values are plotted in Figure 5.5. It is shown that the higher the dispersion value the wider the pulse width becomes.

The dependence of the pulse width on the cavity dispersion is presented in Figure 5.6. It shows a symmetric dependence of the pulse width on the cavity dispersion over the zero-dispersion value. Either the normal dispersion or anomalous dispersion has the same broadening effect on the pulse. The pulse width increases from 13.6 ps for zero dispersion to up to 32.8 ps for an absolute normalized cavity dispersion value of 1.

The pulse-broadening effect of dispersion on the mode-locked pulse can be explained due to the velocity difference between different frequency

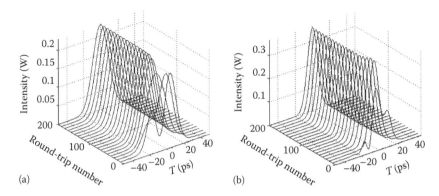

FIGURE 5.4
Pulse buildup in an actively MLFL with (a) normalized cavity dispersion of 0.2 and (b) zero dispersion.

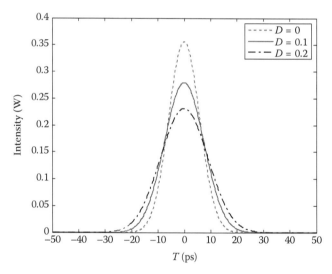

FIGURE 5.5
Steady-state pulses with different cavity dispersions; increasing of the cavity dispersion broadens the mode-locked pulse due to the velocity mismatch of different frequency components in the pulse.

components in the mode-locked pulse. In the anomalous dispersion cavity ($D > 0$), low-frequency (red-shifted) components of the pulse travel slower than the high-frequency (blue-shifted) ones. This results in the broadening of the pulse with high frequencies at the leading edge and low frequencies at the trailing edge. The pulse is called down-chirped pulse. On the opposite, low-frequency components of the pulse travel faster than the high-frequency components in a normal dispersion cavity ($D < 0$). This also causes the pulse

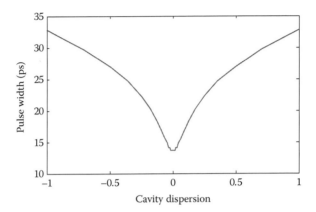

FIGURE 5.6
Symmetric dependence of the pulse width on the normalized cavity dispersion over the zero-dispersion value.

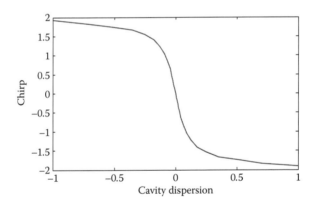

FIGURE 5.7
Up-chirp pulses are generated in a normalized dispersion cavity MLL while down-chirp pulses are generated in an anomalous dispersion cavity.

to broaden but the low frequencies are at the leading edge while the high frequencies are at the trailing edge. And hence an up-chirped pulse is obtained. This is clearly confirmed in Figure 5.7, which shows that up-chirp pulses are generated in a normal dispersion cavity MLL while down-chirp pulses are generated in the anomalous dispersion cavity.

5.3.2 Dispersion Effects in Detuned Actively Mode-Locked Fiber Lasers

As discussed in Chapter 4, when the modulation frequency is slightly detuned from the harmonic of the fundamental frequency, the laser is still under mode-locking states. However, the pulse position is no longer at the transmission peak of the modulation function but shifted linearly with the detuning frequency. The position shift causes a pushing (pulling) effect on the pulse when it passes through the modulator and hence compensates for the detuning. In a dispersion cavity, the shifting of the pulse when detuned is also observed, as shown in Figure 5.8. The figure shows the pulse buildup processes of 10 GHz actively MLFLs when the modulation frequency is detuned by an amount of $\delta f_m = 8 \times 10^{-6}$ ($\Delta f_m = 80$ kHz) for the case of zero dispersion and dispersion equal to 0.1. Compared to the zero-dispersion cavity, the position shifting in a dispersion cavity is less severe.

Figures 5.9 and 5.10 show the effect of dispersion on the steady-state pulse and on its position when the laser is detuned. Both the normal dispersion

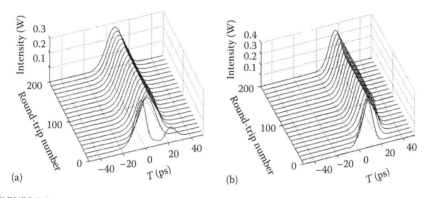

(a)   (b)

FIGURE 5.8
Pulse buildup in an actively MLFL with a detuning frequency of $\delta f_m = 8 \times 10^{-6}$; (a) normalized cavity dispersion $D = 0.1$ and (b) zero dispersion.

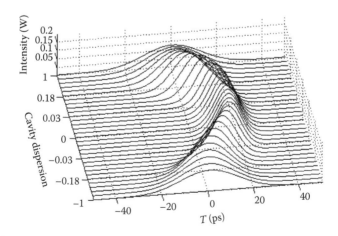

FIGURE 5.9
Steady-state pulses of a detuned MLFL ($\delta f_m = 8 \times 10^{-6}$) with different normalized cavity dispersion values; large dispersion causes the pulses to broaden but reduces the pulse position–shifting effect when detuned.

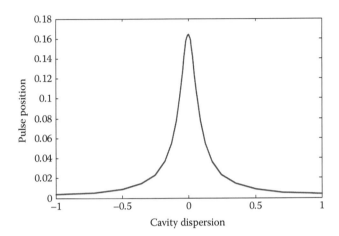

FIGURE 5.10
Increasing of the normalized cavity dispersion (absolute value) reduces the pulse-shifting effect when detuned.

and anomalous dispersion cause the same effects on the detuned laser. The dispersion broadens the pulse but keeps it from shifting far away from the transmission peak. It can be understood that as the pulse becomes wider, it will experience a stronger shaping effect by the modulation, which pushes (or pulls) it back to the transmission peak. Therefore, the pulse is shifted less in the dispersion cavity than in the zero dispersion although the detuning frequency is the same.

Besides the pulse position–shifting effect on the detuned MLFLs, the dispersion also shifts the optical carrier toward a faster/slower wavelength component to compensate for the delay/advance caused by detuning. In the zero-dispersion cavity, all wavelengths travel at the same speed. Therefore, the only way to keep the pulse synchronized with the modulation frequency is to shift the pulse position so that the pushing/pulling effect of the modulator transmission function compensates for the detuning. However, in a dispersion cavity, different wavelengths travel at different velocities. Instead of lasing at the central wavelength, the laser can shift the wavelength to a slower or faster one so that the change of propagation speed cancels the frequency detuning. Figure 5.11 shows the shifting of the lasing wavelength in a positive detuned MLFL when the dispersion varies from the normal dispersion to the anomalous dispersion. It can be seen that the lasing wavelength is shifted to a lower frequency for a normal dispersion cavity ($D < 0$). This is because a lower-frequency component travels faster in a normal dispersion cavity and hence it compensates for the delay caused by the positive detuning. On the other hand, the lasing wavelength is shifted to a higher-frequency component when the cavity dispersion is anomalous ($D > 0$).

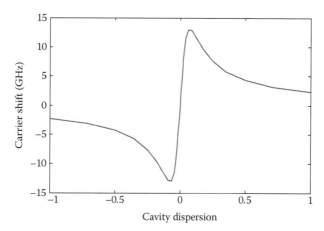

FIGURE 5.11
Change of carrier frequency when the normalized cavity dispersion increases in a detuned actively MLFL ($\Delta f_m = 80\,\text{kHz}$).

5.3.3 Locking Range

Since the dispersion assists the compensation in the detuning by shifting the lasing wavelength, the locking range of the laser is increased with the increasing of dispersion, as shown in Figure 5.12. The locking range has been enhanced from only 9×10^{-6} for zero dispersion to 2.2×10^{-4} for the normalized dispersion of one. This improvement is achieved with the degradation (broadening) of the pulse width. The pulse width has been increased to 28.5 ps due to the dispersion of the single-mode fiber. However,

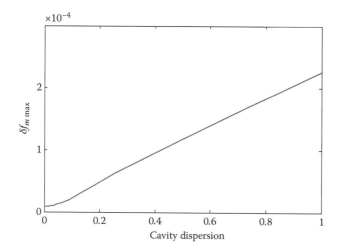

FIGURE 5.12
Locking range vs. the normalized cavity dispersion.

it is noted that the pulse is heavily chirped and hence it can be compressed by propagating through an appropriate length of a dispersive fiber outside the cavity [12].

5.4 Nonlinear Effects in Actively Mode-Locked Fiber Lasers

5.4.1 Zero Detuning

The cavity nonlinearity is determined as

$$\gamma = \frac{2\pi L n_2}{\lambda_0 A_{eff}} \tag{5.27}$$

where
L is the fiber length
n_2 is the fiber nonlinear-index coefficient
λ_0 is the operating wavelength
A_{eff} is the effective area of the fiber

For convenience, we normalize the cavity nonlinearity using

$$\gamma_n = \frac{\gamma P_{sat}}{1.5\tau_G f_m k_{sat}} \tag{5.28}$$

where
P_{sat} is the saturation power of the EDFA
f_m is the modulation frequency
k_{sat} is the saturation coefficient
τ_G is the pulse width given in (5.26)

Figure 5.13 shows the pulse buildup process in an actively MLL with a normalized cavity nonlinearity of 0.2. A steady state pulse is obtained after about 200 round trips and its pulse width is shorter than that without the nonlinearity effect. Comparison between the steady-state pulses of different nonlinearity values and the dependence of pulse width on the cavity nonlinearity are given in Figures 5.14 and 5.15, respectively. As the nonlinearity increases, the pulse becomes shorter and shorter. Since the nonlinearity introduces a nonlinear phase variation to the optical pulse, and the phase change generates new frequencies, the spectrum of the optical pulse is broadened. When the spectrum becomes wider, more modes can be locked and hence the pulse becomes shorter. This is verified in Figure 5.16, which

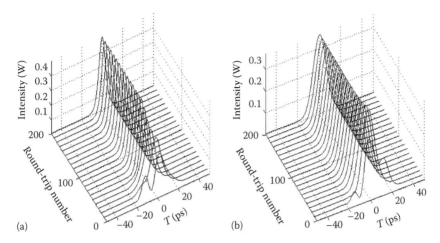

FIGURE 5.13
Pulse buildup in an actively MLFL with (a) normalized cavity nonlinearity of 0.2 and (b) zero nonlinearity.

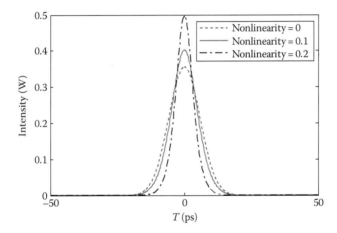

FIGURE 5.14
Pulses with different normalized cavity nonlinearity values.

shows that the spectrum of the pulse with high nonlinearity is wider than that without the nonlinearity effect.

It can be seen from Figures 5.14 and 5.16 that when the normalized non-linearity increases from 0 to 0.2, the pulse width is shortened by half, while the pulse bandwidth is increased four times. This indicates that the pulse is chirped when the nonlinearity increases.

The dependence of the pulse chirp on the cavity nonlinearity is shown in Figure 5.17. The pulse is up chirped with low-frequency components at the leading edge and high-frequency components at the trailing edge. Recalled

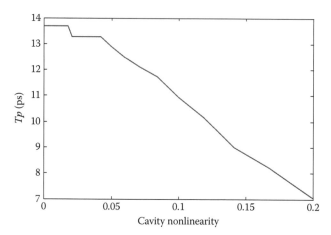

FIGURE 5.15
Pulse shortening when the normalized nonlinearity increases.

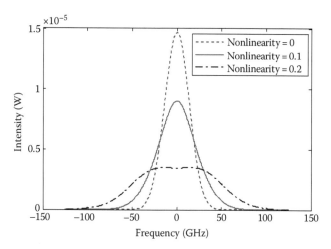

FIGURE 5.16
Pulse spectra with different normalized cavity nonlinearity values.

from Equation 5.13 for the new frequency generated due to the nonlinearity effect of the fiber

$$\delta\omega(T) = -\frac{\partial \phi_{NL}}{\partial T} = -\left(\frac{L}{L_{NL}}\right)\frac{\partial |U(0,T)^2}{\partial T} \tag{5.29}$$

For the leading edge $\partial U^2/\partial T > 0$ gives $\partial \omega < 0$, and thus the low-frequency components are generated. While for the trailing edge, $\partial \omega > 0$, high-frequency components are generated. Therefore, the pulse is up chirped.

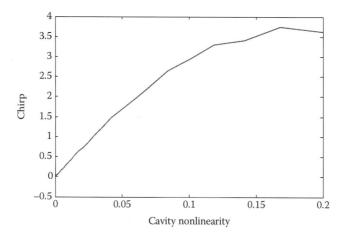

FIGURE 5.17
The nonlinearity causes the pulse up-chirped and the chirp increases as the nonlinearity increases.

5.4.2 Detuning in an Actively Mode-Locked Fiber Laser with Nonlinearity Effect

Figure 5.18a shows the steady-state pulses of an actively MLFL having a cavity nonlinearity of 0.1 with different frequency detuning values. Similar to the case of the linear cavity, the pulse position is shifted away from the transmission peak (the center) of the modulation function. For positive detune ($f_m > 0$), the pulse is delayed and the leading edge experiences less loss than

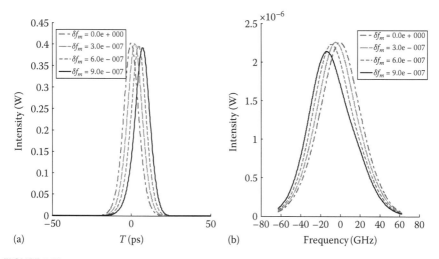

FIGURE 5.18
Steady-state pulses and spectra of an actively MLFL having a normalized cavity nonlinearity of 0.1 with different frequency detune values. (a) Temporal pulses. (b) Spectra.

the trailing edge when passing through the modulator. This pushes the pulse toward the center and hence compensates for the delay caused by the detune. And hence a steady-state pulse is formed. The position shifting also increases linearly with the detune, as shown in Figure 5.19a.

The spectra of the pulses are plotted in Figure 5.18b. It is seen that the lasing wavelength is shifted toward the lower frequency for a positive detune as in the case of normal dispersion. Although the pulse is also up chirped as in the case of normal dispersion but the principle of wavelength shifting is different. The wavelength shifting in the nonlinear case is the result of the filtering effect of the modulator. Since the pulse is up chirped, the low frequency is at the leading edge and hence it experiences lower loss than the high frequency when passing the modulator. Therefore, the spectrum is shifted toward the low-frequency region, as shown in Figure 5.19b. On the other hand, when the detuning is negative, the pulses lag behind and its trailing edge, where the high frequency is distributed, passes through the modulator at the transmission peak. The modulator now acts as a filter, which is in favor of the high-frequency components and, hence, shifts the spectrum toward the high-frequency region.

Although similar to the normal dispersion case where the pulse is up chirped and the spectrum is shifted toward the low-frequency region with a positive detune and toward the high-frequency region with a negative detune; detune in the case of the nonlinear effect is different. While the wavelength shifting in a dispersive cavity helps to compensate and balance for the detuning, the wavelength shifting in the nonlinear cavity does not give any help in compensating for the detune. The wavelength shifting is just the result of the modulator filtering effect on the up-chirped distribution

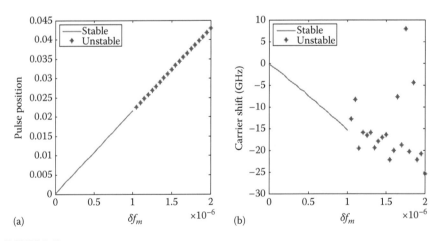

(a)

(b)

FIGURE 5.19

Shifting of the pulse position (a) and lasing wavelength (b) in a detuned actively MLFL with the nonlinear effect.

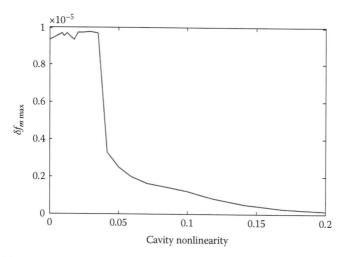

FIGURE 5.20
Detuning range vs. the normalized cavity nonlinearity.

of the frequency components of the pulse. The only effect that reacts and compensates for the detuning is the pushing or pulling effect when the pulse position is shifted. However, these pushing/pulling effects are weaker for the narrower pulse under the nonlinear effect. Therefore, the laser is more sensitive to the detuning. Figure 5.20 shows the locking range of the laser when the nonlinearity is changed. The locking range starts dropping dramatically as the nonlinearity reaches 0.04 where the nonlinearity starts taking effect on shortening the pulse, as shown in Figure 5.15.

5.4.3 Pulse Amplitude Equalization in a Harmonic Mode-Locked Fiber Laser

In harmonic MLLs, the modulation frequency is not equal to the fundamental frequency of the cavity but is a harmonic of it. There is not only one pulse but N pulses traveling together in the cavity. This results in an N times increase of the repetition rate. However, the pulse amplitude may be different from one to the other due to the super-mode noise [26–28], which causes the laser modes to jump from one group to the other group.

The amplitude difference between two pulses in a harmonic MLL without nonlinearity is clearly seen in Figure 5.21a but their amplitudes become equalized when the nonlinearity is added in the cavity, as shown in Figure 5.21b. Nonlinearity helps to equalize the pulse amplitude in harmonic MLLs through the SPM and filtering effect. The higher-energy pulse experiences a higher SPM effect and hence its spectrum becomes wider. Therefore it will experience more loss when passing through the filter. This results in the equalization of the pulse energy.

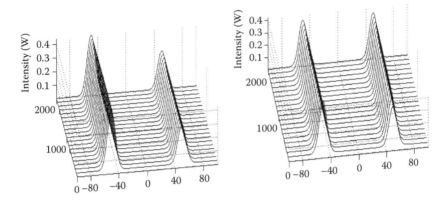

FIGURE 5.21
Harmonic mode-locked pulses without (a) and with (b) nonlinearity effect. The axes are the round-trip number and the time.

5.5 Soliton Formation in Actively Mode-Locked Fiber Lasers with Combined Effect of Dispersion and Nonlinearity

5.5.1 Zero Detuning

As seen in Sections 5.2 and 5.4, nonlinearity shortens and introduces up chirp to the pulse. The output pulse is up chirped, i.e., the low-frequency components are at the leading edge, while the high-frequency components are at the trailing edge. If the anomalous dispersion effect is also introduced into the cavity, it will slow down the leading edge, where the low-frequency components are, and hence may compensate for the up chirp caused by the nonlinearity and compress the pulse even more. Moreover, the coexistence of the anomalous dispersion and nonlinearity may lead to the formation of solitons whose pulse shape is preserved during propagation through the fiber.

As discussed in Sections 5.2 and 5.4, nonlinearity introduces up chirp to the pulse while the anomalous dispersion causes the pulse down chirp. The combination of nonlinearity and dispersion would balance those effects and generate a soliton pulse, which is what happens in the propagation of pulses in a single-mode fiber [29]. Figure 5.22 shows the dependence of the pulse width on the nonlinearity in the laser with a cavity dispersion of 0.1. The pulse width is broadened and increases from 13.5 to 16.8 ps in a linear cavity with a normalized dispersion of 0.1. However, the pulse width decreases dramatically as the nonlinearity increases in the laser and obtains 5.8 ps, less than half of that of the linear cavity, when the normalized nonlinearity is 0.2. A further increase of the nonlinearity makes the laser unstable due to the strong nonlinear phase shift of the pulse. When the dispersion is doubled,

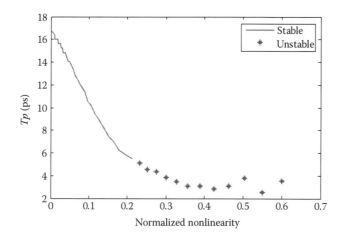

FIGURE 5.22
Pulse width shortens with an increase of nonlinearity, $D = 0.1$.

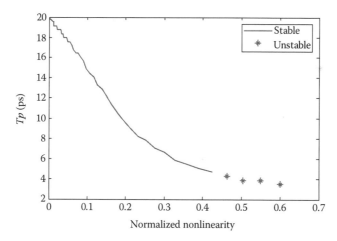

FIGURE 5.23
Pulse width shortens with an increase of nonlinearity, $D = 0.2$.

the required nonlinearity for shortening the pulse is also nearly double, as shown in Figure 5.23.

The buildup process and steady-state pulse of the laser with the normalized cavity dispersion and nonlinearity of 0.1 and 0.2, respectively, are shown in Figure 5.24. The pulse shape is well fitted with a sech² profile. This confirms the theoretical prediction that a soliton can be formed in the laser when the anomalous dispersion and nonlinearity have taken effect together.

Figure 5.25 gives a detailed view on the effects of dispersion and nonlinearity on the pulse width of the mode-locked pulse. A pulse width contour is plotted on a plan of dispersion and nonlinearity. The unstable area is also

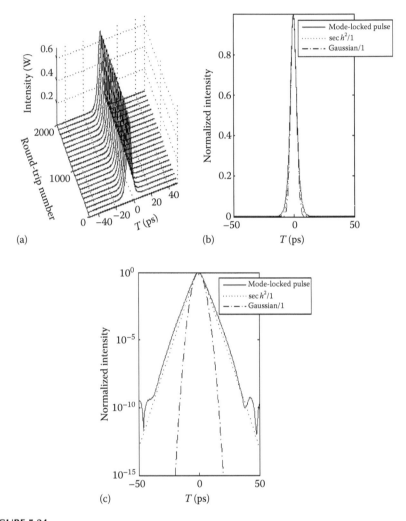

FIGURE 5.24
Soliton pulse formed in the active MLL when the nonlinearity and anomalous dispersion effects are introduced into the cavity. (a) Pulse buildup, (b) steady-state pulse with profile fitting, and (c) replot of (b) in log scale.

presented. The left side of the figure is the combined area of the normal dispersion and nonlinearity. It is shown that the nonlinearity enhances the normal dispersion effect and hence causes the pulse to broaden more. On the opposite, on the right side where the cavity dispersion is anomalous, the pulse is compressed when the nonlinearity is increased. The larger the dispersion value, the higher the nonlinearity required to get the pulse shorter and, thus, a good balance between the dispersion and nonlinearity in the cavity is required to maintain the pulse shape.

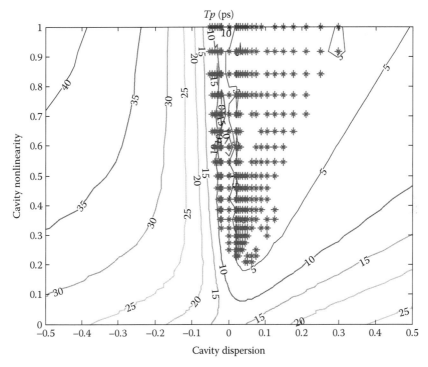

FIGURE 5.25
Pulse width contour with cavity dispersion and cavity nonlinearity as the parameters; * area is an unstable area where the steady-state pulse is not obtained.

5.5.2 Detuning and Locking Range in a Mode-Locked Fiber Laser with Nonlinearity and Dispersion Effect

Similar to the other cases, when the modulation frequency is detuned, the pulse position is also shifted away from the center, as shown in Figure 5.26. The pushing/pulling effect caused by this shifting compensates for the detuning and, hence, keeps the pulse stable. However, as the pulse becomes shorter and shorter under the soliton effect, the pushing/pulling effect becomes weaker and hence the amount of detuning it can compensate for is reduced. This results in a smaller locking range for shorter pulses.

Figure 5.27 plots the locking range contour on the plane of cavity nonlinearity and dispersion. Combining with the pulse width contour in Figure 5.25, we see that the large locking range contours correspond to the wide pulse width contours while the small locking range contours correspond to the narrow pulse width contours. There is a trade-off between the short pulse and the large locking range. If the laser is designed to give a short pulse, then its locking range would be sacrificed and its stability against the change of environment is reduced. On the other hand, if the laser is designed to have a large

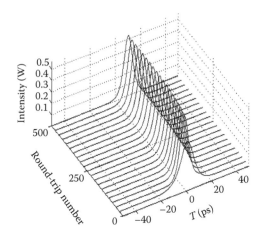

FIGURE 5.26
Pulse buildup in a detuned actively MLFL with both dispersion and nonlinear effects.

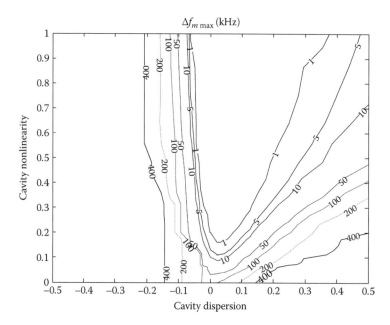

FIGURE 5.27
Locking range contour with cavity dispersion and cavity nonlinearity as the parameters.

locking range, then the shortest pulse that can be obtained is limited and depends on the cavity dispersion. Examples are given in Figures 5.28 and 5.29 where the dependence of the shortest pulse width on the cavity dispersion is plotted for locking ranges of $\delta f_m = 10^{-6}$ ($\Delta f_m = 10$ kHz) and $\delta f_m = 10^{-7}$ ($\Delta f_m = 1$ kHz). It is shown that a shorter pulse is obtained not only when the cavity dispersion is increased, but also when the required nonlinearity is increased.

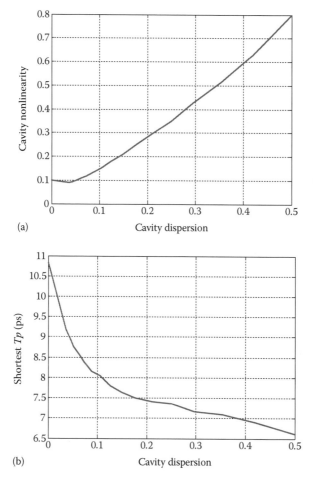

FIGURE 5.28

(a) Shortest pulse width and (b) the required cavity nonlinearity as a function of the cavity dispersion for the locking range of $\delta f_m = 10^{-6}$ ($\Delta f_m = 10\,kHz$).

5.6 Detuning and Pulse Shortening

5.6.1 Experimental Setup

Figure 5.30 shows the experimental setup of an actively MLFL using an EDF as a gain medium. The EDF is 5.6 m long and pumped by a 980 nm laser diode through a WDM coupler. An isolator integrated in the WDM ensures unidirectional lasing. A LiNbO$_3$ Mach–Zehnder intensity modulator (MZIM) is inserted in the ring for mode locking. Mode locking is obtained by introducing

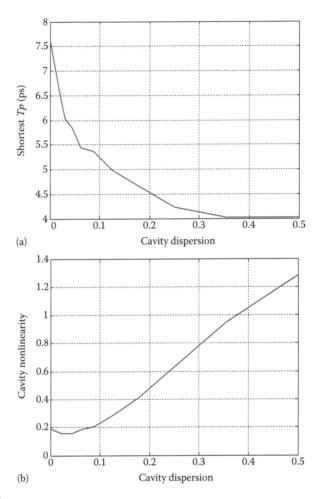

FIGURE 5.29
(a) Shortest pulse width and (b) the required cavity nonlinearity as a function of the cavity dispersion for the locking range of $\delta f_m = 10^{-7}$ ($\Delta f_m = 1\,\text{kHz}$).

a periodic loss to the signal when driving the modulator by a sinusoidal signal extracted from an RF signal generator. The polarization controller (PC) is used to adjust the polarization state of the lightwave signal traveling in the ring. The lasing wavelength can be tuned by adjusting the central wavelength of the thin-film tunable optical BPF with a 3 dB bandwidth of 1.2 nm. The optical signal in the ring is coupled to the output port through a 70:30 coupler.

5.6.2 Mode-Locked Pulse Train with 10 GHz Repetition Rate

The total length of the cavity is about 20.7 m, which corresponds to a fundamental frequency of 9.859 MHz. When the modulation frequency is set at 10.006885 GHz, which corresponds to the 1015th harmonic of the fundamental

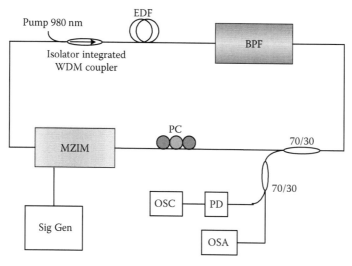

FIGURE 5.30
Experimental setup of an actively MLFL. EDF, erbium-doped fiber; BPF, bandpass filter; PC, polarization controller; MZIM, Mach–Zehnder intensity modulator; PD, photodiode; OSC, oscilloscope; OSA, optical spectrum analyzer; Sig Gen, signal generator.

frequency, a stable mode-locked pulse train is obtained, as shown in Figure 5.31. The FWHM of the pulse measured from the oscilloscope is about 21 ps. It is noted that this value includes the rise times of the photodiode and the oscilloscope. Using an autocorrelator for measuring the pulse width to get rid of those rise times due to the electrical equipment and components, we obtain the pulse width of 14.3 ps.

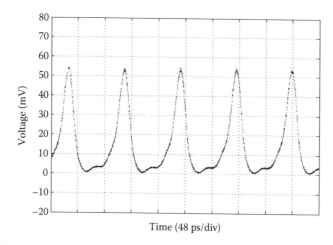

FIGURE 5.31
10 GHz repetition rate pulse train generated from the actively MLFL when the modulator is driven by a sinusoidal RF signal.

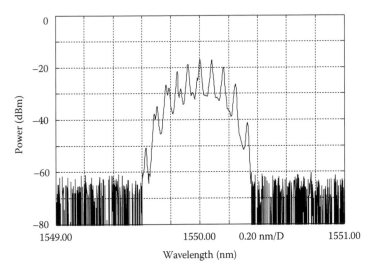

FIGURE 5.32
Spectrum of the mode-locked pulse train; the 10 GHz (0.08 nm) spacing modes are clearly observed.

The optical spectrum of the pulse train is shown in Figure 5.32. It can be seen from Figure 5.32 that the pulse train spectrum has side lobes with a separation of 0.08 nm, which corresponds to a longitudinal mode separation of 10 GHz. This spectrum profile is typical for the MLL and usually referred to as the mode-locked structure spectrum [30–32].

The above spectrum profile can be explained by taking the Fourier transform of the periodic Gaussian pulse train:

$$P(t) = e^{-t^2/2T_0^2} * \left|\left|\left| \left(\frac{t}{T} \right) \right.\right.\right. \tag{5.30}$$

where
T_0 is the width of the Gaussian pulse
T is the pulse repetition period
$\left|\left|\left|(t/T)\right.\right.\right.$ is the comb function given by

$$\left|\left|\left| \left(\frac{t}{T} \right) = \sum_{n=-\infty}^{\infty} \delta(t - nT) \right.\right.\right. \tag{5.31}$$

$$P(\omega) = F(p(t)) = T_0 \sqrt{2\pi}\, e^{-T_0^2 \omega^2/2} \sum_{n=-\infty}^{\infty} \delta(\omega - n\Omega) \tag{5.32}$$

where $\Omega = 2\pi/T$. $P(\omega)$ has the profile of a train of Dirac function pulses separated by the repetition frequency with a Gaussian envelope. This theoretical pulse spectrum $P(\omega)$ is consistent with the experimental one shown in Figure 5.32.

The laser wavelength is tunable over the whole C-band by tuning the central wavelength of the thin-film bandpass filter inserted in the loop.

5.6.3 Wavelength Shifting in a Detuned Actively Mode-Locked Fiber Laser with Dispersion Cavity

To study the dispersion effect in the actively MLFL, we inserted 200 m of SMF fiber into the cavity in the experimental setup shown in Figure 5.33. The total cavity dispersion was about 3.302 ps/nm or equivalent to a normalized dispersion of 0.07. As low as 40 mW of 980 nm signal was pumped to the EDF so that the laser had enough gain for lasing but the mode-locked pulse power was kept at a low level. Therefore, the nonlinear effect could be negligible.

Figure 5.34 shows the mode-locked pulse train of the laser when the modulation frequency was tuned to 10.006900 GHz, corresponding to the 1083th harmonic of the fundamental frequency of 924 kHz (the total length of the cavity is 220.9 m). The pulse was slightly broadened due to the dispersion of the cavity. The pulse width was measured to be about 16.2 ps. The pulse spectrum shown in Figure 5.35 also exhibits the mode locking structure with 0.08 nm spacing peaks. The bandwidth of the signal is about 0.26 nm, which gives a time-bandwidth product (TBP) of 0.526. This is slightly higher

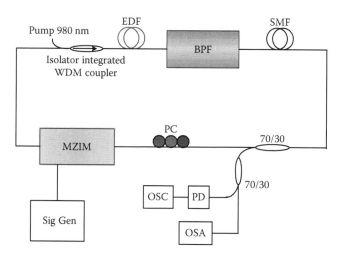

FIGURE 5.33
Experimental setup of an actively MLFL with dispersion cavity enhanced by 200 m of SMF. EDF, erbium-doped fiber; BPF, bandpass filter; PC, polarization controller; MZIM, Mach–Zehnder intensity modulator; PD, photodiode; OSC, oscilloscope; OSA, optical spectrum analyzer; Sig Gen, signal generator.

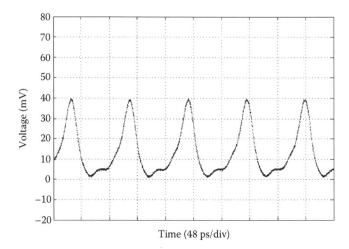

FIGURE 5.34
The mode-locked pulse is broadened due to the dispersion effect when the dispersion of the cavity increases.

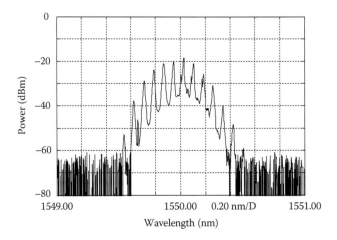

FIGURE 5.35
Optical spectrum of the pulse train shown in Figure 5.34 when the modulation frequency is set at 10.006900 GHz, the 1083 harmonic of the f_R.

than 0.44, the TBP of an unchirped Gaussian pulse. Therefore, the pulse is slightly chirped.

When the modulation frequency is detuned by 10 kHz to 10.006910 GHz, the spectrum is shifted to the shorter wavelength region, as shown in Figure 5.36. The central wavelength is now moved to 1549.9 nm. This confirms the above simulation result where the lasing wavelength is shifted to higher frequency (shorter wavelength) when the abnormal dispersion cavity MLL is positively detuned ($\Delta f_m > 0$). On the other hand, as seen in Figure 5.37, the

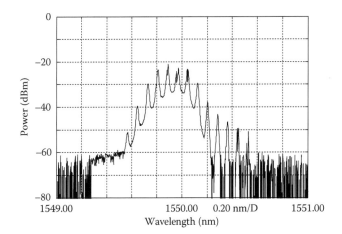

FIGURE 5.36
The spectrum is shifted to the shorter wavelength region when the modulation frequency is positively detuned by 10 kHz from 10.006900 GHz.

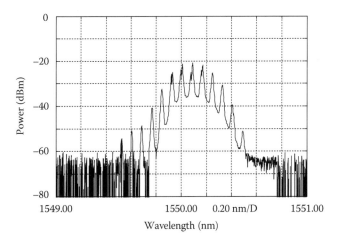

FIGURE 5.37
The spectrum is shifted to the longer wavelength region when the modulation frequency is negatively detuned by 10 kHz from 10.006900 GHz.

lasing wavelength is shifted to longer wavelength when the modulation frequency is decreased by 10 kHz.

5.6.4 Pulse Shortening and Spectrum Broadening under Nonlinearity Effect

The nonlinearity effect is explored by increasing the optical power pumped into the EDF. As the power is increased, the nonlinear effect is enhanced and, thus, contributes to the formation of the mode-locked pulse. Figure 5.38 shows

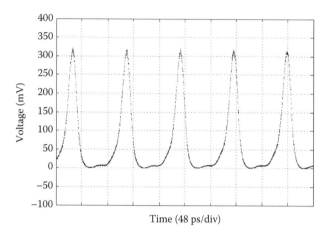

FIGURE 5.38
Oscilloscope trace of the mode-locked pulse train when the pump power increases to 190 mW; pulse is shortened when the pump power increases.

the oscilloscope trace of the pulse train when the pump power is increased to 190 mW. The pulse width has been shortened to 12.3 ps, a 24% decrease compared to that of the dispersion cavity without the nonlinear effect. The pulse will be compressed more if higher optical power is pumped into the EDF.

The optical spectrum of the pulse is shown in Figure 5.39. As discussed in Sections 5.2 and 5.4, the nonlinear effect generates new frequencies when it introduces nonlinear phase change to the pulse and, thus, broadens its spectrum. The 3 dB bandwidth of the pulse has been increased from 0.26 to 0.5 nm when the pump power increases from 40 to 190 mW.

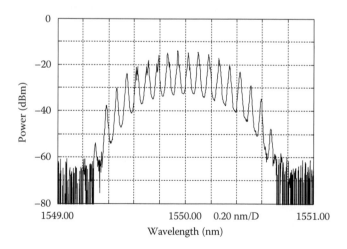

FIGURE 5.39
Optical spectrum of the generated pulse sequence. The nonlinearity causes the pulse spectrum to broaden when the pump power increases to 190 mW.

5.7 Conclusions

We have presented the dispersion and nonlinearity effects on the actively MLFLs. A laser model that includes the optical amplifier, BPF, modulator, and optical fiber has been developed for studying the dispersion and nonlinear effects. It is found that dispersion causes the broadening and chirping of the pulse. The sign of the chirp depends on the cavity dispersion: positive chirp (down chirp) for normal dispersion while negative chirp for anomalous dispersion. Similarly to the zero-dispersion case, the pulse in the dispersive MLFL is also shifted from the transmission peak of the modulation function when the laser is detuned. However, the shifting is reduced as the dispersion increases. The position shifting reduction is the result of moving the lasing wavelength toward a faster/slower wavelength to compensate for the delay/advance of the pulse when detuning. Therefore, the cavity dispersion enhances the locking range and thus improves the stability of the laser against the environmental perturbation.

In contrast to the dispersion effect, the nonlinearity compresses the pulse and reduces the locking range. The mode-locked pulse is positively chirped due to the generation of new frequencies through the SPM effect and thus the pulse bandwidth is increased. When combining the nonlinear effect with the anomalous dispersion effect, we are able to generate the soliton pulse directly from the actively MLFL. The pulse width of the soliton pulse is shorter than that of Gaussian mode-locked pulse. However, there is a trade-off between the short pulse width and the large locking range. If the laser is designed to have a shorter pulse width, the locking range of the laser is narrower and vice verse. Therefore, our laser model can be used to optimize the laser parameters to obtain the required pulse width while maintaining the laser stability over a certain locking range.

References

1. H.A. Haus and W.S. Wong, Solitons in optical communications, *Rev. Mod. Phys.*, 68(2), 423–444, 1996.
2. D.S. Seo, D.E. Leaird, A.M. Weiner, S. Kamei, M. Ishii, A. Sugita, and K. Okamoto, Continuous 500 GHz pulse train generation by repetition-rate multiplication using arrayed waveguide grating, *Electron. Lett.*, 39(15), 1138–1140, 2003.
3. N.H. Seong and D.Y. Kim, A new figure-eight fiber laser based on a dispersion-imbalanced nonlinear optical loop mirror lumped dispersive elements, *IEEE Photon. Technol. Lett.*, 14(4), 459–461, 2002.
4. A.E. Siegman, *Lasers*, University Science Books, Mill Valley, CA, 1986.
5. X. Shan and D.M. Spirit, Novel method to suppress noise in harmonically modelocked erbium fiber lasers, *Electron. Lett.*, 29(11), 979–981, 1993.

6. Y. Shi, M. Sejka, and O. Poulsen, A unidirectional Er^{3+}-doped fiber ring laser without isolator, *IEEE Photon. Technol. Lett.*, 7(3), 290–292, 1995.

7. Y. Shiquan, L. Zhaohui, Z. Chunliu, D. Xiaoyi, Y. Shuzhong, K. Guiyun, and Z. Qida, Pulse-amplitude equalization in a rational harmonic mode-locked fiber ring laser by using modulator as both mode locker and equalizer, *IEEE Photon. Technol. Lett.*, 15(3), 389–391, 2003.

8. J.J.E. Slotine and W. Li, *Applied Nonlinear Control*, Prentice-Hall, Englewood Cliffs, NJ, 1991.

9. C.S. Gardner, J.M. Greene, M.D. Kruskal, and R.M. Miura, Method for solving the Korteweg-deVries equation, *Phys. Rev. Lett.*, 19, 1095 LP–1097 LP, 1967.

10. A. Hasegawa and F. Tappert, Transmission of stationary nonlinear optical pulses in dispersive dielectric fibers. I. Anomalous dispersion, *Appl. Phys. Lett.*, 23, 142–144, 1973.

11. L.F. Mollenauer, R.H. Stolen, and J.P. Gordon, Experimental observation of picosecond pulse narrowing and solitons in optical fibers, *Phys. Rev. Lett.*, 45, 1095 LP–1098 LP, 1980.

12. G.P. Agrawal, *Nonlinear Fiber Optics*, Academic Press, San Francisco, CA, 2001.

13. R.A. Fisher and W. Bischel, The role of linear dispersion in plane-wave self-phase modulation, *Appl. Phys. Lett.*, 23, 661–663, 1973.

14. H.A. Haus and Y. Silberberg, Laser mode-locking with addition of nonlinear index, *IEEE J. Quantum Electron.*, 22, 325–331, 1986.

15. F.X. Kartner, D. Kopf, and U. Keller, Solitary-pulse stabilization and shortening in actively mode-locked lasers, *J. Opt. Soc. Am. B*, 12, 486–496, 1995.

16. B. Bakhski and P.A. Andrekson, 40 GHz actively modelocked polarisation-maintaining erbium fibre ring laser, *Electron. Lett.*, 36, 411–413, 2000.

17. M. Nakazawa, E. Yoshida, and Y. Kimura, Ultrastable harmonically and regeneratively modelocked polarisation-maintaining erbium fibre ring laser, *Electron. Lett.*, 30, 1603–1605, 1994.

18. M. Nakazawa, Ultrafast optical pulses and solitons for advanced communications, in *Lasers and Electro-Optics, 2003. CLEO/Pacific Rim 2003. The 5th Pacific Rim Conference*, Taipei, Taiwan, 2003.

19. F.X. Kartner, U. Morgner, S.H. Cho, J. Fini, J.G. Fujimoto, E.P. Ippen, V. Scheuer, M. Tilsch, and T. Tschudi, Advances in short pulse generation, in *Lasers and Electro-Optics Society Annual Meeting, 1998, LEOS '98*, IEEE, Orlando, FL, 1998.

20. D. Foursa, P. Emplit, R. Leners, and L. Meuleman, 18 GHz from a σ-cavity Er-fibre laser with dispersion management and rational harmonic active mode-locking, *Electron. Lett.*, 33, 486–488, 1997.

21. B. Bakhshi, P.A. Andrekson, and X. Zhang, 10 GHz modelocked, dispersion-managed and polarisation-maintaining erbium fibre ring laser with variable output coupling, *Electron. Lett.*, 34, 884–885, 1998.

22. M. Horowitz, C.R. Menyuk, T.F. Carruthers, and I.N. Duling III, Theoretical and experimental study of harmonically modelocked fiber lasers for optical communication systems, *J. Lightwave Technol.*, 18, 1565–1574, 2000.

23. M. Nakazawa, M. Yoshida, and T. Hirooka, Ultra-stable regeneratively mode-locked laser as an opto-electronic microwave oscillator and its application to optical metrology, *IEICE Trans. Electron.*, E90C, 443–449, 2007.

24. K. Tamura and M. Nakazawa, Dispersion-tuned harmonically mode-locked fiber ring laser for self-synchronization to an external clock, *Opt. Lett.*, 21, 1984–1986, 1996.

25. G.P. Agrawal, *Applications of Nonlinear Fiber Optics*, Academic Press, San Diego, CA, 2001.
26. O. Pottiez, O. Deparis, R. Kiyan, M. Haelterman, P. Emplit, P. Megret, and M. Blondel, Supermode noise of harmonically mode-locked erbium fiber lasers with composite cavity, *IEEE J. Quantum Electron.*, 38, 252–259, 2002.
27. M. Nakazawa, K. Tamura, and E. Yoshida, Supermode noise suppression in a harmonically mode locked fibre laser by self phase modulation and spectral filtering, *Electron. Lett.*, 32, 461–463, 1996.
28. Y.H. Li, C.Y. Lou, J. Wu, B.Y. Wu, and Y.Z. Gao, Novel method to simultaneously compress pulses and suppress supermode noise in actively mode-locked fiber ring laser, *IEEE Photon. Technol. Lett.*, 10, 1250–1252, 1998.
29. G.P. Agrawal, *Fiber-Optic Communication Systems*, 3rd edn., Wiley-Interscience, New York, 2002.
30. M. Nakazawa and E. Yoshida, A 40-GHz 850-fs regeneratively FM mode-locked polarization-maintaining erbium fiber ring laser, *IEEE Photon. Technol. Lett.*, 12, 1613–1615, 2000.
31. H. Takara, S. Kawanishi, M. Saruwatari, and K. Noguchi, Generation of highly stable 20 GHz transform-limited optical pulses from actively mode-locked Er^{3+}-doped fibre lasers with an all-polarization maintaining ring cavity, *Electron. Lett.*, 28, 2095–2096, 1992.
32. T. Pfeiffer and G. Veith, 40 GHz pulse generation using a widely tunable all-polarization preserving erbium fibre ring laser, *Electron. Lett.*, 29, 1849–1850, 1993.

6

Actively Mode-Locked Fiber Lasers with Birefringent Cavity

This chapter presents the locking of a fiber laser in which the fiber is of a birefringent type whose refractive index depends on the polarization of the lightwaves. The principles of operation and application to switching and very fast tunable fiber lasers are described. Both theoretical and experimental works are discussed.

6.1 Introduction

In theory, an SMF can have a perfect cylindrical symmetry and can indiscriminately propagate a polarized light along any direction. The light emitted from fiber lasers, hence, can have any polarization. With this property, fiber lasers are usually modeled using scalar equations in which the polarizations of the fields are ignored [1–5]. These models have successfully explained and predicted several characteristics of the lasers such as lasing threshold, lasing wavelength, lasing modes, mode locking, pulse width, steady state. In addition, they can also be applied to analyze lasers that employ polarization-maintaining (PM) fibers to keep the light polarization unchanged along the laser cavity [6]. In such lasers, since the light polarization is fixed along an eigenstate of the fiber, normally the X-axis, they can be analyzed using the scalar models.

However, in practice, the laser cavity is not an ideal isotropic medium. Different polarized components may undergo different processes due to the cavity's anisotropies. In fact, several phenomena related to the polarization dynamics of the lasers have been reported [7]. In 1992, Bielawski et al. presented an anti-phase behavior in the Nd-doped fiber laser [8]. By simultaneously measuring the two orthogonal polarizations, the authors observed that the maximum output intensity in one polarization corresponds to a minimum peak in the other one. The variation of the threshold pump power due to gain and loss anisotropies in a Nd-doped fiber laser was also observed in 1994 [9]. The authors showed that the pump power threshold can be varied by controlling the anisotropy of the gain medium. Those phenomena were supposed to relate to the pump anisotropies or the asymmetries of the lasing medium.

Besides the pump and lasing medium anisotropies, the birefringence of the cavity may also be the cause of polarization dynamics in the lasers. Laser frequency splitting due to cavity birefringence was reported [9,10]. In Ref. [3] two quarter-wave plates were inserted into the cavity to separate the longitudinal modes. By adjusting the rotation angle of the two wave plates, dual-frequency lasing was observed. Similarly, Wey et al. [11] used a KD*P crystal to induce the birefringence in the cavity and also obtained similar results.

A passive MLFL based on the nonlinear polarization rotation (NPR) effect is one of the mode-locked fiber models that considers the polarization of the light inside the cavity [3,12]. The mode-locking mechanism is that the pulse peak polarization is rotated owing to the nonlinear phase change so that it experiences less loss than the wings with low intensity. Thus the pulse can become shorter and shorter after every round trip in the cavity till reaching equilibrium at which a mode-locked pulse train is formed. However this model just considers the nonlinear phase change and ignores the velocity mismatch between two orthogonal components, which is caused by the birefringence of the fiber.

Therefore, in this chapter, we present the study on an MLFL with a birefringence cavity in which the polarization of the optical field inside the cavity is considered. By introducing the birefringence into the laser cavity, we can reveal several lasers' dynamic behaviors, including polarization switching and dual-locking range operation.

6.2 Birefringence Cavity of an Actively Mode-Locked Fiber Laser

As presented in Chapters 2 through 4, an actively MLFL is normally characterized by its fundamental frequency f_R, which is inversely proportional to the round-trip traveling time. The mode-locking condition is obtained when the modulation frequency is equal to a harmonic of f_R [11]. A small amount of detuning from the ideal locking modulation frequency can be used. It can also be shown that the laser is still locked into one particular mode if the detuning is within the locking range. However, this condition is no longer valid when the birefringence of the cavity is large enough that it cannot be ignored. Since the lightwaves traveling in the X-axis see an effective refractive index n_x different from n_y when traveling in the Y-axis, the traveling round-trip time of the ring is thus different from X-axis to Y-axis. The laser can thus resonate at two different fundamental frequencies, f_{Rx} for the X-axis and f_{Ry} for the Y-axis, instead of a single fundamental frequency as reported in [13–15]. When the modulation frequency is equal to a multiple integer of f_{Rx} the laser is excited and mode locked in the X-axis, resulting in the output pulse having an X-polarized state. On the other hand, when the modulation frequency is equal to a multiple integer of f_{Ry}, harmonic mode locking in the Y-axis is obtained. The locking ranges can be estimated. The difference between two modulation frequencies can be derived as follows.

In the fiber ring laser, the birefringence is induced by integrating a length of a hi-birefringent (HiBi) fiber into the ring cavity. The refraction index difference between the fast and slow axes of the fiber can be calculated as [16]

$$\Delta n = \frac{\lambda}{L_b} \tag{6.1}$$

where
λ is the operating wavelength
L_b is the beat length of the fiber

The round-trip traveling time difference between the two locking regimes is thus given by

$$\Delta T = L_{PM} \times \frac{\Delta n}{c} \tag{6.2}$$

where
L_{PM} is the length of the PM fiber
c is the speed of light in vacuum

Therefore, the difference between the two fundamental frequencies f_{Rx} and f_{Ry} is

$$\Delta f_R = f_{Ry} - f_{Rx} = \frac{1}{T_{Ry}} - \frac{1}{T_{Rx}} = \frac{\Delta T}{T_{Ry} T_{Rx}} = \Delta T \times f_R^2 \tag{6.3}$$

where
T_{Rx}, T_{Ry} are the round-trip times of the light traveling in the X-axis and Y-axis, respectively
f_R is the average fundamental frequency

$$f_R = \sqrt{f_{Rx} f_{Ry}} \tag{6.4}$$

Substituting (6.1) through (6.3) into (6.4) one can obtain

$$\Delta f_R = \frac{L_{PM} \lambda f_R^2}{L_b c} \tag{6.5}$$

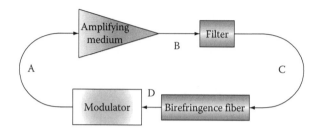

FIGURE 6.1
Model of an actively MLFL with a birefringence cavity.

When the laser is locked to the Nth order harmonic mode of the fundamental frequency, the difference between the two modulation frequencies, f_{mx} and f_{my}, for locking the laser ring into the two orthogonal regimes is

$$\Delta f_m = f_{mx} - f_{my} = N\Delta f_R = N\frac{L_{PM}\lambda f_R^2}{L_b c} \tag{6.6}$$

6.2.1 Simulation Model

Figure 6.1 shows the model of an MLFL with a birefringence cavity. The model consists of an amplifier, a BPF, a piece of birefringence fiber, and an intensity modulator.

The optical signal is modeled by an electric-field vector (E_x, E_y), where E_x and E_y are the envelopes of the electric fields on the X-axis and Y-axis, respectively. Initially, E_x and E_y are a set of random complex numbers representing a random noise. After passing through the amplifier, the signal becomes

$$\begin{pmatrix} E_x(t) \\ E_y(t) \end{pmatrix}_B = \begin{pmatrix} E_x(t) \\ E_y(t) \end{pmatrix}_A \times \sqrt{G} + \begin{pmatrix} E_{ASE,x} \\ E_{ASE,y} \end{pmatrix} \tag{6.7}$$

where $E_{ASE,x}$ and $E_{ASE,y}$ are the ASE noises of the amplifier and their spectral densities are given by

$$\frac{dP_{ASE,y}}{df} = \frac{dP_{ASE,y}}{df} = \frac{10^{NF/10}(G-1)hf}{2} \tag{6.8}$$

where
  NF is the noise figure
  $h = 6.626 \times 10^{-34}$ is the Planck's constant
  f is the optical frequency
  $G = P_{out}/P_{in}$ is the intensity gain of the amplifier

The saturation of the amplifier is also taken into account and its effect on the gain factor G is given as

$$G = \frac{g_{ss}}{1 + P/P_{sat}} \tag{6.9}$$

where
g_{ss} is the small-signal gain factor
P is the signal power
P_{sat} is the saturation power

The signal is then passed through a bandpass filter. The filter is modeled as a first-order transmission transfer function with a Gaussian profile

$$T(f) = \exp\left(-2\ln 2 \times \frac{f^2}{\Delta f^2}\right) \tag{6.10}$$

where Δf is the 3 dB bandwidth

The signal in the frequency domain at the output of the filter is

$$\begin{pmatrix} E_x(f) \\ E_y(f) \end{pmatrix}_C = \begin{pmatrix} E_x(f) \\ E_y(f) \end{pmatrix}_B \times \exp\left(-2\ln 2 \times \frac{f^2}{\Delta f^2}\right) \tag{6.11}$$

After being filtered, the signal propagates in a HiBi fiber, which can be modeled by an NLSE [17].

$$\frac{\partial A_x}{\partial z} + \beta_{1x}\frac{\partial A_x}{\partial t} + \frac{i\beta_{2x}}{2}\frac{\partial^2 A_x}{\partial t^2} + \frac{\alpha}{2}A_x = i\gamma\left(\left|A_x\right|^2 + B\left|A_y\right|^2\right)A_x$$

$$\frac{\partial A_y}{\partial z} + \beta_{1y}\frac{\partial A_y}{\partial t} + \frac{i\beta_{2y}}{2}\frac{\partial^2 A_y}{\partial t^2} + \frac{\alpha}{2}A_y = i\gamma\left(\left|A_y\right|^2 + B\left|A_x\right|^2\right)A_y \tag{6.12}$$

where
$i = \sqrt{-1}$
A_x and A_y are the slowly varying envelops of the electric fields in the two orthogonal polarizations
z is the propagation distance along the fiber
β_{1x}, β_{1y} and β_{2x}, β_{2y} are the first-order and second-order mode propagation constants in the two orthogonal polarizations, respectively
α is the fiber attenuation
γ is the nonlinear coefficient of the fiber

The coefficient B is defined as the constant related to the nonlinear effects, the self-phase modulation effect, which is also material dependent.

For simplicity, only the birefringence term in the NLSE is kept while the loss, nonlinear, and dispersive terms are ignored so that only the effect of birefringence on pulse propagation in the cavity can be isolated and studied. A simple equation governing the propagation of the signal in a birefringence fiber is given as

$$\left(\frac{\partial}{\partial z} + \beta_{1x,y}\frac{\partial}{\partial t}\right)E_{x,y}(z,t) = 0 \tag{6.13}$$

where $\beta_{1x,y}$ is the propagation constant on the X-axis and Y-axis. Hence the X-axis and Y-axis fields, after passing through the birefringence fiber, experience different delays due to the birefringence, and the output signal is

$$\begin{pmatrix} E_x(t) \\ E_y(t) \end{pmatrix}_D = \begin{pmatrix} E_x(t - \beta_{1x}L) \\ E_y(t - \beta_{1y}L) \end{pmatrix}_C \tag{6.14}$$

where L is the length of the birefringence fiber. This birefringence delay leads to two different fundamental frequencies f_{Rx} and f_{Ry} for the two orthogonal fields E_x and E_y, respectively. The signal is then modulated by the MZI modulator. After one round trip, the signal returns to point A as

$$\begin{pmatrix} E_x(t) \\ E_y(t) \end{pmatrix}_{A'} = \begin{pmatrix} E_x(t) \\ E_y(t) \end{pmatrix}_D \times \Gamma \cos\left\{\frac{\pi}{4}\left[1 - m\cos(\omega_m t)\right]\right\} \tag{6.15}$$

where
   m is the modulation index
   ω_m is the modulation frequency
   Γ represents the total loss of the cavity

The signal is fed in the amplifier again and completes another round trip. The process is repeated until equilibrium is reached.

6.2.2 Simulation Results

The parameters used in the simulation are as follows: Small-signal gain $g_{ss} = 18\,dB$, $NF = 8\,dB$, output saturation power $P_{sat} = 5\,dBm$, filter bandwidth $BW = 1.2\,nm$, fundamental frequencies of the X-axis and Y-axis $f_{Rx} = 12.00059\,MHz$, $f_{Ry} = 12.00127\,MHz$, harmonic order $N = 834$, modulation index $m = 0.5$, total cavity loss $\Gamma = 10\,dB$, and the operating wavelength $\lambda = 1550\,nm$.

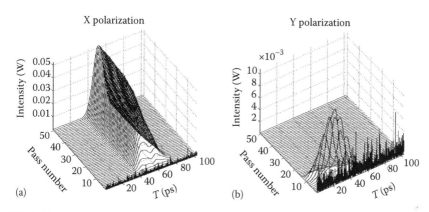

FIGURE 6.2
Pulse evolution in the cavity when the modulation frequency is set to 10.00849 GHz. (a) In the X polarization direction. (b) In the Y polarization direction.

Figure 6.2 shows the simulation results when the modulation frequency is set to the multiple integer of the X-axis fundamental frequency ($f_m = f_{mx} = Nf_{Rx} = Nf_{Ry} - N\Delta f_R = 10.008490$ GHz). It is clearly shown that initiating from the *noise* in the X-axis component gradually forms the mode-locked pulse while the Y-axis component is suppressed to zero after several round trips. This results in the X-polarized mode-locked pulse. The laser is thus locked in the X-axis.

On the other hand, when the modulation frequency is increased by $\Delta f_m = 570$ kHz to be equal to a multiple of f_{Ry} ($f_m = f_{my} = Nf_{Ry} = Nf_{Rx} + N\Delta f_R = 10.009060$ GHz), the X-axis component is suppressed and the Y-axis component is built up, as shown in Figure 6.3. Moreover, in Figure 6.4, there is no mode-locked pulse formed in either the X-axis or Y-axis when the modulation frequency is set in the middle of f_{mx} and f_{my}.

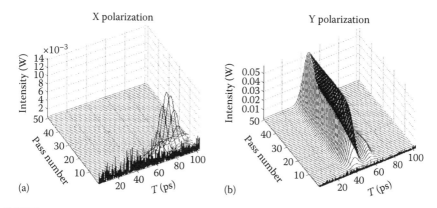

FIGURE 6.3
Pulse evolution in the cavity when the modulation frequency is set to 10.00906 GHz. (a) In the X polarization direction. (b) In the Y polarization direction.

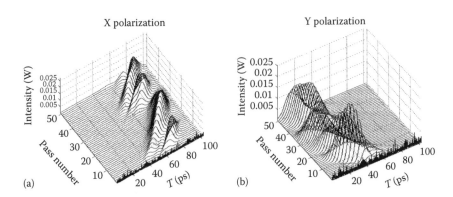

FIGURE 6.4
Pulse evolution in the cavity when the modulation frequency is set to 10.00876 GHz. (a) In the X polarization direction. (b) In the Y polarization direction.

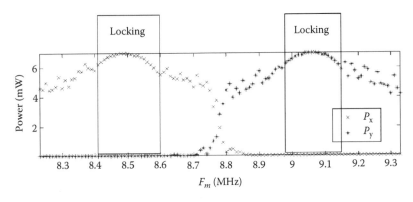

FIGURE 6.5
Dependence of the output power of the X-axis and Y-axis components on the detuning of modulation frequency f_m (with Δf_m =100 kHz); F_m is the offset to 10 GHz: $F_m=f_m - 10$ GHz.

Figure 6.5 plots the output power of the X-axis and Y-axis components when the modulation frequency is detuned. It can be seen that there are two locking ranges. One is located at f_{mx} and the other at f_{my}. When the modulation frequency is in the locking range of f_{mx}, the laser is mode locked and the output pulse is polarized on the X-axis. We call this the H-mode regime. In contrast, the laser is mode locked in the V-mode regime, in which the output pulse is Y polarized, if the modulation is in the locking range of f_{my}. Outside the locking ranges, mode locking is not obtained.

It is noted that when the locking range of H-mode and V-mode is overlapped, the two locking ranges are merged. This occurs when Δf_R is small and the simulation result is plotted in Figure 6.6. It can be seen that there is a transition point at which the output pulse polarization is switched from X-axis to Y-axis and vice versa.

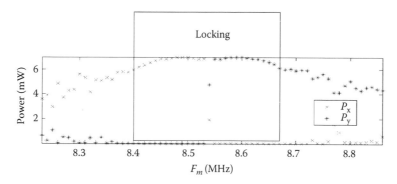

FIGURE 6.6
Dependence of the output power of the X-axis and Y-axis components on the detuning of modulation frequency f_m when Δf_m is reduced to 100 kHz; F_m is the offset to 10 GHz: $F_m = f_m - 10$ GHz.

6.3 Polarization Switching in an Actively Mode-Locked Fiber Laser with Birefringence Cavity

6.3.1 Experimental Setup

Figure 6.7 shows the experimental setup of a semiconductor optical amplifier (SOA)-based actively MLFL. The laser has a ring configuration incorporating an isolator to ensure unidirectional lasing. The optical gain is provided by an SOA

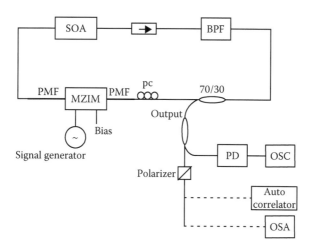

FIGURE 6.7
Experimental setup of an SOA-based harmonic MLFL. SOA, semiconductor optical amplifier; BPF, bandpass filter; pc, polarization controller; MZIM, Mach–Zehnder intensity modulator; PMF, polarization maintaining fiber; PD, photodiode; OSC, oscilloscope; OSA, optical spectrum analyzer. The component with the arrow is an optical isolator.

with a small signal gain of 18.2 dB and a saturation power of 11.3 dBm while a thin-film tunable optical BPF with a 3 dB bandwidth of 1.2 nm is used to tune the lasing wavelength. The mode locking of the lasing lightwave is obtained through the amplitude modulation process by incorporating an LiNbO$_3$ MZIM in the ring with the insertion loss of 3 dB. As large as 30% of the optical power in the ring can be coupled to the output port through a 70:30 coupler.

The output signal is split into two paths for simultaneous monitoring and characterization. One is amplified and fed into a high-speed photodiode (45 GHz 3 dB bandwidth) for conversion into the electrical signal and then monitored by a high-speed 50 GHz sampling oscilloscope. The other path is analyzed by a polarizer followed by an autocorrelator and/or an optical spectrum analyzer.

6.3.2 Results and Discussion

The fundamental frequency of the ring, estimated as the inverse of the round-trip period of the lightwave traveling in the ring, is about 12 MHz. Figure 6.8a shows the generated harmonic mode-locked pulse sequence at a repetition rate of about 10 GHz when the modulation frequency is set at 10.008426 GHz, according to the 834th harmonic of the fundamental frequency. The optical spectrum of the pulse sequence shown in Figure 6.8b clearly shows the lasing longitudinal mode spacing of 0.08 nm. The 3 dB bandwidth of the spectrum envelop has been measured as large as 0.77 nm. The locking range is about 260 kHz. However, when the modulation frequency is tuned to 10.009106 GHz, or 680 kHz away from 10.008426 GHz, a stable well-defined mode-locked pulse train appears again, as shown in Figure 6.9. The 3 dB bandwidth is estimated to be 0.36 nm under this condition. It is interesting that the polarization state of the pulse sequence in this case is orthogonal to that of the previous one, which is mode locked at 10.008426 GHz. Hence we denote the two polarization locking states as the V- and the H-mode regimes, respectively.

6.3.2.1 H-Mode Regime

Figure 6.10a and b show the autocorrelation traces of the X and Y polarization components, respectively, of the H-mode pulse sequence. Higher power on the X-axis demonstrates that the pulse has the major polarization state aligning along this axis. This is confirmed by the optical spectra of the two orthogonal polarization components shown in Figure 6.11. The central wavelength modes of the pulse polarize along the X-axis and, hence, pass through the X-axis polarizer without attenuation while suffering a loss of as much as 10 dB when passing through the Y-axis polarizer. It is noted that the sideband modes far from the central wavelength do not have a strong X-polarized state as the central wavelength. They have equal powers on both axes.

The pulse width of the Y-axis component is slightly larger than that of the X-axis. The pulse FHWM of 24.8 and 19.8 ps have been achieved, assuming a Gaussian shape, for the Y and X components, respectively.

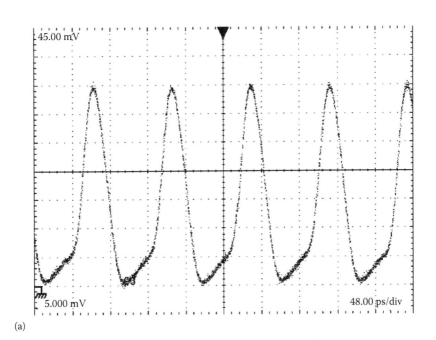

(a)

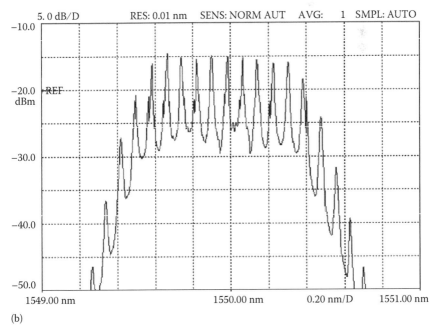

(b)

FIGURE 6.8
(a) H-mode pulse trace and (b) its optical spectrum.

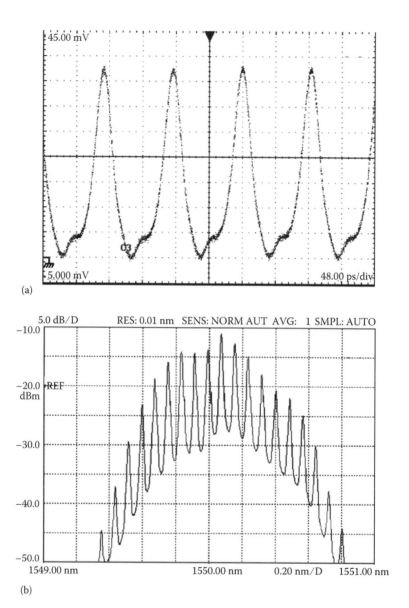

(a)

(b)

FIGURE 6.9
(a) V-mode pulse trace and (b) its optical spectrum.

6.3.2.2 V-Mode Regime

The polarization state of the mode-locked pulse is now switched to the other orthogonal state, the V-mode regime, by tuning the modulation frequency to 10.009106 GHz. The autocorrelation traces shown in Figure 6.12 indicate that the pulse power is more distributed on the Y-axis than on the X-axis,

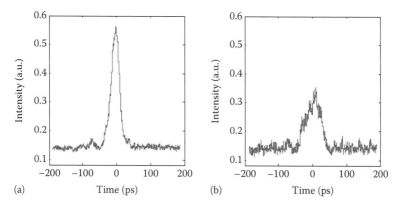

FIGURE 6.10
Autocorrelation traces of H-mode pulses: (a) X-axis component and (b) Y-axis component.

hence contrasting to that observed in the H-mode regime. In addition, the optical spectra presented in Figure 6.13 show that the central wavelength is polarized along the Y-axis and has not been affected by the Y-axis polarizer. However, there is a slight difference of the polarization state of the sideband modes from the H-mode regime. The sideband modes are now polarized along the axis orthogonal to the polarization axis of the central wavelength and significantly attenuated when passing through the Y-axis polarizer.

Similar to the H-mode regime, the pulse width of the major axis (the higher power axis) component is slightly narrower than that of the minor axis. The pulse widths are 29.4 and 19.1 ps for the X-axis and Y-axis components, respectively. This can be easily understood that the minor axis component is detuned while the major axis component is tuned to its locking frequency.

It is also noted that the TBP are very large in both locking regimes. This indicates that the pulses are heavily chirped and can be linearly compressed externally [16].

In the experiment, $L_{PM} = 3.2$ m, $L_b = 3.5$ mm, $\lambda = 1550$ nm, $f_R = 12$ MHz, $N = 834$, and the fundamental frequency difference and the modulation frequency difference between the two axes are thus estimated as $\Delta f_R = 680$ Hz and $\Delta f_m = 567$ kHz. These values agree well with the experimental results, where the modulation frequencies for X- and Y-polarized mode-locking states are $f_{mx} = 10.008426$ GHz and $f_{my} = 10.009106$ GHz, respectively. The difference is $\Delta f_m = 680$ kHz.

6.3.3 Dual Orthogonal Polarization States in an Actively Mode-Locked Birefringent Fiber Ring Laser

6.3.3.1 Experimental Setup

Figure 6.14 shows the experimental setup of an actively MLFL incorporating a birefringent cavity. The laser has a ring configuration with 15 m of

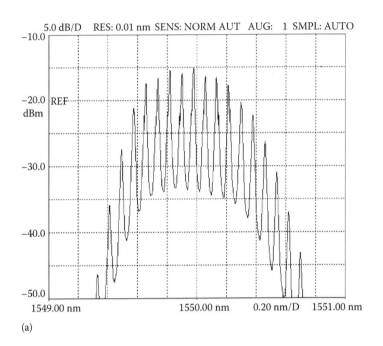

(a)

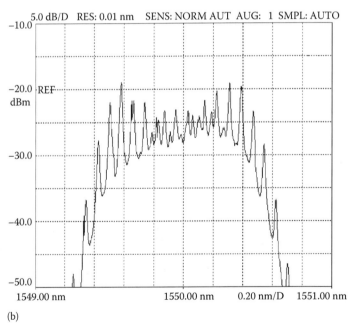

(b)

FIGURE 6.11
Optical spectra of H-mode pulses: (a) X-axis component and (b) Y-axis component.

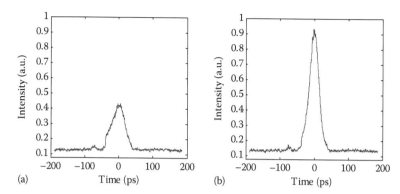

FIGURE 6.12

Autocorrelation traces of V-mode pulses: (a) X-axis component and (b) Y-axis component.

polarization-maintaining erbium-doped fiber (PM-EDF) serving as a gain medium. This PM-EDF also causes the laser cavity to become birefringent. The PM-EDF is pumped by a 980 nm laser diode through a WDM coupler. An isolator integrated in the WDM ensures unidirectional lasing. A thin-film tunable optical BPF with a 3 dB bandwidth of 1.2 nm is used to tune the lasing wavelength. A LiNbO₃ MZIM is incorporated in the cavity to stimulate mode locking through the amplitude modulation process. The modulator is driven by a microwave signal extracted from a signal generator. The polarization controllers (PC1 and PC2) are used to adjust the polarization state of the lightwave signal traveling in the ring. The optical signal in the ring is coupled to the output port through a 70:30 fiber coupler.

The output signal is detected by a photodiode (45 GHz 3 dB bandwidth) and then monitored by a high-speed sampling oscilloscope (50 GHz 3 dB bandwidth). A polarizer is also inserted in front of the photodiode when the polarization state of the output signal is analyzed. An optical spectrum analyzer with a resolution of 0.01 nm is used to record the pulse spectrum.

6.3.3.2 Results and Discussion

The total length of the cavity is about 29.5 m, which corresponds to the fundamental frequency f_R of 6.923 MHz. The modulator is biased at the quadrature point. The central wavelength of the filter is 1550 nm. The modulation frequency is 2.997668 GHz, which corresponds to the resonance of the 433th harmonic frequency component. When the polarization controller is adjusted at an appropriate position, we observed a well-defined mode-locked pulse train on the oscilloscope, as shown in Figure 6.15a. It is interesting to note that the pulse sequence has two amplitude levels. This indicates that the pulse train actually consists of low- and high-amplitude pulses, which can be denoted as the X- and Y-pulse sequences, respectively. It is noticed that the X-pulse sequence appears brighter than that of the Y pulse. This means

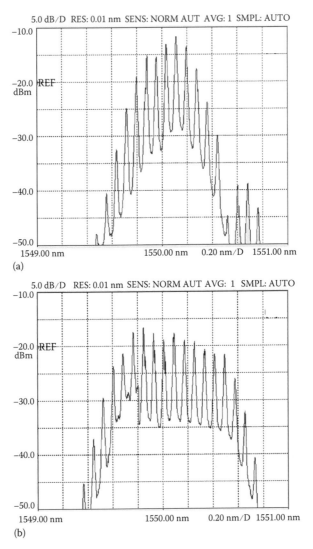

FIGURE 6.13
Optical spectra of V-mode pulses: (a) X-axis component and (b) Y-axis component.

that there are more X pulses than Y pulses in the pulse train. The sampling oscilloscope samples the pulse train for every sampling interval and each sample gives a dot on the screen. Therefore, more samples obtained for the high population of X pulses will result in a brighter trace of the X pulse on the screen.

Figure 6.15b shows the optical spectrum of the pulse train. The spectrum is totally different from a normal mode-locked pulse train spectrum. It does not follow a Gaussian shape with the peak at the filter center wavelength;

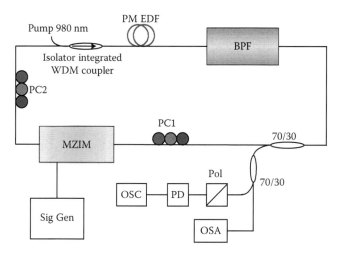

FIGURE 6.14
Experimental setup of an actively MLFL with a birefringence cavity. PM EDF, Polarization maintaining erbium-doped fiber; BPF, bandpass filter; PC, polarization controller; MZIM, Mach–Zehnder intensity modulator; Pol, polarizer; PD, photodiode; OSC, oscilloscope; OSA, optical spectrum analyzer; Sig Gen, signal generator.

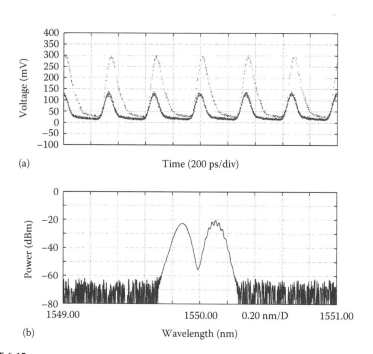

FIGURE 6.15
Oscilloscope trace of the output pulses (a) shows dual-amplitude property of the pulse train; the pulse spectrum (b) also shows dual wavelengths.

instead there exists two peaks located on the two sides of the filter center wavelength. The separation is about 0.25 nm. The longer wavelength peaks at 1550.12 nm and has the mode-locked profile of several longitudinal modes with a spacing of about 0.02 nm that corresponds to the modulation frequency of 2.998 GHz. The shorter wavelength peaks at 1549.87 nm and has a smoother structure. It thus demonstrates that the mode-locked pulses do not only have dual amplitudes in the time domain but also dual wavelengths in the frequency domain. It is questionable whether the two wavelengths are from the two types of pulses. One corresponds to the X pulses and the other corresponds to the Y pulses.

To answer this question, we operated the laser in two different modes by adjusting the polarization controllers. Figure 6.16 shows the oscilloscope trace of the pulse train and its spectrum when the polarization controller is adjusted to support only the X pulse. There is now only the low-amplitude component in the pulse trace. The clear underfoot pulse trace indicates that there is no pulse dropping. Every time slot in the pulse train is filled with the X pulse. This results in the mode locking as observed in its spectrum in Figure 6.16b. The spectrum shows that the X pulse corresponds to the long wavelength component of the dual-pulse spectrum.

Furthermore, the polarization controllers are also adjusted to lock the laser to the high-amplitude level pulse sequence. The low-amplitude pulses no

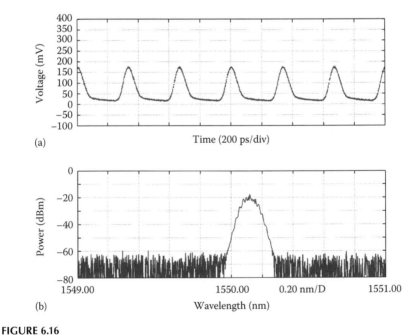

FIGURE 6.16
(a) Pulse trace and (b) its spectrum when the polarization controllers are adjusted so that only low-amplitude pulses exist in the loop; clear underfoot shows no pulse dropping.

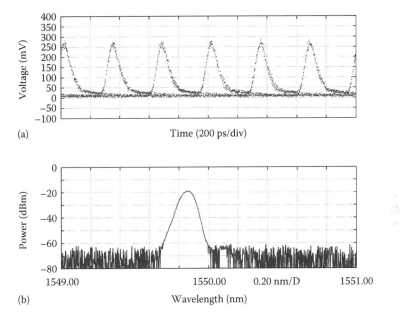

(a)

(b)

FIGURE 6.17
(a) Pulse trace and (b) its spectrum when the polarization controllers are adjusted so that only high-amplitude pulses exist in the loop; underfoot shows that there are some pulses dropped in the loop.

longer exist, as shown in Figure 6.17a. In contrast to the X pulse–locking case, the pulse trace here has an underfoot. This means that not every time the slot in the pulse train is filled. It is easily understood that the pulse dropping is caused by the energy limitation of the laser. High energy of the Y pulse requires high power from the pump to fill the slots all the time. Failure to this requirement results in pulse dropping [4]. It is noted that the high-amplitude pulse is brighter than that of the dual-pulse case. This means there are more Y pulses in the loop than before. The energy of the X pulses transfers to the Y pulses and hence more Y pulses are formed, but not all slots of the pulse train circulating in the loop are filled. Pulse dropping causes unequal spacing of the pulse. Therefore, the Fourier-transformed pulse sequence no longer has a smooth mode-locking structure as the spectrum shown in Figure 6.17b. This spectrum corresponds to the short-wavelength component of the dual-pulse spectrum.

Dual-pulses locking can be explained as follows: With a proper setup of polarization controller in the cavity, the laser can support two orthogonal modes that can be considered as the two eigen-modal solutions of the coupled NLSEs.

In the Nth order harmonic MLFL, there is not just one pulse but there are N pulses simultaneously traveling in the loop. The pulse of any solution can occupy one time slot in the pulse train. Therefore, there is a mixture of pulses from the two modes in the pulse train, which appears as dual-amplitude pulses overlapping on the oscilloscope screen.

Moreover, the slow-axis and fast-axis modes are under different detuning due to the birefringence of the cavity. The slow-axis mode is negatively detuned while fast-axis mode is positively detuned. It should be noted that the laser cavity has a normal dispersion with an average value of −28.5 ps/nm/km. In order to keep synchronizing, the wavelength of the slow-axis mode is shifted to the longer wavelength so that it travels faster than the fast-axis mode in the normal dispersion cavity. In contrast, the fast-axis mode is shifted to a shorter wavelength and, hence, is slowed down by the normal dispersion of the fiber cavity [18,19]. The wavelength shifting of the two modes is clearly observed in Figure 6.17. The X-pulse spectrum is in the longer wavelength region while the Y-pulse spectrum is in the shorter wavelength region.

To verify that the pulse train consists of pulses on two orthogonal axes, we use a polarizer to analyze the output pulse train. The polarizer is rotated into two orthogonal positions. Figure 6.18a is the pulse trace when the polarizer is at position 1. The high-amplitude pulses are attenuated and appear as the underfoot while the low-amplitude pulses reach the maximum value. In contrast, as shown in Figure 6.18, the low-amplitude pulses are highly attenuated and the high-amplitude pulses reach their maxima when the polarizer is set at position 2.

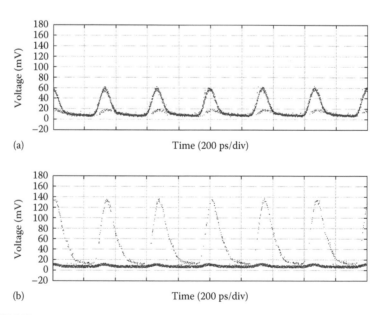

FIGURE 6.18
(a) X-axis component and (b) Y-axis component of the dual-pulse train; high-amplitude pulses and low-amplitude pulses are orthogonally polarized.

It is noted that although no stabilization technique was applied, stable pulses were obtained for several hours when the laser was operated to generate either the X pulse or Y pulse alone; dual-pulse operation was stable for more than 1 h. Besides, the phenomenon was also observed when the modulation frequency was increased to 10 GHz, but the stability was an issue. Moreover, adjustments need to be carefully handled since the higher the frequency, the larger the bandwidth of the MLL and, hence, the PMD effect will cause some polarization fading and coupling.

6.3.4 Pulse Dropout and Sub-Harmonic Locking

In addition to the generation of dual-polarization state pulses, we have also recorded the pulse dropout and sub-harmonic locking when the polarization controllers were set to obtain coupling between the two orthogonal modes. The experimental setup is the same as described in Section 6.3.3. When the polarization controller was adjusted at an appropriate position, we observed a well-defined mode-locked pulse train on the oscilloscope, as shown in Figure 6.2a. Starting from this setting, the evolution of the pulse train can be observed by rotating the middle plate of the polarization controller.

Remarkable frames extracted from the video file are presented in Figure 6.19a through d. It can be seen from Figure 6.19b that the underfoot appears at every alternative pulse position as the polarization controller is rotated. Pulses in the pulse train are no longer unified but divided into two groups, denoted group A and group B, respectively. Pulses in group B are gradually dropped out when the PC is rotated. And finally, all pulses of group B are dropped and the pulse train consists of pulses from group A only (see Figure 6.19c). The repetition rate of the pulse train is half of the modulation frequency, and thus sub-harmonic locking occurs. It is noted that the amplitude of the pulse in this case is nearly double that of the normal locking case since the energy of the dropped pulses was transferred to the remaining pulses. Further rotation of the PC results in the dropping of pulses in group A, as shown in Figure 6.19.

Pulse dropout and sub-harmonic locking were also observed when the modulation frequency was increased to 6.009125 GHz. However, the pulses in this case are classified into four groups and have the pattern of ABCDABCD, as shown in Figure 6.20. The pulses were also gradually dropped as the PC was rotated. The evolution of the pulse train is as follows: ABCDABCD, _BCD_BCD, _ _CD_ _CD, _ _ _ D_ _ _D. This results in the sub-harmonic locking of the laser to a quarter of the modulation frequency.

This phenomenon can be explained due to the beating between different modes simultaneously oscillating in the laser cavity. In a birefringent cavity, there are two cavities, one with a fundamental frequency of f_{Rx}, and the other with f_{Ry}. Thus, there are two sets of modes oscillating inside the laser cavity. With a proper setting of the PC, these two sets may oscillate independently or mutually and couple energy from one to another and beat with each other

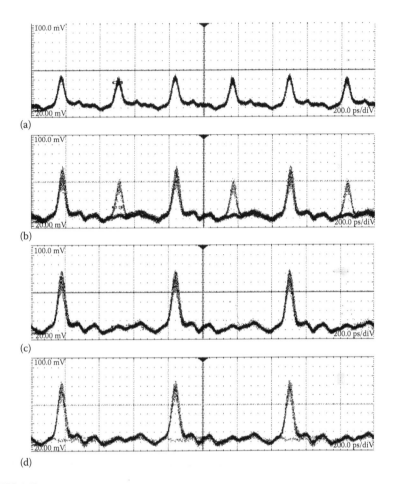

FIGURE 6.19
Oscilloscope trace of the output pulses: (a) full rate pulse train with clear pedestral, (b) pulses of group B are dropped in some slots, (c) All group-B pulses are dropped leaving a sub-harmonic mode-locked pulse with repetition rate at half of the modulation frequency, and (d) pulses of group A are dropped when the polarization controller was further rotated.

inside the cavity. The independent oscillation of the two sets causes dual-polarization state locking, which has been reported in [20]. In contrast, the beating of the two sets may change the pattern of the pulse train through pulse dropping and causes the laser to operate at a repetition rate lower than the modulation frequency, i.e., sub-harmonic locking.

6.3.5 Concluding Remarks

By employing a birefringent ring cavity, we have generated a dual-amplitude pulse train from an actively MLFL. The pulse train consists of pulses polarized on both X-axis and Y-axis. The X-polarized pulses have a lower

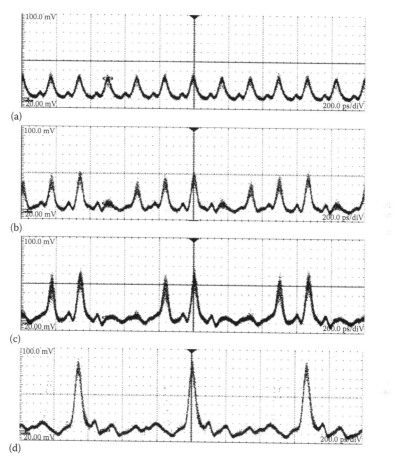

FIGURE 6.20
Oscilloscope trace of the output pulses (a) full rate pulse train with clear pedestral, (b) group-A pulses are dropped, (c) group-B pulses are also dropped, and (d) pulses from groups A, B, and C are dropped, leaving a sub-harmonic mode-locked pulse with a repetition rate that is one quarter of the modulation frequency.

amplitude and lasing at a longer wavelength than that of the Y-polarized mode. Therefore, we observed the dual-amplitude pulse trace and dual wavelengths on the optical spectrum analyzer. Besides that, locking to only the X-axis pulse or the Y-axis pulse has also been obtained by controlling the coupled polarization state of the generated lightwave.

In addition, the switching of the polarization state of the output pulse sequence of an actively MLFL as a result of birefringence in the ring has been reported. Since there are two different fundamental frequencies according to the resonances of the two orthogonally polarized regimes, the polarization state of the laser can be switched between the two orthogonal states by tuning the modulation frequency. Theoretical equations for predicting the tuning frequency has been developed and show good agreement with the experimental results.

Moreover, sub-harmonic locking, where the output pulse's train repetition rate is half or a quarter of the modulation frequency, has been obtained by the proper control of the polarization state of the light inside the birefringent cavity. This sub-harmonic locking may be exploited to develop optical frequency dividers or optical clock recovery circuits for fully optical signal processing systems.

6.4 Ultrafast Tunable Actively Mode-Locked Fiber Lasers

6.4.1 Introduction

Tunable optical filters are essential elements for selecting the lasing wavelength of actively MLFLs. Numerous techniques have been developed for tuning the filter's central wavelength. The simplest tuning technique is mechanically rotating the angular position of the substrate in the commonly used tunable thin-film filter. This rotation changes the light path length and hence setting a different resonant condition at a particular wavelength at which the maximum transmission occurs. Another technique is the mechanical or piezoelectrical stretching of the resonant cavity in a Fabry–Perot filter [21]. Thermal tuning has also been introduced in thin-film filters [22], arrayed waveguide grating [23], or fiber Bragg grating [24]. In addition, the redirection property of the liquid crystal under an electric field has been employed to make an electrically tunable filter [25]. In this technique, the physical etalon filter thickness is unchanged but the reflective index is altered under the redirection of the liquid crystals inside the cavity, hence leading to the change of the effective thickness of the etalon filter. Another technique involving the interaction between the lightwave and surface acoustic waves (SAW) in an anisotropic $LiNbO_3$ to convert a transverse electric guided mode to a transverse magnetic mode has also been applied to implement acousto-optic tunable filters [26]. The central wavelength is tuned by changing the frequency of the SAW for satisfying the phase-matching condition of different guided modes.

However, besides the acousto-optic filters, all those tuning techniques involve the movement or rotation of the material: moving of the substrate, lengthening of the cavity by stretching or heating, and rotating of the liquid crystals. Such mechanical tuning causes the slow response time in the order of μs or ms. The tuning time of the acoustic filter is also in the order of μs.

In this chapter, we present an electrically tunable filter with picosecond-scale response time. By introducing a phase modulator into a Lyot birefringence fiber filter, we are able to control the phase shift of the lightwave inside the retarder section and hence the resonant wavelength can be easily tuned. Since only the phase of the signal is changed and no mass movement is required, our proposed scheme can yield an ultrafast tuning speed, in the order of 10 ps. The fast tuning speed of the proposed filter is demonstrated by the implementation of a frequency shift keying (FSK) modulation at 10 Gbps.

We then apply the filter to design an ultrafast wavelength tunable actively MLFL. The lasing wavelength is easily tuned with a DC voltage applied to a LiNbO$_3$ phase modulator. Moreover, the modulator in our scheme is also used to modulate the signal, thus reducing the requirement of an additional modulator for mode locking. This helps to simplify the setup and reduce the cost as well. Since the tuning mechanism is based on the electro-optic effect of the LiNbO$_3$ crystal with response time of tens of picoseconds, the tuning speed is much faster than that based on the liquid crystal or mechanical rotation of the angle-tuned etalon filter [26–30]. Therefore, our scheme is promising for applications that require fast tuning of the wavelength such as wavelength switching, remote sensor, and optical spread spectrum communications.

6.4.2 Birefringence Filter

Figure 6.21 shows the setup of a Lyot filter using a HiBi fiber as a retarder. The filter consists of a piece of HiBi fiber sandwiched between two polarizers. The HiBi fiber is spliced with a 45° angle rotated of its slow axis to the polarizer. Using the Jones matrix representation, we obtain the transmission function of the filter as

$$T(\lambda) = \cos^2\left(\frac{\pi L \Delta n}{\lambda}\right) \qquad (6.16)$$

where
  L is the length of the fiber
  $\Delta n = \lambda_0/L_B$ is the birefringence of the fiber, and L_B is the beat length at the reference wavelength λ_0

This is a comb filter with a free spectral range (FSR) of $\lambda_0 L_B/L$. The calculated transmission function of a Lyot filter with a length of 0.7 m of Panda fiber is shown in Figure 6.22. The FSR of the filter is 7.75 nm. The FSR of the filter can be increased while keeping the same filter bandwidth by cascading

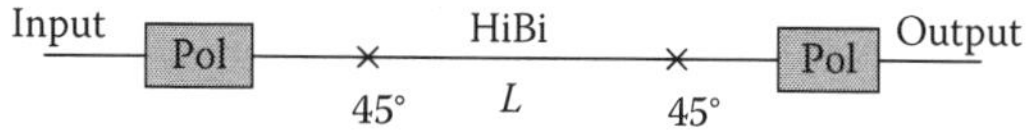

FIGURE 6.21
Setup of a Lyot filter with HiBi fiber serving as a retarder. Pol, Polarizer.

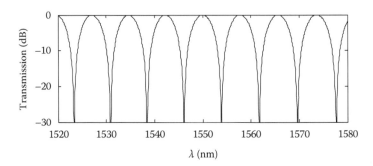

FIGURE 6.22
Transmittance of a Lyot filter with a 0.7 m of HiBi fiber serving as a retarder. The free spectral range: FSR = 7.75 nm; bandwidth: $BW = 3.875$ nm.

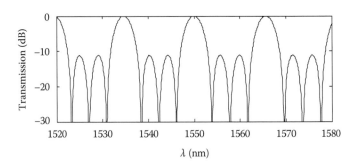

FIGURE 6.23
The FSR of the filter is enlarged by cascading two first-order Lyot filters to form a second-order Lyot filter. The FSR is twice that of the first-order Lyot filter while keeping the same bandwidth.

several Lyot filters to form a higher-order filter. This is achieved using the Vernier effect by cascading several Lyot filters, with each filter having a slightly different FSR, resulting in a larger overall FSR of the cascaded filters. The lengths of individual filters' Hibi fiber are $L, L/2, L/4, \ldots, L/2^N$. Figure 6.23 shows the transmission function of a second-order Lyot filter. The filter FSR has been doubled to 15.5 nm while the bandwidth is kept unchanged.

6.4.3 Ultrafast Electrically Tunable Filter Based on Electro-Optic Effect of LiNbO$_3$

6.4.3.1 Lyot Filter and Wavelength Tuning by a Phase Shifter

The central wavelength of the filter can be tuned by varying the length of the Hibi fiber through stretching or heating the fiber. However, the tuning speeds of those methods are slow and the response times are limited to ms. It is thus proposed to insert a phase shifter into the Hibi fiber section to introduce an ultrafast phase shift to the lightwave signal. The transmission function now becomes

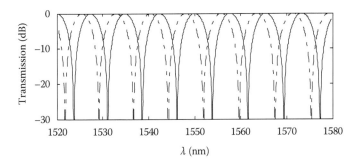

FIGURE 6.24
Central wavelength of the filter can be tuned by a phase shifter. (Solid line—zero phase shift; dash-dotted line—$\pi/2$ phase shift.)

$$T(\lambda) = \cos^2\left(\frac{\pi L \Delta n}{\lambda} + \varphi\right) \tag{6.17}$$

The central wavelength can now be easily tuned by varying the shifted phase φ. Figure 6.24 shows the shifting of the central wavelength when the phase is shifted by $\pi/2$.

6.4.3.2 Experimental Results

An experimental setup of an electrically tunable filter is shown in Figure 6.25. The filter consists of two polarizers sandwiching an optical phase modulator. All the fibers in the setup are HiBi Panda fibers. The modulator pigtails with lengths of $L_1 = 7.6\,\text{m}$ and $L_2 = 2.8\,\text{m}$, respectively, are spliced with a rotated angle of $45°$ to the slow axes of the fiber pigtail of the polarizer. By applying an electrical voltage to the phase modulator, we can easily shift the phase of the signal and hence tune the central wavelength of the filter.

The transmission function of the filter is given as

$$T(\lambda) = \cos^2\left(\frac{\pi B}{\lambda} + \frac{\pi V}{V_\pi}\right) \tag{6.18}$$

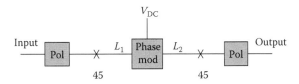

FIGURE 6.25
Setup of an electrically tunable HiBi-fiber Lyot filter with a phase modulator acting as a phase shifter. Pol, Polarizer; Phase Mod, phase modulator.

where

V is the applied voltage

V_π is the π-phase-shift voltage of the modulator, $B = \Delta n_{PMF} (L_1 + L_2) \pm \Delta n_c L_c$,
Δn_{PMF} is the birefringence of the PM fiber, and L_c and Δn_c are the length and birefringence of the phase modulator crystal, respectively

The shifted wavelength is given by a linear function of the applied voltage:

$$\Delta\lambda = \frac{FSR}{2V_\pi} V \qquad (6.19)$$

Figure 6.26 shows the transmission function of the setup filter. The FSR of the filter is about 16.06 nm. The central wavelength has been tuned from 1550.14 to 1543.52 nm when the applied voltage increased from 0 to 0.9 V, respectively. We obtain a shifting of FSR/2 when the applied voltage is equal to V_π as expected. However, in the experiment, the wavelength has been tuned by FSR/2 while the applied voltage is less than V_π(5 V). This enhancement of wavelength tuning is due to the increase of the temperature through the heating effect. As the phase modulator used in the experiment is a traveling waveguide type, it consumes the electrical power of the applied voltage and heats the modulator. This causes the thermal expansion and pyro-optic effect on the LiNbO$_3$ crystal [31–34], hence contributing to a larger shifting of the central wavelength.

It is suggested that the modulator's temperature should be stabilized by using a temperature controller. Another possible solution would be the use of isolated electrodes instead of the traveling waveguide electrodes for controlling the phase modulator. With the isolated electrodes, the driving current is nullified and, thus, the heating effect can be avoided.

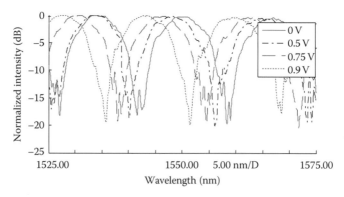

FIGURE 6.26
Central wavelength of the filter is shifted when a DC voltage is applied to the phase modulator.

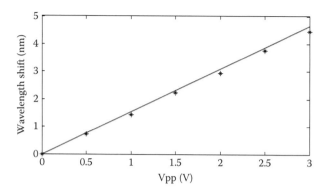

FIGURE 6.27
Shifting of the central wavelength as a function of applied square wave voltage (peak-to-peak).

To eliminate the thermal effect on the measurement of the wavelength shift, we applied a square wave to the phase modulator and measured the wavelength shift when the voltage was switched from one peak to the other. The results are shown in Figure 6.27. The wavelength shift increases linearly with an increase of the applied voltage. This is consistent with the theoretical estimation.

The fast tuning response of the filter has been tested with the FSK modulation setup shown in Figure 6.28. Signals from two lasers at 1543 and 1550 nm are combined by a 50/50 fiber coupler and fed into the filter. The phase modulator is driven by a bit stream of 1100001011000000 from a bit pattern generator (BPG) at 10 Gbps. The BPF is tuned to either the 1543 or 1550 nm to monitor the signal at the correspondent wavelength. The results in Figure 6.29 show that the filter alternately passes the signals at the two wavelengths when driven by the bit stream. The rise time of about 40 ps has been measured.

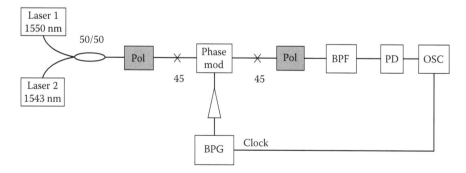

FIGURE 6.28
Experimental setup of a 10 Gbps FSK modulation using the high-speed tunable filter. BPG, bit pattern generator; BPF, bandpass filter; PD, photo diode; OSC, oscilloscope.

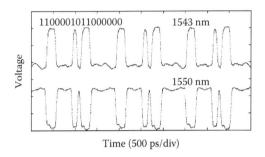

FIGURE 6.29
Lights at 1543 and 1550 nm alternately pass through the filter when the phase modulator is driven by a bit stream of 1100001011000000; switching up to 10 Gbps has been obtained.

Since the tuning mechanism is based on the electro-optic effect of the LiNbO$_3$ crystal, we can obtain a very fast tuning with sub-ps-scale time response. The tuning speed is only limited due to the slow rise time of the modulator's electrodes. Given this limitation, we still achieve the tuning speed in the order of 10 ps using the current 40 GHz phase modulator technology [35]. This is faster by an order of 2 compared to ns- or ms-scale tuning speed of the liquid crystal or thermal-based tunable filters [22,25].

6.4.4 Ultrafast Electrically Tunable MLL

6.4.4.1 Experimental Setup

The electrically tunable filter is then used in an actively MLFL, as shown in Figure 6.30 The laser also includes 5 m of EDF pumped by a 980 nm laser diode through an isolator integrated with the WDM coupler, 50 m of dispersion shifted fiber (DSF), two polarization controllers, and a 90/10 output coupler. A bias DC voltage and a microwave signal are combined together by a bias tee to drive the phase modulator. The modulator here plays dual roles: a mode locker and a wavelength tuner. The lasing wavelength can be tuned by controlling the bias DC voltage. As the bias DC voltage changes, the phase modulator shifts the phase of the signal by an amount proportional to the DC voltage and hence changes the central wavelength of the filter. When the modulator is driven by a microwave signal, it introduces phase modulation to the optical signal passing through it. If the microwave frequency is equal to the harmonic of the fundamental frequency of the laser cavity, the laser is mode locked and generates pulse train with the repetition rate at the microwave frequency.

The signal is coupled out through a 90/10 fiber coupler and detected by a 45 GHz bandwidth photodiode. The detected signal is then monitored using a high-speed oscilloscope with 70 GHz sampling head. The optical signal tapped out from the WDM coupler was monitored using an optical spectrum analyzer with a resolution of 0.02 nm.

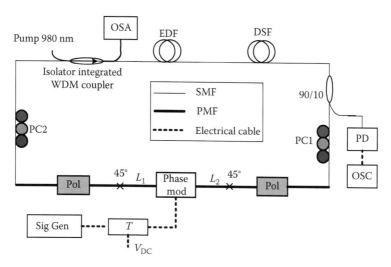

FIGURE 6.30

Setup of an actively MLFL using the electrically tunable Lyot filter described in Section 6.3; the phase modulator serves as a mode locker as well as a wavelength selector. Pol, Polarizer; Phase Mod, phase modulator; EDF, erbium-doped fiber; DSF, dispersion shifted fiber; PC, polarization controller; PD, photodiode; OSA, optical spectrum analyzer; OSC, oscilloscope; T, bias tee; Sig Gen, signal generator.

6.4.4.2 Experimental Results

Figure 6.31a shows the oscilloscope trace of the mode-locked pulse train when a zero DC voltage and a 10.002223 GHz microwave signal are applied to the modulator. The pulse width measured from the oscilloscope is about 21 ps. It is noticed that this includes the rise time of the photodiode and the sampling head. The pulse train is stable for more than 2 h without any adjustment of the polarization controllers or modulation frequency. The optical spectrum of the pulse train in Figure 6.31b clearly shows the modes with a separation of 0.08 nm, which corresponds to the 10 GHz repetition rate of the pulses. The pulse has a 0.41 nm bandwidth spectrum centered at 1558.85 nm. The TBP of 1.07 indicates that the pulse is highly chirped. This is due to the chirp caused by the phase modulator and the dispersion, assuming that it is operating below the nonlinear threshold. When the signal power level becomes large, i.e., above the nonlinear threshold and high pump power, the chirp comes from the nonlinearity of the fiber. It seems from the description here that there are no nonlinear effects in the cavity of the fiber. It is noticed that the DSF and the right-hand side polarizer in the setup can be removed to simplify the laser design. However, employing these components helps to reduce the super-mode noise and equalize the pulse amplitude through the additive pulse limiting (APL) effect [36].

The electrical spectrum and phase noise power density of the pulse train are plotted in Figure 6.32. The root mean square (RMS) timing jitter of the

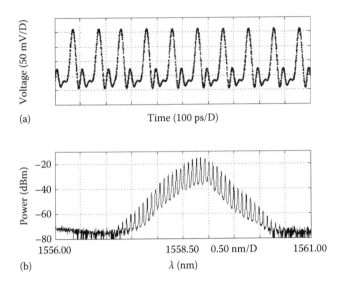

FIGURE 6.31
Pulse train generated from (a) the electrical wavelength tunable actively MLL and (b) its optical spectrum.

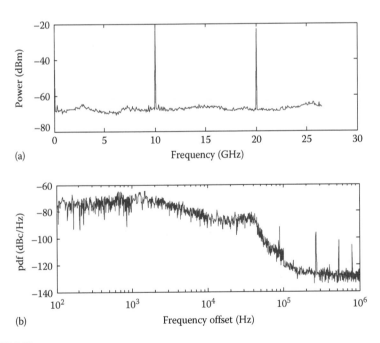

FIGURE 6.32
(a) Electrical intensity spectrum and (b) phase noise spectrum density of the mode-locked pulse train.

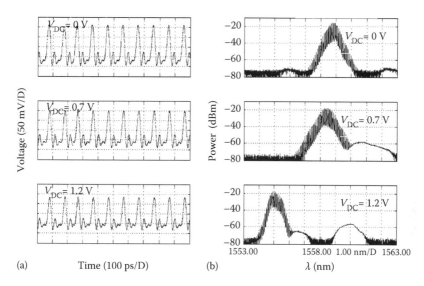

(a)   Time (100 ps/D)   (b) λ (nm)

FIGURE 6.33
Shifting of the wavelength when changing the bias voltage applied to the modulator: (a) oscilloscope traces and (b) optical spectra.

pulse train is 630 fs, which is obtained by performing integration on the electrical spectrum from 100 Hz to 1 MHz. This is consistent with the timing jitter of 530 fs calculated from the intensity spectrum using the Vonderlinde method [37].

The tuning of wavelength by applying an electrical bias voltage to the modulator is shown in Figure 6.33. The lasing wavelengths are 1558.85, 1558.45, and 1554.98 nm for the DC voltage of 0, 0.7, and 1.2 V, respectively. The lasing wavelength does not shift linearly with the DC voltage. This is due to the thermal effect as explained above. Using a phase modulator with isolated electrical electrodes for the DC bias would avoid such thermal induced shifting effect.

6.4.5 Concluding Remarks

We have proposed and demonstrated an electronically tunable optical filter with fast tuning characteristic. The filter consists of a PM-pigtail phase modulator spliced between two polarizers with a 45° rotating angle of the principal axes. Tuning of 6.62 nm for a 16.06 nm FSR filter has been obtained with an applied voltage of 0.9 V. Fast tuning speed with a rise time of 40 ps has been demonstrated.

We have also demonstrated the principles and operation of an actively MLFL with a fast electrical wavelength-tuning ability. A phase modulator inserted in a Lyot birefringence filter has been used as both a mode locker and a wavelength selector. Since the wavelength tuning is based on the electro-optic effect of the LiNbO$_3$ crystal with the response time of about tens of picoseconds, our scheme is promising for a very fast electrically tunable pulsed laser source. Possible applications of such a laser include wavelength switching, remote sensing, and optical spread spectrum.

6.5 Conclusions

So far, in Chapters 2 through 5 we have proposed and demonstrated a novel analysis method for studying the characteristics and performance of actively MLLs. Instead of using the self-consistence condition to obtain the steady-state pulse parameters as in the classical analysis methods, we have derived a set of mathematical series to trace the evolution of the signal in the laser cavity. Using this approach, we not only obtained the steady-state pulse parameters but also observed the transient evolution of the pulse. Round-trip-by-round-trip shaping of the pulse from the ASE noise under the modulation and filtering effects has been observed. In addition, the detuning of the modulation frequency of the actively MLLs was investigated by using this method. It is shown that mode locking still occurs when the modulation frequency is detuned within a certain locking range. This locking range is limited due to the shifting of the pulse position when the laser is detuned. Except for the shifting of the pulse position, the other steady-state pulse parameters when detuned are the same as those when exactly tuned.

Another advantage of our proposed series method is that it can be used for analyzing the laser model with ASE noise, which is normally ignored in classical analyses. Signal-to-noise ratio (SNR) of the laser can be calculated using the series method and its dependence on the cavity design parameters such as the cavity loss and amplifier noise has been discussed. It is found that the laser's SNR can be improved by minimizing the cavity loss and amplifier noise. On the other hand, detuning would decrease the laser's SNR and the presence of ASE noise limits the detuned frequency to a narrower locking range compared to the case in which an ideal noiseless amplifier is employed.

Besides the theoretical analysis method, we have also developed a laser model for the numerical simulation of the actively MLFLs. Dispersion and nonlinear effects play very important roles in the pulse formation and they have been studied in detail. Chirped and broadened pulses are attributed to the dispersion of the laser cavity. The sign of the chirp depends on the cavity dispersion: positive chirp (down chirp) for normal dispersion and negative chirp (up chirp) for anomalous dispersion. It is interesting that the lasing wavelength is no longer at the center of the intra-cavity filter but shifted away due to the dispersion, when the laser is detuned. The wavelength shifting toward the faster/slower wavelength compensates for the delay/advance caused by detuning and, hence, enhances the locking range of the laser. This improves the stability of the laser against any environmental perturbations.

In contrast, the laser is more sensitive to detuning when a large nonlinearity is introduced into the cavity. Although pulse shortening is obtained with a nonlinear cavity, the locking range is dramatically reduced. The pulse bandwidth is wider due to new frequencies generated through the SPM effect. Moreover, the mode-locked pulse is positively chirped. By combining the anomalous dispersion effect with the nonlinear effect, a soliton pulse can be directly generated from the laser, and a shorter pulse width is obtained from this soliton-forming effect. However, pulse is shortened with the trade-off for the locking range. Therefore, the laser parameters must be optimized to obtain the required pulse width while the laser stability is still maintained over a certain locking range.

Furthermore, we have also investigated techniques to improve the tunability of the lasers. The lasers are normally tuned by using mechanically tunable optical filters whose tuning speeds are rather slow, in the order of µs or ms. We have proposed and designed an ultrafast electrically tunable optical filter. The filter is based on a Lyot birefringence filter setup where the birefringence wave plates are replaced by polarization maintaining fibers. We introduced a phase modulator to shift the phase of the signal and thus tune the filter wavelength. Fast tuning speed with a rise time in the order of picoseconds has been demonstrated.

The proposed ultrafast filter has been applied into our design to obtain an actively MLFL with a fast electrical wavelength-tuning ability. Moreover, the phase modulator, which is used to tune the wavelength in the filter, has been also used to mode lock the laser. This simplifies the design and thus lowers the cost of the laser. Since our scheme uses the $LiNbO_3$ crystal's electro-optic effect with the response time of about tens of picoseconds for wavelength tuning, it is very promising for the development of ultrafast electrically tunable pulse laser sources.

An actively MLFL with a birefringent cavity has also been investigated. By introducing birefringence into the cavity, we have generated a pulse train with dual amplitudes from an active MLFL. Different to the normal mode-locked pulses with identical amplitude and polarization state, pulses polarized on both X-axis and Y-axis simultaneously exist in the output pulse train. The two orthogonal pulse sequences have different amplitudes and lase at different wavelengths. Dual wavelengths are the results of red shift and blue shift of the X and Y polarization states of the generated pulses, respectively, due to the detuning phenomena. Locking to the individual X-polarized or Y-polarized pulse is also obtained by adjusting the polarization controllers inside the cavity.

In addition, switching of the polarization state of the output pulse train generated from an active harmonic MLFL by exploiting the birefringence of the fiber ring has been reported for the first time. Two polarization modes have different fundamental frequencies and, hence, the laser can be forced to the mode-locking states of the two orthogonal polarization regimes by tuning or detuning the modulation frequency.

Moreover, by properly controlling the polarization state of the light inside the birefringent cavity laser, we have generated a pulse train with a repetition rate at half or a quarter of the modulation frequency. This sub-harmonic locking may find such applications as in the design and development of optical frequency dividers and optical clock recoveries, which are very important components in fully optical signal processing systems.

References

1. D. Kuizenga and A. Siegman, FM and AM mode locking of the homogeneous laser—Part I: Theory, *IEEE J. Quantum Electron.*, 6, 694–708, 1970.
2. H. Haus, A theory of forced mode locking, *IEEE J. Quantum Electron.*, 11, 323–330, 1975.
3. G.P. Agrawal, *Applications of Nonlinear Fiber Optics*, Academic Press, San Diego, CA, 2001.
4. M. Horowitz, C.R. Menyuk, T.F. Carruthers, and I.N. Duling III, Theoretical and experimental study of harmonically modelocked fiber lasers for optical communication systems, *J. Lightwave Technol.*, 18, 1565–1574, 2000.
5. E. Marti-Panameno, J.J. Sanchez-Mondragon, and V.A. Vysloukh, Theory of soliton pulse forming in an actively modelocked fiber laser, *IEEE J. Quantum Electron.*, 30, 822–826, 1994.
6. M. Nakazawa and E. Yoshida, A 40-GHz 850-fs regeneratively FM mode-locked polarization-maintaining erbium fiber ring laser, *IEEE Photon. Technol. Lett.*, 12, 1613–1615, 2000.
7. R. Leners, P.L. Francois, and G. Stephan, Simultaneous effects of gain and loss anisotropies on the thresholds of a bipolarization fiber laser, *Opt. Lett.*, 19, 275–277, 1994.
8. S. Bielawski, D. Derozier, and P. Glorieux, Antiphase dynamics and polarization effects in the Nd-doped fiber laser, *Phys. Rev.*, 46, 2811–2822, 1992.
9. S.L. Zhang, M. Lu, G.F. Jin, and M.X. Wu, Laser frequency split by an electro-optical element in its cavity, *Opt. Commun.*, 96, 245–248, 1993.
10. R.J. Oram, I.D. Latimer, S.P. Spoor, and S. Bocking, Longitudinal mode separation tuning in 633 nm helium-neon lasers using induced cavity birefringence, *J. Phys. D: Appl. Phys.*, 26, 1169–1172, 1993.
11. J.S. Wey, J. Goldhar, and G.L. Burdge, Active harmonic modelocking of an erbium fiber laser with intracavity Fabry-Perot filters, *J. Lightwave Technol.*, 15, 1171–1180, 1997.
12. H.A. Haus, Mode-locking of lasers, *IEEE J. Sel. Top. Quantum Electron.*, 6, 1173–1185, 2000.
13. K.K. Gupta, N. Onodera, K.S. Abedin, and M. Hyodo, Pulse repetition frequency multiplication via intracavity optical filtering in AM mode-locked fiber ring lasers, *IEEE Photon. Technol. Lett.*, 14, 284–286, 2002.

14. H. Takara, S. Kawanishi, M. Saruwatari, and K. Noguchi, Generation of highly stable 20 GHz transform-limited optical pulses from actively mode-locked Er^{3+}-doped fibre lasers with an all-polarisation maintaining ring cavity, *Electron. Lett.*, 28, 2095–2096, 1992.

15. N. Onodera, Supermode beat suppression in harmonically mode-locked erbium-doped fibre ring lasers with composite cavity structure, *Electron. Lett.*, 33, 962–963, 1997.

16. G.P. Agrawal, *Fiber-Optic Communication Systems*, 3rd edn., Wiley-Interscience, New York, 2002.

17. G.P. Agrawal, *Nonlinear Fiber Optics*, Academic Press, San Francisco, CA, 2001.

18. G. Zhu and N.K. Dutta, Dispersion effects on the detuning properties of actively harmonic mode-locked fiber lasers, *Opt. Express*, 13, 2688, 2005.

19. K. Tamura and M. Nakazawa, Dispersion-tuned harmonically mode-locked fiber ring laser for self-synchronization to an external clock, *Opt. Lett.*, 21, 1984–1986, 1996.

20. H.Q. Lam, P. Shum, L.N. Binh, and Y.D. Gong, Polarization-dependent locking in SOA harmonic mode-locked fiber laser, *IEEE Photon. Technol. Lett.*, 18, 2404–2406, 2006.

21. J. Stone and L.W. Stulz, Pigtailed high-finesse tunable fiber Fabry-Perot interferometers with large, medium and small free spectral ranges, *Electron. Lett.*, 23, 781–783, 1987.

22. M. Lequime, R. Parmentier, F. Lemarchand, and C. Amra, Toward tunable thin-film filters for wavelength division multiplexing applications, *Appl. Opt.*, 41, 3277–3284, 2002.

23. S. Suzuki, A. Himeno, and M. Ishii, Integrated multichannel optical wavelength selective switches incorporating an arrayed-waveguide grating multiplexer and thermooptic switches, *J. Lightwave Technol.*, 16, 650–655, 1998.

24. A. Iocco, H.G. Limberger, and R.P. Salathe, Bragg grating fast tunable filter, *Electron. Lett.*, 33, 2147–2148, 1997.

25. M.W. Maeda, J.S. Patel, C.L. Lin, J. Horrobin, and R. Spicer, Electronically tunable liquid-crystal-etalon filter for high-density Wdm systems, *IEEE Photon. Technol. Lett.*, 2, 820–822, 1990.

26. L.N. Binh, J. Livingstone, and D.H. Steven, Tunable acoustooptic Te-Tm mode converter on a diffused optical-waveguide, *Opt. Lett.*, 5, 83–84, 1980.

27. B. Bakhski and P.A. Andrekson, 40 GHz actively modelocked polarisation-maintaining erbium fibre ring laser, *Electron. Lett.*, 36, 411–413, 2000.

28. G.T. Harvey and L.F. Mollenauer, Harmonically mode-locked fiber ring laser with an internal Fabry–Perot stabilizer for soliton transmission, *Opt. Lett.*, 18, 107, 1993.

29. E. Yoshida and M. Nakazawa, 80~200 GHz erbium doped fibre laser using a rational harmonic mode-locking technique, *Electron. Lett.*, 32, 1370–1372, 1996.

30. K.K. Gupta and N. Onodera, Regenerative mode locking via superposition of higher-order cavity modes in composite cavity fiber lasers, *Opt. Lett.*, 30, 2221–2223, 2005.

31. H. Nagata, Y.G. Li, W.R. Bosenberg, and G.L. Reiff, DC drift of x-cut LiNbO$_3$ modulators, *IEEE Photon. Technol. Lett.*, 16, 2233–2235, 2004.

32. M.L. Dennis, I.N. Duling, and W.K. Burns, Inherently bias drift free amplitude modulator, *Electron. Lett.*, 32, 547–548, 1996.

33. N.A. Whitaker, R.J. Lustberg, M.C. Gabriel, and H. Avramopoulos, Low-drift modulator without feedback, *IEEE Photon. Technol. Lett.*, 4, 855–857, 1992.
34. A.R. Beaumont, C.G. Atkins, and R.C. Booth, Optically induced drift effects in lithium-niobate electrooptic wave-guide devices operating at a wavelength of 1.51-Mu-M, *Electron. Lett.*, 22, 1260–1261, 1986.
35. E.L. Wooten, K.M. Kissa, A. Yi-Yan, E.J. Murphy, D.A. Lafaw, P.F. Hallemeier, D. Maack et al., A review of lithium niobate modulators for fiber-optic communications systems, *IEEE J. Sel. Top. Quantum Electron.*, 6, 69–82, 2000.
36. C.R. Doerr, H.A. Haus, E.P. Ippen, L.E. Nelson, M. Shirasaki, and K. Tamura, Additive-pulse limiting, *Opt. Lett.*, 19, 31, 1994.
37. D. Vonderlinde, Characterization of the noise in continuously operating mode-locked lasers, *Appl. Phys. B*, 39, 201–217, 1986.

7

Ultrafast Fiber Ring Lasers by Temporal Imaging

The generation of ultrashort pulses with high repetition rates is very important for future ultrahigh bit-rate optical communication systems. An active mode-locked fiber laser is a potential source of such pulses. However, the pulse repetition rate is usually limited by the bandwidth of the intracavity modulator. Hence, some techniques have to be applied to increase the repetition frequency of the pulses generated. In this chapter, we focus on the generation of high-repetition-frequency optical pulses using the fractional temporal Talbot effect and briefly on some other repetition rate multiplication techniques. The stability studies using the phase plane analysis (which is commonly employed to study the stability of nonlinear control systems) will also be discussed. The phase plane analysis is commonly used in nonlinear control systems and, to the best of our knowledge, this is the first time that the technique is being applied in the repetition rate multiplication laser system.

7.1 Repetition Rate Multiplication Techniques

In order to increase the system's line rate, many repetition rate multiplication techniques have been proposed and demonstrated, out of which rational harmonic detuning [1], fractional temporal Talbot effect [2], intracavity optical filtering [3], higher-order FM mode locking [4], optical time domain multiplexing [5,6], etc. Longhi et al. [7] had shown a 16-times bit-rate multiplication, achieving 40 GHz, 10 ps pulse train from a base rate of 2.5 GHz mode-locked optical pulses using a 100 cm long linearly chirped fiber grating (LCFG). More recently, Azana et al. [8] demonstrated the use of short superimposed fiber Bragg grating (FBG) structures for generating ultrahigh repetition rate pulse bursts, i.e., a 170 GHz optical pulse train from a 10 GHz mode-locked fiber laser. The superimposed FBGs are made of LCFGs operating in the reflection mode. Besides FBG, a dispersive fiber has also been used to achieve the same objective [9].

7.1.1 Fractional Temporal Talbot Effect

The self-imaging effect of a plane periodic grating, when the grating is illuminated by a monochromatic light beam, was reported by Talbot in 1836 [10], and is known as the Talbot effect. This effect has been widely used in several optics fields, such as holography, image processing and synthesis, photolithography, optical testing, optical computing, and optical metrology. Fractional Talbot effect refers to the superimposition of shifted and complex weighted replicas of the original object resulted from the diffraction of a periodic object at fractions of the Talbot distance [11] in the spatial domain. The fractional temporal Talbot effect refers to the interference effect between the dispersed pulses in the optical fiber and in the time domain. By using this interference effect, one can multiply the system's repetition rate. The essential element of this technique is the dispersive medium, such as a LCFG and a dispersive single-mode fiber, out of which, LCFG can be designed to provide the required bandwidth and dispersion characteristics in a more compact form.

When a pulse train is transmitted through the optical fiber, the phase shift of the kth individual lasing mode due to the group velocity dispersion (GVD) is given by [9]

$$\varphi_k = \frac{\pi\lambda^2 D z k^2 f_r^2}{c} \tag{7.1}$$

where
  λ is the center wavelength of the mode-locked pulses
  D is the fiber's GVD
  z is the fiber length
  c is the speed of light in vacuum
  k is the propagation constant in vacuum
  f_r is the repetition frequency

This phase shift generally induces pulse broadening and distortion. When φ_k is $2\pi k^2$ (no phase shift between the lasing modes) or its multiple, the initial pulse shape is restored even after transmission through a long dispersive fiber as long as the coherence length of each lasing mode is maintained. This corresponds to

$$z_T = \frac{2c}{\lambda^2 |D| f_r^2} = \frac{2}{\Delta\lambda f_r |D|} \tag{7.2}$$

where
  $\Delta\lambda = f_r \lambda^2/c$ is the spacing between spectral components of the pulse train in the wavelength domain
  f_r is the spacing between spectral components of the pulse train in the frequency domain

For example, if $\Delta\lambda = 1\,\text{nm}$ then, correspondingly, $f_r = 125\,\text{GHz}$ at $\lambda = 1550\,\text{nm}$. Thus, for lower repetition frequency there is a need to use wide passband optical filter associated with the ring laser. In contrast, when the fiber length is equal to $z_T/(2m)$ (where $m = 2, 3, 4, \ldots$), every mth lasing mode oscillates in phase then the oscillation waveform maximums accumulate. However, the phases of other modes are mismatched, weakening their contributions to the formation of pulse waveform. This leads to the generation of a pulse train with a multiplied repetition frequency with m times. The highest repetition rate obtainable is limited by the duration of the individual pulses; as the pulses start to overlap when the pulse duration becomes comparable to the pulse train period, i.e., $m_{\text{max}} = \Delta T/\Delta t$, where ΔT is the pulse train period and Δt is the pulse duration.

The pulse duration does not change much even after the multiplication, because every mth lasing mode dominates in the pulse waveform formation of m times multiplied pulses. The pulse waveform therefore becomes identical to the one generated from the mode-locked laser, with the same spectral envelope. The optical spectrum of the signal does not change after the multiplication process, because this technique utilizes only the change-of-phase relationship between lasing modes and does not use the fiber's nonlinearity. The effect of higher-order dispersion might degrade the quality of the multiplied pulses, i.e., pulse broadening, appearance of pulse wings, and pulse-to-pulse intensity fluctuation. In this case, any medium to compensate for the higher-order dispersion would be required in order to complete the multiplication process. To achieve higher multiplications, the input pulses must have a broad spectrum and the fractional Talbot length must be very precise in order to receive high-quality pulses. If the pulses exist in the nonlinear regime and experience the anomalous dispersion along the fiber, solitonic action occurs and prevents the linear Talbot effect from occurring. Thus, the power of the input signal must be below that required to induce the nonlinear fiber effects such as the self-phase modulation in order for the Talbot effect to work properly to increase the pulse's repetition rate.

7.1.2 Other Repetition Rate Multiplication Techniques

In this section, we give a brief description on other repetition rate multiplication techniques available to date, such as intracavity filtering and optical time division multiplexing (OTDM). Rational harmonic detuning is another simple technique, which has been widely adopted to increase the repetition rate. However, we will only discuss it in Chapter 8.

Intracavity optical filtering uses modulators and a high-finesse Fabry–Perot filter (FFP) within the laser cavity to achieve a higher repetition rate. In a conventional AM harmonic mode locking, an RF modulation frequency $f_m = qf_c$ (where f_c is the cavity resonance frequency) is applied to the MZM modulator placed inside the ring cavity to lock the phase of the longitudinal modes with frequency spacing equal to f_m. In a mode-locked fiber ring laser, when intracavity optical filtering via an FFP with a free spectral range (FSR) equal to f_m is introduced, the phase-locked longitudinal modes with frequency spacing equal to the FSR are synchronized to the transmission peaks of FFP and are passed through the filter, while the intermediate cavity resonance modes are blocked by the FFP. This results in the generation of highly stable optical pulse trains with repetition rate equal to the FSR.

When the modulation signal frequency, a sub-harmonic multiple of the FSR of FFP, is applied to the MZM, the generated optical pulses will have the repetition rate equal to FSR. It is important that a modulation signal frequency f_m must not only be a harmonic multiple of the cavity resonance frequency f_c, but also a sub-harmonic multiple of the FSR of FFP, such that $f_m = \text{FSR}/m$, where m is the sub-harmonic number. When the FSR of FFP is a multiple integer m of one of the longitudinal modes of the cavity, the intermediate modes are suppressed, resulting in an optical pulse train with a repetition frequency of FSR or $f_r = mf_m$, hence increasing the repetition rate of the optical pulse train. In addition, selective filtering of randomly oscillating intermediate longitudinal modes via the intracavity FFP also leads to enhanced pulse stability in the generated optical pulse train, e.g., lower supermode noise, amplitude noise, phase noise, or timing jitter.

Another method used to increase the repetition frequency of the system is by using the OTDM technique. It is done by creating certain time delay in each multiplexed optical path, so as to fill up the gap between the original optical pulses. Daoping et al. [5] have shown an eight-times repetition rate multiplication utilizing three loop-connecting fiber couplers, achieving a 20 GHz optical pulse train. Yamada has also demonstrated 640 GHz–1.28 THz pulse trains by using the OTDM technique [6]. They were using a seven-step Mach–Zehnder interferometer in their setup, which was formed on a silica-based planar lightwave circuit. It has [12] also been demonstrated that 500 GHz optical pulse train generation using an arrayed waveguide grating (AWG) under this principle. However, owing to the coherent interference between the OTDM channels, the maximum feasible increase in the system's line rate using a practical multiplexer is limited. Furthermore, the multiplied pulses suffer from severe amplitude modulation and each burst of pulse lies at a different wavelength. Furthermore, this technique requires a precise adjustment of the optical path delay.

7.1.3 Experimental Setup

The input to the fractional temporal Talbot multiplier is the output pulse train obtained from the 10 GHz active harmonically mode-locked fiber ring laser as described in Chapter 4. The multiplier is made up of a rim of about 3 km of dispersion compensating fiber (DCF), with the dispersion value of −98 ps/nm/km at 1550 nm. The schematic of the experimental setup is shown in Figure 7.1. The variable optical attenuator is used to reduce the optical power of the pulse train generated by the mode-locked fiber ring

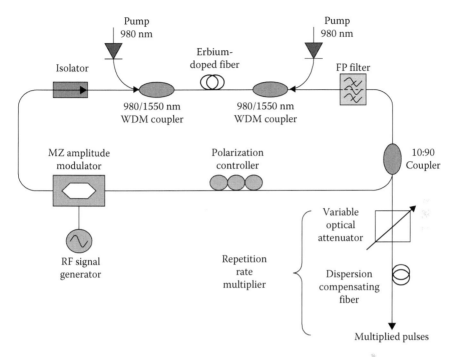

FIGURE 7.1
Fractional temporal Talbot-based repetition rate multiplication setup.

laser and, hence, to remove the nonlinear fiber effect on the pulse before entering the multiplier. Based on the calculation, the required length of DCF ($D = -98$ ps/nm/km at 1550 nm) for a six times multiplication on a 10 GHz signal is 3.185457 km.

7.1.4 Results and Discussion

The original and multiplied pulse trains and their spectra are shown in Figures 7.2 and 7.3. The input to the multiplier is a 10.217993 GHz pulse train, obtained from the active harmonically mode-locked fiber ring laser, operating at 1550.2 nm, as shown in Figure 7.2. The spectra of the un-multiplied and multiplied signals are the same. This is simply because the Talbot effect is using only the linear phase difference to achieve the repetition rate multiplication and not the fiber nonlinearity, hence, no change in the spectra should be observed. The linewidth separation for both cases is about 0.08 nm.

The stability of the high repetition rate pulse train is one of the main concerns for practical high-speed optical communications systems. Conventionally, the stability analyses of such laser systems are based on the linear behavior of the laser in which we can analytically analyze the system behavior in both time and frequency domains. However, when the

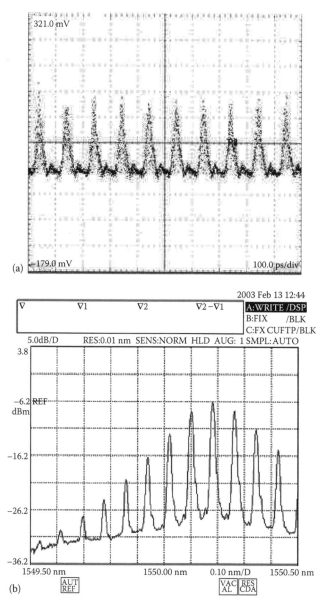

FIGURE 7.2
Original 10 Gbps un-multiplied (a) pulse train and (b) its spectrum.

mode-locked fiber laser is operating under the nonlinear regime, none of these standard approaches can be used, since a direct solution of the nonlinear different equation is generally impossible. Hence, the frequency domain transformation is not applicable. Although multiplying a system's repetition rate using the fractional temporal Talbot is based on the linear behavior of

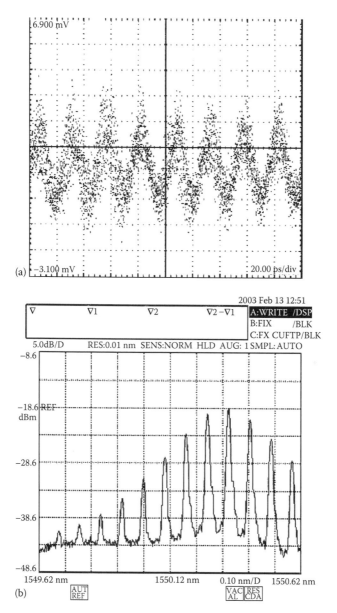

FIGURE 7.3
40 Gbps multiplied (a) pulse train and (b) its spectrum.

the laser, there are still some inherent nonlinearities affecting its stability, such as the saturation of the embedded gain medium, nonquadrature biasing of the modulator, fiber nonlinearities, etc., hence, a nonlinear stability approach must be adopted. Here, we focus on the stability and transient analyses of the Talbot-based multiplied pulse trains using the nonlinear control analytical technique—the phase plane analysis [13]. This is the first time, to the best of our knowledge, that the phase plane analysis is being used to study the stability and transient performances of a system whose repetition rate is multiplied using the fractional temporal Talbot effect.

The phase plane analysis is a graphical method of studying second-order nonlinear systems. The result is a family of system motion of trajectories on a two-dimensional plane, which allows us to visually observe the motion patterns of the system. Nonlinear systems can display more complicated patterns in the phase plane, such as multiple equilibrium points and limit cycles. In the phase plane, a limit cycle is defined as an isolated closed curve. The trajectory has to be both closed, indicating the periodic nature of the motion, and isolated, indicating the limiting nature of the cycle [13].

The system modeling for our fractional temporal Talbot multiplier is done based on the following assumptions: (1) perfect output pulse from the mode-locked fiber ring laser without any timing jitter; (2) the multiplication is achieved under ideal conditions (i.e., exact fiber length for a certain dispersion value); (3) no fiber nonlinearity is included in the analysis of the multiplied pulse; (4) no other noise sources are involved in the system; and (5) the non-Gaussian lasing modes have equal amplitude, and the Gaussian lasing modes also have equal amplitude.

7.2 Uniform Lasing Mode Amplitude Distribution

A uniform amplitude distribution of the lasing modes is assumed at the first instance, i.e., ideal mode-locking condition. The simulation is done based on the 10 GHz pulse train, centered at 1550 nm, with a fiber dispersion factor of −98 ps/km/nm, a 1 nm flat-top passband filter is used in the cavity of the mode-locked fiber laser. The estimated Talbot distance is 25.484 km.

The original pulse (obtained directly from the mode-locked laser) propagation behavior and its phase plane are shown in Figures 7.4a and 7.5a. From the phase plane obtained, one can observe that the origin is a stable node and the limit cycle around that vicinity is a stable limit cycle. This agrees very well with our first assumption: an ideal pulse train at the input of the multiplier. Also, we present the pulse propagation behavior and its corresponding phase plane for two-times, six-times, and eight-times the GVD-based bit-rate multiplication system in Figures 7.4 and 7.5. The shape of the phase graph exposes the phase between the displacement and its derivative. As for the phase planes shown in the following sections, the X-axis is $E(t)$ – signal energy and Y-axis is its derivative; solid and dotted lines represent the real and imaginary parts of the energy, respectively.

As the multiplication factor increases, the system trajectories are moving away from the origin. As for the six-times and eight-times multiplications,

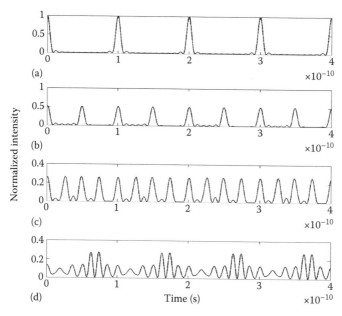

FIGURE 7.4
Pulse propagation of (a) original pulse, (b) 2×, (c) 4×, and (d) 8× multiplication with 1 nm filter bandwidth using the analysis of equal amplitude of the lasing modes.

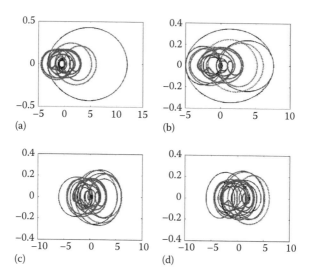

FIGURE 7.5
Phase plane of (a) original pulse, (b) 2×, (c) 4×, and (d) 8× multiplication with 1 nm filter bandwidth and using the analysis of equal amplitude of the lasing modes.

there is neither a stable limit cycle nor a stable node on the phase planes even with the ideal multiplication parameters. Here we see the system trajectories spiral out to an outer radius and back to the inner radius. The change in the radius of the spiral is the transient response of the system. Hence, with the increase in the multiplication factor, the system trajectories become more sophisticated. Although the GVD-based repetition rate multiplication method uses only the phase change effect in the multiplication process, the inherent nonlinearities still affect its stability indirectly. Despite the reduction in the pulse amplitude, we observe uneven pulse amplitude distribution in the multiplied pulse train. The percentage of unevenness increases with the multiplication factor in the system.

7.2.1 Gaussian Lasing Mode Amplitude Distribution

This set of the simulation models the practical filter used in the system. It gives us a better insight on the behavior of the repetition rate multiplication system–based on the fractional temporal Talbot effect. The parameters used in the simulation are exactly the same, except that the flat-top filter has been changed to a Gaussian-like passband filter. The spirals of the system trajectories and the uneven pulse amplitude distribution are more severe than those in the uniform lasing mode amplitude analysis. The simulation results are shown in Figures 7.6 and 7.7.

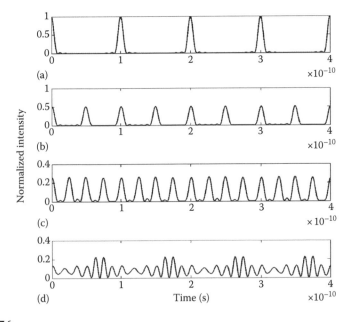

FIGURE 7.6

Pulse propagation of (a) original pulse, (b) 2×, (c) 4×, and (d) 8× multiplication with 1 nm filter bandwidth using the analysis of Gaussian amplitude of the lasing modes.

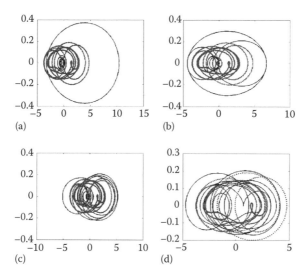

FIGURE 7.7
Phase plane of (a) original pulse, (b) 2×, (c) 4×, and (d) 8× multiplication with 1 nm filter bandwidth and using the analysis of Gaussian amplitude of the lasing modes.

7.2.2 Filter Bandwidth Influence

The bandwidth of the filter used in the MLFRL affects the stability of the repetition rate multiplication system using the fractional temporal Talbot. The earlier analyses are based on a 1 nm filter bandwidth. The number of modes locked within a laser system increases with the filter bandwidth, which gives us a better quality of the mode-locked pulses. The simulations shown below are based on the Gaussian lasing mode's amplitude distribution, 3 nm filter bandwidth, and other parameters remain unchanged. With a wider filter bandwidth, the pulse width and the percentage of the pulse amplitude fluctuation will decrease. This suggests a better stability condition. Instead of spiraling away from the origin, the system trajectories move inward to the stable node. However, this leads to a more complex pulse formation system (Figures 7.8 and 7.9).

7.2.3 Nonlinear Effects

When the input signal enters the nonlinear region, the Talbot multiplier loses its multiplication capability as predicted earlier. The additional nonlinear phase shift due to the high input power is added to the total phase shift and destroys the phase change condition of the lasing modes required by the multiplication condition. The additional nonlinear phase shift also changes the pulse shape and the phase plane of the multiplied pulses. The nonlinear phase shift of the kth lasing mode is related to the input power as follows [14]:

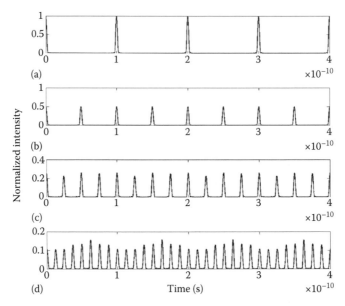

FIGURE 7.8
Pulse propagation of (a) original pulse, (b) 2×, (c) 4×, and (d) 8× multiplication with 3 nm filter bandwidth using the analysis of Gaussian amplitude of the lasing modes.

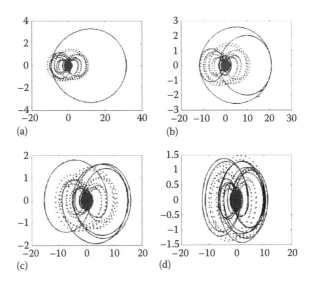

FIGURE 7.9
Phase plane of (a) original pulse, (b) 2×, (c) 4×, and (d) 8× multiplication with 3 nm filter bandwidth using the analysis of Gaussian amplitude of the lasing modes.

$$\theta_k = \frac{2\pi}{\lambda} \cdot \frac{n_2 n_0}{2\eta_0} |E|^2 L = \frac{2\pi}{\lambda} n_2 \frac{P}{A_{eff}} L \qquad (7.3)$$

where

n_0 and n_2 are the linear and nonlinear refractive indexes of the fiber, respectively

η_0 is the wave impedance in vacuum

E is the optical electric field

P is the optical power

A_{eff} is the effective mode area depending on the modal field profile in the fiber

L is the cavity length

The parameters used in the simulation are $n_2 = 3.0 \times 10^{-20}$ m²/W, $n_0 = 1.45$, $A_{eff} = 50\,\mu m^2$ (Figures 7.10 and 7.11).

7.2.4 Noise Effects

The above simulations are all based on the noiseless situation. However, in practical optical communication systems, noises are always the sources

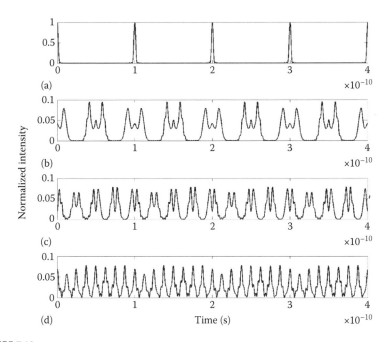

FIGURE 7.10
Pulse propagation of (a) original pulse, (b) 2×, (c) 4×, and (d) 8× multiplication with 3 nm filter bandwidth, input power = 1 W, using the analysis of Gaussian amplitude of the lasing modes.

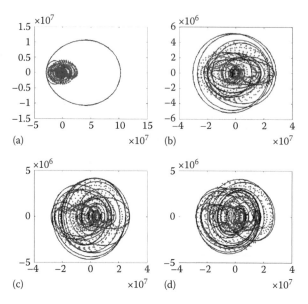

FIGURE 7.11
Phase plane of (a) original pulse, (b) 2×, (c) 4×, and (d) 8× multiplication with 3 nm filter band-width, input power = 1 W, and using the analysis of Gaussian amplitude of the lasing modes.

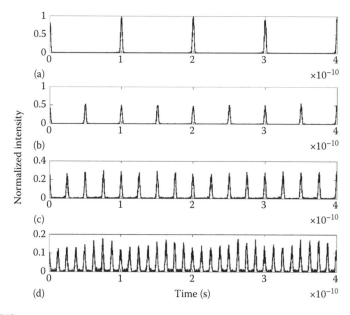

FIGURE 7.12
Pulse propagation of (a) original pulse, (b) 2×, (c) 4×, and (d) 8× multiplication with 3 nm filter band-width and 0 dB signal-to-noise ratio using the analysis of Gaussian amplitude of the lasing modes.

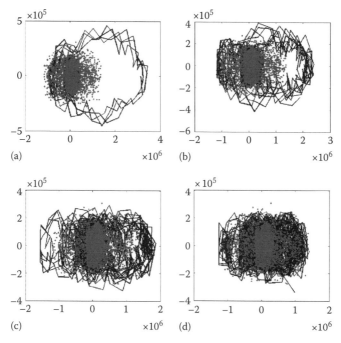

FIGURE 7.13

Phase plane of (a) original pulse, (b) 2×, (c) 4×, and (d) 8× multiplication with 3 nm filter bandwidth, and 0 dB signal-to-noise ratio using the analysis of Gaussian amplitude of the lasing modes.

of nuisance, which can cause system instability; therefore, it must be taken into consideration for the system stability studies. Figures 7.12 and 7.13 show the results in which the signal level is equal to the noise level (a very noisy system), that is, the signal-to-noise ratio is 0 dB. Since the optical intensity of the *m* times multiplied pulse is *m* times less than the original pulse, it is more vulnerable to noise. The signal is difficult to differentiate from the noise within the system if the power of multiplied pulse is too small. The phase plane of the multiplied pulse is distorted due to the presence of noise, which leads to poor stability performance. It shows a total random fashion of the system trajectories.

7.3 Conclusions

We have demonstrated a six-times repetition rate multiplication by using the fiber-based fractional temporal Talbot effect; and hence 40 GHz pulse train is obtained from 10 GHz mode-locked fiber laser source. However, its stability is of great concern for practical use in the optical communications systems. Although the repetition rate multiplication based on the fractional temporal Talbot technique is linear in nature, the inherent nonlinearities in such a system may disturb its stability. Therefore, any linear approach may not be suitable in deriving the system stability. Stability analysis for this multiplied pulse train has been studied by using the nonlinear control stability theory—the phase plane analysis. Surprisingly, from the analysis model, the stability of the multiplied pulse train can hardly be achieved even under perfect multiplication conditions. Furthermore, we observed uneven amplitude distribution of the bit-rate-multiplied pulse using the fractional temporal Talbot, which was due to the energy variations between the pulses that caused some energy beating between them. Another possibility is the divergence of the pulse's energy variation in the vicinity around the equilibrium point that leads to instability.

The pulse's amplitude fluctuation increases with the multiplication factor. A better stability condition can be achieved with a wider filter bandwidth. The nonlinear phase shift due to the SPM effect and noises in the system challenge the system stability of the multiplied pulses. They not only change the pulse shape of the multiplied pulses, but also distort the phase plane of the system. Hence, the system stability is greatly affected by the SPM effect as well as system noises.

References

1. G.R. Lin, Y.C. Chang, and J.R. Wu, Rational harmonic mode-locking of erbium-doped fiber laser at 40 GHz using a loss-modulated Fabry-Perot laser diode, *IEEE Photon. Tech. Lett.*, 16(8), 1810–1812, 2004.
2. J. Azana and M.A. Muriel, Temporal self-imaging effects: Theory and application for multiplying pulse repetition rates, *IEEE J. Quantum Electron.*, 7(4), 728–744, 2001.
3. K.K. Gupta, N. Onodera, K.S. Abedin, and M. Hyodo, Pulse repetition frequency multiplication via intracavity optical filtering in AM mode-locked fiber ring lasers, *IEEE Photon. Tech. Lett.*, 14(3), 284–286, 2002.
4. K.S. Abedin, N. Onodera, and M. Hyodo, Repetition-rate multiplication in actively mode-locked fiber lasers by higher-order FM mode-locking using a high finesse Fabry Perot filter, *Appl. Phys. Lett.*, 73(10), 1311–1313, 1998.
5. W. Daoping, Z. Yucheng, L. Tangjun, and J. Shuisheng, 20 Gb/s optical time division multiplexing signal generation by fiber coupler loop-connecting configuration, in *Fourth Optoelectronics and Communications Conference*, Vol. 1, Beijing, China, pp. 364–366, 1999.
6. E. Yamada, E. Yoshida, T. Kitoh, and M. Nakazawa, Generation of terabit per second optical data pulse train, *Electron. Lett.*, 31(16), 1342–1344, 1995.

7. S. Longhi, M. Marano, P. Laporta, O. Svelto, M. Belmonte, B. Agogliati, L. Arcangeli, V. Pruneri, M.N. Zervas, and M. Ibsen, 40-GHz pulse train generation at 1.5 μm with a chirped fiber grating as a frequency multiplier, *Opt. Lett.*, 25(19), 1481–1483, 2000.

8. J. Azana, R. Slavik, P. Kockaert, L.R. Chen, and S. LaRochelle, Generation of customized ultrahigh repetition rate pulse sequences using superimposed fiber Bragg gratings, *J. Lightwave Technol.*, 21(6), 1490–1498, 2003.

9. S. Arahira, S. Kutsuzawa, Y. Matsui, D. Kunimatsu, and Y. Ogawa, Repetition frequency multiplication of mode-locked using fiber dispersion, *J. Lightwave Technol.*, 16(3), 405–410, 1998.

10. H.F. Talbot, Facts relating to optical science No. IV, *Philos. Mag. Ser.*, 3(9), 401–407, 1836.

11. H. Hamam, Simplified linear formulation of Fresnel diffraction, *Opt. Commun.*, 144, 89–98, 1997.

12. D.S. Seo, D.E. Leaird, A.M. Weiner, S. Kamei, M. Ishii, A. Sugita, and K. Okamoto, Continuous 500 GHz pulse train generation by repetition-rate multiplication using arrayed waveguide grating, *Electron. Lett.*, 39(15), 1138–1140, 2003.

13. J.J.E. Slotine and W. Li, *Applied Nonlinear Control*, Prentice-Hall, Englewood Cliffs, NJ, 1991.

14. K. Ogusu, Dynamic behavior of reflection optical bistability in a nonlinear fiber ring resonator, *IEEE J. Quantum Electron.*, 32(9), 1537–1543, 1996.

8

Terahertz Repetition Rate Fiber Ring Laser

Although some repetition rate multiplication techniques have been proposed and demonstrated, as discussed in Chapter 7, the frequencies and repetition rates achieved are still within the GHz range, which is still far below the optical fiber transmission capacity predicted by Shannon. This leaves plenty of room for us to explore the possibilities of the generation of ultrahigh-speed optical pulse trains, which is a key requirement in the ultrahigh bit-rate optical communication systems.

In this chapter, the modulating signal integrated in a conventional actively mode-locked laser system is modified. The stability of the generated pulses is studied. The novel fiber laser structure in generating terahertz operation is also described with its theoretical model.

8.1 Gaussian Modulating Signal

Other than the conventional rational harmonic detuning technique, we propose to investigate a similar system, but with a different modulation signal, i.e., a Gaussian modulation signal. The theory of mode locking has been well developed by Haus [1], with the assumption that the modulating signal is a conventional co-sinusoidal periodic signal. However, for modulating signals other than that, the model may not be appropriate in determining the system behavior. Hence, a more fitting mode-locking model needs to be developed.

In this section, we develop a set of mode-locking equations for the mode-locked pulses with a Gaussian modulating signal, which is often generated by step recovery diodes (SRDs) and used in many applications.

The electric field amplitude of a Gaussian pulse train can be described as follows [2]:

$$f(t) = \sum_{n=-\infty}^{+\infty} a_n \exp(jn\Omega_m t) \tag{8.1}$$

with

$$a_n = \frac{1}{T} \int_{-T/2}^{T/2} M \exp\left(-\frac{t^2}{2\sigma^2}\right) \exp\left(-jn\Omega_m t\right) dt \tag{8.2}$$

where

$j^2 = -1$

T is the period

M is the peak amplitude of the Gaussian pulse

σ is the half duration of the pulse at the $1/e$ point

Ω_m is the modulation frequency

and the Fourier coefficients a_n can be solved as follows [3]:

$$a_n = M\sqrt{\frac{\pi}{2}\frac{\sigma}{T}} \exp\left[-2\left(n\pi\frac{\sigma}{T}\right)^2\right]$$

$$\cdot \left[erf\left(\frac{T}{2\sqrt{2}\sigma} + j\sqrt{2}n\pi\frac{\sigma}{T}\right) - erf\left(-\frac{T}{2\sqrt{2}\sigma} + j\sqrt{2}n\pi\frac{\sigma}{T}\right) \right] \tag{8.3}$$

and $a_n = a_{-n}$.

A laser is formed by introducing a gain medium. Generally, several axial modes will be lasing if the gain level is higher than the threshold required. By following the procedures proposed by Haus [1], the mode-locking equation can be modified as follows, with $\Omega_m = \Delta\Omega$ being the modulation frequency that produces sidebands at $\omega_0 \pm \Delta\Omega$, where ω_0 is the central carrier frequency. These injections lock the adjacent modes, which in turn lock their neighbors. Let A_n be the amplitude of the axial mode of frequency $\omega_0 \pm n\Delta\Omega$. The amplitude changes within each pass through the amplifier of loss $(1-l)$ and peak gain $(1+g)$, where $l \ll 1$ and $g \ll 1$, and is given by

$$\Delta A_n = \left[\frac{g}{1+\left(\frac{n\Delta\Omega}{\Omega_g}\right)^2} - l\right] A_n + \sum_{m=-\infty}^{+\infty} a_n A_{n+m} - \sum_{n=-\infty}^{+\infty} a_n A_n + a_0 A_n \tag{8.4}$$

where Ω_g is the gain bandwidth of the amplifier within the system. The terms on the right-hand side of the above equation describe the gain and loss of the laser, contributions from the Gaussian side-bands, contributions to the Gaussian side-bands, and the average DC component of the nth mode itself.

This expression can be transformed into a standard operator by the following approximations: (1) the discrete frequency spectrum with Fourier components at $n\Delta\Omega$ is replaced by a continuum spectrum, which is a function of $\Omega = n\Delta\Omega$; (2) the frequency-dependent gain can be expanded up to the second order in $\Omega = n\Delta\Omega$.

$$\frac{1}{1+\left(\Omega/\Omega_g\right)^2} = 1 - \left(\frac{\Omega}{\Omega_g}\right)^2 + \left(\frac{\Omega}{\Omega_g}\right)^4 - \left(\frac{\Omega}{\Omega_g}\right)^6 + \cdots = 1 - \left(\frac{\Omega}{\Omega_g}\right)^2 + O\left(\Omega^4\right) \quad (8.5)$$

Hence, arriving at the following:

$$\Delta A(\Omega) = (g-1)A(\Omega) - g\left(\frac{\Omega}{\Omega_g}\right)^2 A(\Omega) + \sum_{n=-\infty}^{+\infty} a_n A_n(\Omega) - \sum_{n=-\infty}^{+\infty} a_n A(\Omega) + a_0 A(\Omega) \quad (8.6)$$

where

$A(\Omega)$ is the amplitude of the axial mode frequency of Ω

$\Omega_m = 2\pi f_m$ is the modulation frequency

a_0 is the DC value of the modulating signal

$A_n(\Omega)$ is the amplitude of the nth mode at a frequency of $n\Omega_m$ away from the central frequency

We approximate the above equation with a second-order differential equation with the assumption that the contributions of higher-order components are small and negligible. The steady-state solution to the equation is a Gaussian pulse whose pulse width τ and net gain g are related as follows:

$$\tau^4 = \frac{g}{a_1 \Omega_m^2 \Omega_g^2} \quad (8.7)$$

$$g - 1 + a_0 = a_1 \Omega_m^2 \tau^2 \quad (8.8)$$

where [3],

$$a_0 = M\sqrt{2\pi}\frac{\sigma}{T}erf\left(\frac{T}{2\sqrt{2}\sigma}\right) \quad (8.9)$$

$$a_1 = C \left\{ \begin{array}{l} \left| 2erf(x) + \dfrac{\exp(-x^2)}{\pi x} \left[1 - \cos(2xy) \right] \right. \\[2em] \left. + \dfrac{4}{\pi} \exp(-x^2) \sum_{p=1}^{+\infty} \dfrac{\exp\left(-\dfrac{1}{4} p^2 \right)}{p^2 + 4x^2} f_p(x,y) + \varepsilon(x,y) \right| \end{array} \right. \qquad (8.10)$$

$$C = M \sqrt{\frac{\pi}{2} \frac{\sigma}{T}} \exp\left[-2 \left(\pi \frac{\sigma}{T} \right)^2 \right] \qquad (8.11)$$

$$x = \frac{T}{2\sqrt{2}\sigma} \qquad (8.12)$$

$$y = \sqrt{2}\pi \frac{\sigma}{T} \qquad (8.13)$$

$$f_p(x,y) = 2x - 2x \cosh(py)\cos(2xy) + p \sinh(py)\sin(2xy) \qquad (8.14)$$

$$\varepsilon(x,y) \approx 10^{-16} \, | \, erf(x, jy) \, | \approx 0 \qquad (8.15)$$

where $\varepsilon(x, y)$ is the allowable error of the system, which is very small and is thus neglected in our simulation. From the obtained equations, the system gain condition (Equation 8.8) has been improved by the DC component a_0 of the Gaussian modulating signal.

For the rational harmonic mode-locking case, the generated pulse width should be modified to include the effect of the detuning factor m, and it is proportional to $a_1^{-1/4}(q+1/m)^{-1/2}$ for the qth harmonic mode locking. When the detuning factor m is large, the contribution of $1/m$ becomes negligible; thus, the pulse width is very much determined by the a_1 parameter, which is in turn determined by the duty cycle of the modulating signal.

The relationship between the pulse width τ and duty cycle is shown in Figure 8.1, where the duty cycle is defined as $2\sqrt{(\ln 2)}\sigma/T$. For a co-sinusoidal modulating signal, the pulse width does not depend on the duty cycle. In this simulation, only the effect of the duty cycle is considered, and the other system parameters are held constant.

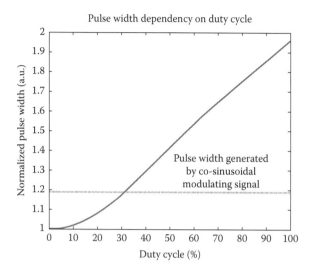

FIGURE 8.1

Relationship between the normalized pulse width and the duty cycle. (Solid line—Gaussian modulating signal; dotted line—co-sinusoidal modulating signal.)

From the obtained result (Figure 8.1), we see that the generated mode-locked pulse width is in fact proportional to the duty cycle of the Gaussian modulating signal and that the narrowest mode-locked pulses are formed at the minimum duty cycle. For practical considerations, at a relatively low (but nonzero) duty cycle, the minimum pulse width is determined by other system parameters, such as the amplifier gain, gain bandwidth, modulation frequency, etc., based on (8.7). For a minimum obtainable pulse width, the modulating signal must first fulfill the gain condition as stated in Equation 8.8. However, when the duty cycle is too large, i.e., a broad modulating pulse, the pulses generated are comparative to the CW signal, since it has lost its mode-locking capability. A better operating region for this Gaussian modulating signal is a duty cycle of less than 30%, whereby the pulses generated are narrower than that of the co-sinusoidal modulating signal. As a matter of fact, this type of signal is readily generated by an SRD, which is used in our experimental setup.

For a conventional mode-locking system, i.e., with a co-sinusoidal modulating signal, the maximum rational harmonic detuning factor, $m_{max,cos} = 2\pi[M/(2g)]^{1/4}(f_g/f_m)^{1/2}$, which is proportional to the square root of the number of modes (i.e., $(f_g/f_m)^{1/2}$) locked within the system. However, with the Gaussian modulating signal, this factor becomes $m_{max,gau} = 2\pi[a_1/g]^{1/4}(f_g/f_m)^{1/2}$, which not only depends on the number of modes locked, but is also inversely proportional to the duty cycle of the modulating signal. The relationship can be easily obtained by inverting the pulse width dependency chart (Figure 8.1). Hence, the maximum allowable multiplication factor, $m_{max,gau}$, relative to the co-sinusoidal modulating signal, can be improved by reducing the duty cycle to less than 30%. This improvement factor can be easily obtained by finding $m_{max,gau}/m_{max,cos} = (2a_1/M)^{1/4}$. Other than the theoretical considerations, the factor m_{max} is also limited by the availability of the modulator driver, i.e., an SRD, and the filter bandwidth, which limits the gain bandwidth, and hence the number of modes locked within the system.

Having known the pulse width generated by the Gaussian modulating signal, we now aim to determine the stable operation condition of this type of modulating signal. In order to generate stable soliton pulses in an active mode-locked laser system, two conditions have to be fulfilled: (1) the soliton pulse must experience a lower loss in the cavity than the Gaussian pulse and (2) the energy fluctuation of the soliton must be damped, mathematically shown in Equations 8.8 and 8.9. When condition 1 is met, the soliton will have sufficient gain to survive; however, the Gaussian noise is not given enough gain to build up. Condition 2 implies that the filter loss must be larger than the modulator loss so as to weaken the soliton energy fluctuation caused by the modulator [4]. A positive soliton energy fluctuation shortens the pulse and broadens the spectrum. On the one hand, the decreased pulse width decreases the modulator loss as the soliton becomes more concentrated around the maximum transmission of the modulator. On the other hand, the filter loss increases as a result of the broadened spectrum. For stability, the filtering effect must be stronger than the modulation effect.

For an active mode-locked laser system with a Gaussian-like modulating signal, the relationship between the pulse width τ and the net gain with cavity dispersion taken into account are as follows:

$$\frac{1}{\tau^4} = \frac{a_1\Omega_m^2}{\dfrac{1}{\Omega_f^2} - jD} \tag{8.16}$$

$$g - l + a_0 = a_1\Omega_m^2\tau^2 \tag{8.17}$$

where
$j = \sqrt{-1}$
a_0 and a_1 are the Fourier coefficients for the Gaussian signal
g and l are the gain and loss of the system, respectively

Physically, a_0 and a_1 are the DC component and the strength of the modulating signal, respectively, which are duty cycle dependent. One thing to note is that the system net gain has been improved by the DC component of the modulating signal. Ω_m is the angular modulation frequency, $D = \beta_2L/2$ with an average group velocity dispersion β_2 and cavity length L, λ is the carrier wavelength, and Ω_f is the bandwidth of the laser cavity. Normally, this bandwidth corresponds to the bandwidth of the optical bandpass filter (BPF) incorporated in the laser.

By considering the timing jitter and frequency perturbation produced by the modulator and filter of the system and following the procedure presented by Haus and Mecozzi [5], the stability conditions stated in Chapter 2 have been modified as follows:

$$\text{Condition 1: } \frac{\pi^2}{12} a_1 \Omega_m^2 \tau^2 + \frac{1}{3\Omega_f^2 \tau^2} < \text{Re} \left[a_1 \Omega_m^2 \left(\frac{1}{\Omega_f^2} - jD \right) \right]^{1/2} \tag{8.18}$$

$$\text{Condition 2: } \frac{\pi^2}{12} a_1 \Omega_m^2 \tau^2 < \frac{1}{3\Omega_f^2 \tau^2} \tag{8.19}$$

It should be stressed that the equations used to deduce the system stability conditions is only valid when the pulse change per round trip is small. This sets the upper limit to the maximum nonlinear phase shift that the soliton may acquire per round trip: stability of a periodically amplified soliton requires the nonlinear phase shift $\ll 4\pi$ [6]. Figure 8.2 shows the stability region for both conventional and Gaussian-like modulating signals in an active mode-locked laser system. The lined regions are the stable soliton operation regions, where both stability conditions are simultaneously met.

From Figure 8.2, condition 1 for both cases requires $(1/\Omega_f^2 D) < 1.53$, i.e., for a given fixed filter bandwidth, a minimum anomalous dispersion is required. This is because the dispersion, which is balanced by the self-phase modulation (SPM) effect for soliton formation, increases the Gaussian loss and thus

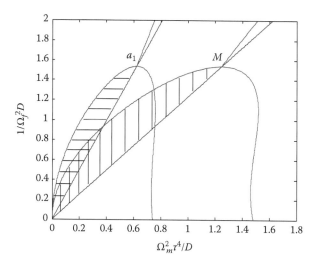

FIGURE 8.2
Stable operating region for a soliton in an active mode-locked laser system with a conventional co-sinusoidal modulating signal (*M*) and a Gaussian-like modulating signal (a_1).

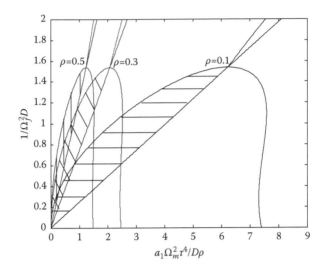

FIGURE 8.3
Stability region for soliton formation in an active mode-locked laser system for various duty cycle parameters.

suppresses the Gaussian noise (condition 1 for both cases). However, the dispersion should not be too large, so as to keep the nonlinear phase shift $\ll 4\pi$. Furthermore, as depicted in the figure, the pulse width of the Gaussian case is much shorter than the conventional case, with the same modulation frequency, which has also been shown in the previous section. By inspection, the stable operating region for a Gaussian-like modulating signal is smaller than that of the conventional one, i.e., more stringent control parameters are required for the stable operation. Steeper slope for condition 2 also reveals a larger modulation fluctuation in the Gaussian modulating case.

By the linear approximation of the a_1 parameter, which is valid for coarse system estimation, we are able to show the stability relation with respect to the duty cycle of the Gaussian modulating signal, as shown in Figure 8.3. For a small duty cycle, one can expect to have a larger stable soliton operation region.

8.2 Rational Harmonic Detuning

Rational harmonically mode-locked fiber ring lasers have attracted much attention due to their ability to generate higher repetition rate optical pulses. By applying a slightly deviated frequency from the multiple of the fundamental cavity frequency, higher rate pulses can be achieved [7,8]. Recently, Lin et al. [8] have achieved the 40th order of rational harmonic mode-locking using a loss-modulated Fabry–Perot laser diode, with a base frequency of 1 GHz. Zhu et al. [9] demonstrated a two-step 80 GHz high-quality pulse train generation involving fourth-order rational harmonic mode-locking on a 10 GHz modulation signal in conjunction with a frequency doubler. The mode-locked fiber laser was stabilized using a regenerative-type baseline extraction feedback technique.

By applying a slight frequency deviation, $\Delta f = f_c/m$, where m is the integer and f_c is the fundamental cavity frequency, the modulation frequency of the harmonically mode-locked laser, f_m becomes

$$f_m = qf_c + \frac{f_c}{m}$$

$$mf_m = (qm + 1)f_c \qquad (8.20)$$

$$f_r = mf_m$$

where
   q is the qth harmonic of the fundamental cavity frequency
   f_r is the repetition frequency of the system, which has been increased to m-times of f_m

However, this technique suffers from the inherent pulse amplitude instability, which includes both amplitude noise and inequality in the pulse amplitude; furthermore, it gives poor long-term stability. The problem is due to the millisecond gain lifetime of erbium because the fiber laser cannot equalize the pulse energies, and hence the output may contain unequal pulse amplitudes or even pulse dropouts.

Some pulse amplitude equalization techniques have been proposed to solve the problem; some of these use a two-photon absorption (TPA) semiconductor mirror [10], a double-pass modulator [11], linear and nonlinear optical filter combinations [12], etc.

8.2.1 Experimental Setup

The experimental setup of the active harmonically mode-locked fiber ring laser is shown in Figure 8.4. The principal element of the laser is an optical closed loop with an optical gain medium, a Mach–Zehnder amplitude modulator (MZM), an optical polarization controller (PC), an optical BPF, optical couplers, and other associated optics.

The gain medium used in our fiber laser system is an EDFA with a saturation power of about 16 dBm. A polarization-independent optical isolator is used to ensure unidirectional lightwave propagation as well as to eliminate back reflections from the fiber splices and optical connectors. A free space filter with a 3 dB bandwidth of 4 nm at 1555 nm is inserted into the cavity to select the operating wavelength of the generated signal and to

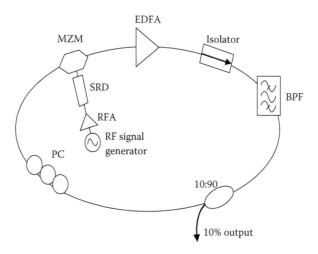

FIGURE 8.4
Experimental setup of an active mode-locked fiber ring laser based on the Gaussian pulse modulated method.

reduce the noise in the system. In addition, it is responsible for the selection of longitudinal modes in the mode-locking process. The birefringence of the fiber is compensated by a PC, which is also used for the polarization alignment of the linearly polarized lightwave before entering the planar structure modulator for better output efficiency. Pulse operation is achieved by introducing a 10 GHz Ti:LiNbO$_3$ MZM based on the asymmetric coplanar traveling wave structure into the cavity with a half-wave voltage V_π of 7.8 V and an insertion loss of ≤7 dB. The modulator is DC biased near the quadrature point and not more than the V_π value such that it operates around the linear region of its characteristic curve. We adopt amplitude modulation in this experiment because it gives better stability than the frequency modulation. The modulator is driven by a 100 ± 5 MHz, ~100 ps SRD, which is in turn driven by a RF signal generator followed by a RF amplifier (RFA). The modulating signal generated by the SRD is a ~1% duty cycle Gaussian pulse train. The output coupling of the laser is optimized using a 10/90 optical coupler. Ninety percent of the optical field power is coupled back into the cavity ring loop, while the remaining portion is taken out as the output of the laser and analyzed using an oscilloscope and an optical spectrum analyzer.

An SRD is also known as a charge storage diode that can be used to generate short electrical pulses from a co-sinusoidal wave drive because its operation depends on the ability to store change [13]. It is simply a PIN junction diode that has a special doping profile, to give a special characteristic curve. Like a normal PN diode, the SRD conducts when forward biased. However, it also shows significant conduction when reverse biased, but only for a short while. The transition from the conducting and nonconducting states of the reverse biased diode occurs in a time frame of typically less than 100 ps. This means the current waveform has a high harmonic content and can be used to generate short pulses [13,14].

8.2.2 Results and Discussion

The theoretical background of the phase plane analysis has been given in Chapter 7. Here, we will focus on the modeling of a rational harmonic mode-locked fiber ring laser system with the following assumptions: (1) the mode-locking condition has been achieved and the detuned frequency is perfectly adjusted according to the fraction number required, m; (2) optical amplifier gain saturation (this is valid because we are operating the laser in a multiple-pass optical amplification situation, where the laser will eventually saturate); (3) a small harmonic distortion; (4) no fiber nonlinearity is included in the analysis; (5) no other noise sources are involved; (6) amplitude distribution analysis of a Gaussian lasing mode; (7) gain bandwidth is determined by the filter bandwidth; and (8) linearly polarized light is used in the system.

We do not consider the initial pulse forming process of the laser. This is simply because it involves the dynamics of the optical amplifier and adds to the complexity of the system, which is not the interest of this study. Furthermore, we are looking at the pulse propagation behavior, once rational harmonic mode-locking has been achieved, as time progresses. Harmonic distortion is applied to the system by introducing small variations to different harmonics up to the m order for an m-fraction detuned system. The overlap between the linewidth of the actual lasing mode (without frequency detuning) and the detuning amount also contributes to the distortion, and is considered in our simulation.

The phase plane of a perfect 10 GHz mode-locked pulse train with a bandwidth of 4 nm and centered at a wavelength of 1550 nm; without any frequency detuning is shown in Figure 8.5, and the corresponding pulse train is shown in Figure 8.6a. From the phase plane obtained, one can easily observe that the origin is a stable node and the limit cycle around that vicinity is a stable limit cycle, hence leading to a stable system trajectory. Four-times multiplication pulse trains, i.e., $m = 4$, without and with 5% harmonic distortion are shown in Figure 8.6b and c, respectively. Their corresponding phase planes are shown in Figure 8.7a and b.

For the case of zero harmonic distortion, which is the ideal case, the generated pulse train is perfectly multiplied with equal amplitude and the phase plane has stable symmetry periodic trajectories around the origin too. However, for a practical case, i.e., with 5% harmonic distortion, it is obvious that the pulse amplitude is unevenly distributed, which agrees very well with many experimental findings obtained in Zhu et al. [9,15], Gupta et al. [16], Wu and Dutta [7]. Its corresponding phase plane shows more complex and asymmetric system trajectories, hence reflecting its undesirable amplitude fluctuations.

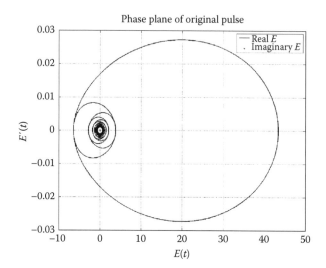

FIGURE 8.5
Phase plane of a 10 GHz rational harmonic mode-locked pulse train.

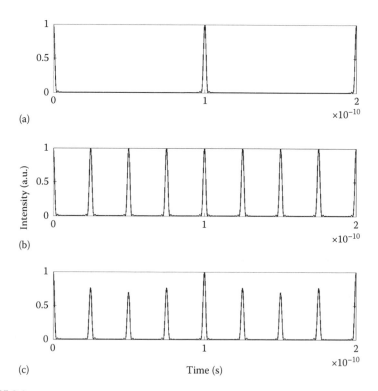

FIGURE 8.6
Normalized pulse propagation of (a) original pulse; detuning fraction of 4, with (b) 0% and (c) 5% harmonic distortion noise in a 10 GHz rational harmonic mode-locked pulse train.

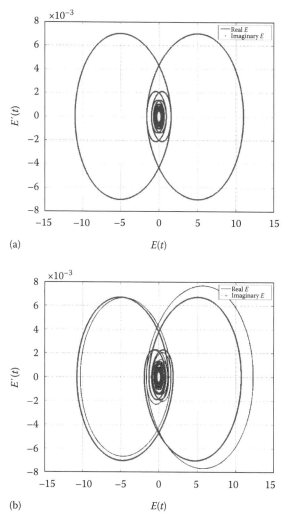

(a)

(b)

FIGURE 8.7
Phase plane of a detuned pulse train, $m=4$, (a) 0% and (b) 5% harmonic distortion in a 10 GHz rational harmonic mode-locked pulse train.

We define the percentage of amplitude fluctuation, %F as follows:

$$\%F = \frac{E_{max} - E_{min}}{E_{max}} \times 100\% \qquad (8.21)$$

where E_{max} and E_{min} are the maximum and minimum peak amplitudes of the generated pulses.

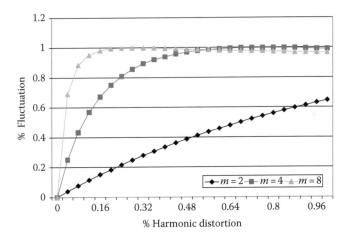

FIGURE 8.8
Relationships between the amplitude fluctuation and the percentage harmonic distortion.

For any practical mode-locked laser system, fluctuations above 50% should be considered as unacceptable. Therefore, this is one of the limiting factors in a rational harmonic mode-locking fiber laser system. The relationships between the percentage fluctuation and harmonic distortion for three multipliers ($m = 2$, 4, and 8) are shown in Figure 8.8. Thus, the obtainable rational harmonic mode-locking is very much limited by the harmonic distortion of the system. For a high multiplier, a small change in the harmonic distortion leads to a large change in the system output, and hence poor system stability. Increasing the system bandwidth, i.e., the bandwidth of the filter used in the system, will allow more modes to be locked in the system, so as to improve the pulse quality. However, this may increase the ASE noise level of the system, hence challenges its stability. Therefore, this is a trade-off between the pulse quality and system stability.

For 100% fluctuation, it means no repetition rate multiplication, but with additional noise components; a typical pulse train and its corresponding phase plane are shown in Figures 8.9 (lower plot) and 8.10, respectively, with $m = 8\%$ and 20% harmonic distortion. The asymmetric trajectories of the phase graph explain the amplitude unevenness of the pulse train. Furthermore, it shows a more complex pulse formation process and undesirable pulse instability. Hence, it is clear that for any harmonic mode-locked laser system, the small side pulses generated are largely due to the improper or not exact tuning of the modulation frequency of the system. An experimental result is depicted in Figure 8.13 for comparison.

By the careful adjustment of the modulation frequency, polarization, gain level, and other parameters of the fiber ring laser, we managed to obtain the 660th and 1230th order of rational harmonic detuning in the mode-locked fiber ring laser with a base frequency of 100 MHz, hence achieving 66 and 123 GHz repetition frequency pulse operations, respectively. The

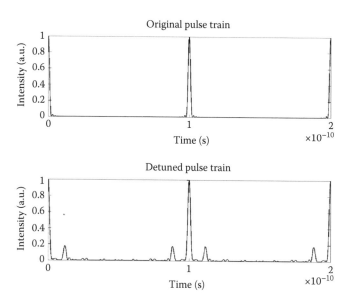

FIGURE 8.9
A 10 GHz pulse train (upper plot), pulse train with $m = 8\%$ and 20% harmonic distortion (lower plot).

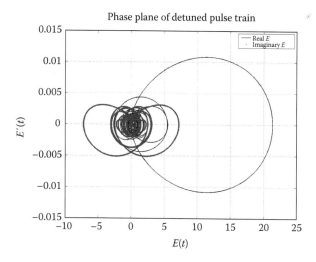

FIGURE 8.10
Phase plane of the pulse train with $m = 8\%$ and 20% harmonic distortion.

autocorrelation traces and optical spectra of the pulse operations are shown in Figure 8.11. With Gaussian pulse assumption, the obtained pulse widths of the operations are 2.5456 and 2.2853 ps, respectively. For the 100 MHz pulse operation, i.e., without any frequency detuning, the generated pulse width is about 91 ps. Thus, not only did we achieve an increase in the pulse

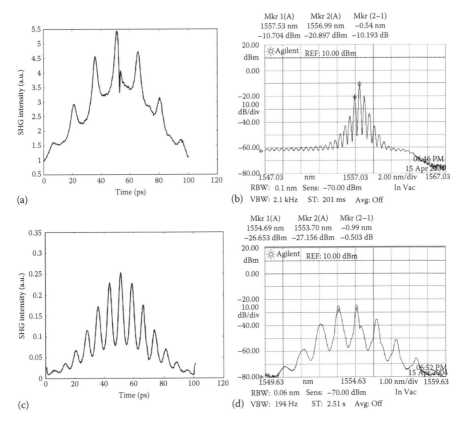

FIGURE 8.11
Autocorrelation traces and optical spectra of (a, b) 66 GHz and (c, d) 123 GHz pulse operations.

repetition frequency, but also a decrease in the generated pulse width. This pulse narrowing effect is partly due to the SPM effect of the system, as observed in the broadened optical spectra. Another reason for this narrow pulse width is the low duty cycle of the modulating signal as described in the previous section. Besides the uneven pulse amplitude distribution, a high level of pedestal noise is also observed in the obtained results.

For 66 GHz pulse operation, a 4 nm bandwidth filter is used in the setup, and it is removed in the 123 GHz operation. It is done so as to allow more modes to be locked during the operation, thus achieving a better pulse quality. In contrast, this increases the level of difficulty significantly in the system tuning and adjustment. As a result, the operation is very much determined by the gain bandwidth of the EDFA used in the laser setup.

The simulated phase planes for the above pulse operation are shown in Figure 8.12. They are done based on the 100 MHz base frequency, 10 round trips condition, and 0.001% of harmonic distortion contribution. There is no stable limit cycle in the phase graphs obtained; hence, the system stability

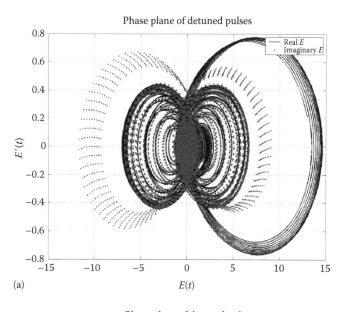

(a)

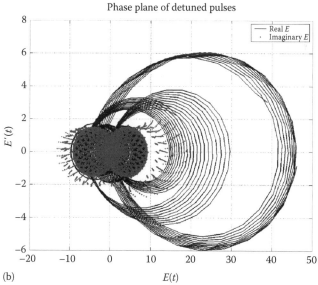

(b)

FIGURE 8.12
Phase plane of the (a) 66 GHz and (b) 123 GHz pulse train with 0.001% harmonic distortion noise.

is hardly achievable, which is a well-known fact in rational harmonic mode locking. Asymmetric system trajectories are observed in the phase planes of the pulse operations. This reflects the unevenness of the amplitude of the pulses generated. More complex pulse formation processes have also been revealed in the phase graphs obtained. Additionally, poor long-term stability

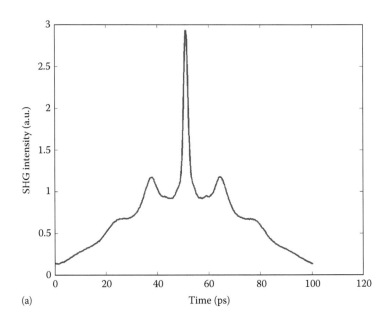

(a)

Mkr 1(A)	Mkr 2(A)	Mkr (2–1)
1554.394 nm	1553.706 nm	−0.688 nm
−34.487 dBm	−35.836 dBm	−1.349 dB

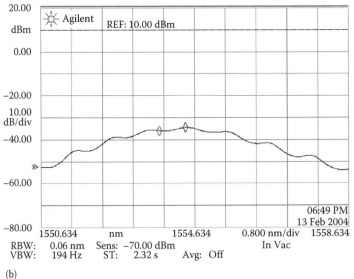

(b)

FIGURE 8.13
(a) Autocorrelation trace and (b) optical spectrum of the pulse train with a 100 MHz frequency deviation in the mode-locked fiber ring laser.

is also observed for both cases, since the trajectories are spiraling unequally with decreasing amplitude around the phase planes.

By a very small amount of frequency deviation, or improper modulation frequency tuning in the general context, we obtain a pulse train with ~100 MHz with small side pulses in between, as shown in Figure 8.13. It is rather similar to Figure 8.9 (lower plot) shown in the earlier section despite the level of pedestal noise in the actual case. This is mainly because we do not consider other sources of noise in our modeling, except the harmonic distortion.

8.3 Parametric Amplifier–Based Fiber Ring Laser

New high-power light sources and optical fibers with high nonlinearity, as well as the need of amplification outside the conventional erbium band has increased the interest of optical fiber parametric amplifiers (PA). It offers a wide gain bandwidth and is similar to the Raman amplifier tailored to operate at any wavelength [17]. As the parametric gain process does not rely on energy transitions between energy states, it enables a wideband and flat gain profile contrary to Raman and EDFA. The underlying process is based on the highly efficient four-photon mixing relying on the relative phase between the four interacting photons. In this section, we will study the behavior of a parametric amplifier in a fiber ring laser.

8.3.1 Parametric Amplification

Parametric amplification is achieved by manipulating the mixing between four lightwaves at three frequencies, based on the fact that the fiber refractive index is intensity dependent. A signal at frequency ω_s and a pump at ω_p will mix and modulate the refractive index of the fiber such that a third lightwave, also at ω_p, will create sidebands at $\omega_p \pm (\omega_s - \omega_p)$, which will result in signal gain and the generation of idler [17].

The parametric gain is very much dependent on the phase condition between the four lightwaves during the transmission through the nonlinear medium, which can be categorized into linear and nonlinear phase shifts. A linear phase shift is due to the group velocity dispersion of the medium, whereas the nonlinear phase shift is dependent on the pump power and the nonlinear coefficient of the medium. Under a perfect phase-matching condition, i.e., no phase mismatch, the maximum obtainable single-pass gain in an optical fiber in dB is given as follows [17]:

$$G_{max} = 10\log\left[\frac{1}{4}\exp(2\gamma P_p L)\right] \tag{8.22}$$

where

γ is the nonlinear coefficient of the fiber
P_p is the pump power
L is the effective length of the fiber

8.3.2 Experimental Setup

An experimental setup of the parametric amplifier–based fiber ring laser is shown in Figure 8.14. A CW distributed feedback (DFB) laser operating at 1547 nm is used as the pump source for the laser. The CW signal is modulated by a MZM, which is driven by a 100 MHz SRD, hence achieving ~100 MHz pulse train with a duty cycle of ~1%. This is the Gaussian-like modulating signal used to modulate the cavity loss of the system.

The signal is then amplified by an EDFA and filtered by a BPF before being coupled into the ring via a 3 dB coupler. The fiber ring consists of an optical closed loop with a 200 m highly nonlinear fiber (HNLF) with a zero dispersion wavelength, λ_0 at 1542 nm, dispersion slope of 0.032 ps/nm²/km, and the nonlinear coefficient, $\gamma = 10/\text{W/km}$, an optical isolator, a 1.5 km SMF, and a PC. The isolator is to ensure unidirectional pulse propagation and to minimize the back reflections from connectors used in the ring. Since four-wave mixing (FWM) is very much polarization dependent, the PC becomes the key player in the loop. The signal and pump are then fed back to the loop via the 3 dB coupler. The total loop loss is estimated to be about 10 dB.

8.3.3 Results and Discussion

8.3.3.1 Parametric Amplifier Action

The pumping wavelength is chosen in such a way that it is near the zero dispersion wavelength of the HNLF and possess a small anomalous dispersion value to give a good gain efficiency. With 200 m of HNLF, peak pump power

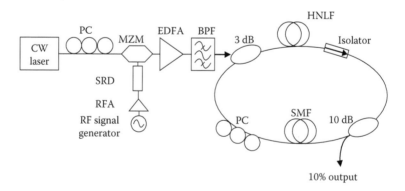

FIGURE 8.14
Experimental setup of a PA-based fiber ring laser.

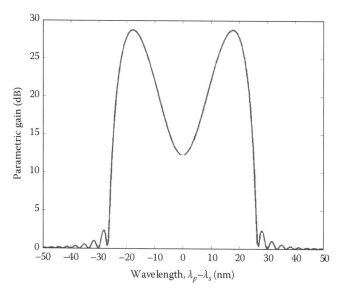

FIGURE 8.15
Simulated PA gain profile.

of 33 dBm, pumping wavelength λ_p of 1547 nm, and nonlinear coefficient $\gamma = 10/W/km$, we obtain a maximum single-pass gain of 28.6 dB theoretically. The simulated PA gain profile is shown in Figure 8.15. We obtained a closed loop gain of about 19.5 dB experimentally (as shown in Figure 8.16), with a total loop loss estimation of about 10 dB, and the result agrees very well with the theoretical calculation. In Figures 8.17 through 8.19, we investigate the effect of the length of HNLF, pump power, and pump wavelength used in affecting the PA gain profile. Gain loop will decrease with a decrease in the length of the HNLF and pump power, however no change in the amplified signal wavelength in the HNLF case. The wavelength of the amplified signal will shift toward the pump wavelength when the pump power is reduced. The gain bandwidth of the amplifier will increase if the pump wavelength is set closer to the zero dispersion wavelength of the HNLF. Furthermore, the signal wavelength will also be shifted away from the pump wavelength.

8.3.3.2 Ultrahigh Repetition Rate Operation

The total cavity loop length is measured to be 1842.8 m, and its corresponding fundamental frequency is 112.2 kHz. Hence, the locking occurs at about the 890th harmonic mode. In a conventional active harmonically mode-locked fiber ring laser, the mode-locker, i.e., a modulator, is placed within the ring cavity. However, in this PA-based fiber ring laser, the mode locking process is controlled by the modulated pump signal. In the following discussion, we assume that the phase condition in the HNLF is matched, so that we obtain

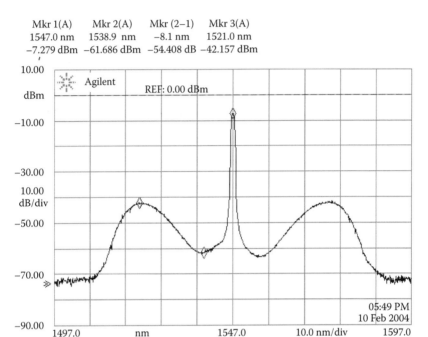

FIGURE 8.16
PA gain curve obtained experimentally.

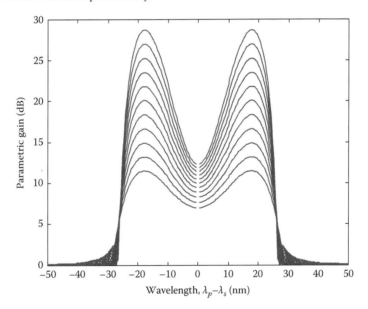

FIGURE 8.17
Effect of the HNLF length on the PA gain profile.

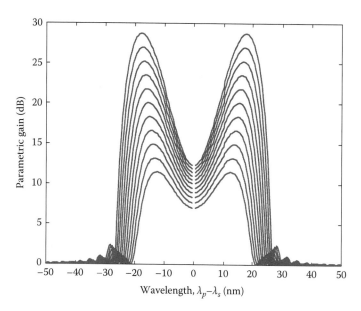

FIGURE 8.18
Effect of the peak power on the PA gain profile.

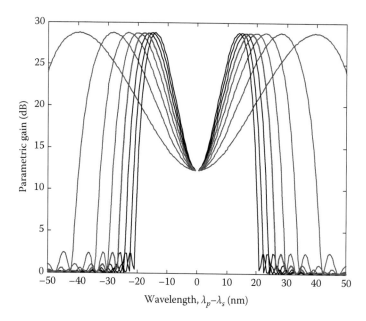

FIGURE 8.19
Effect of the spectral difference between the pump and signal wavelength on the PA gain profile.

a good gain response in the laser. This assumption is valid since the gain curves are obtained as desired.

Under the frequency unmatched situation, i.e., the modulation frequency is not a multiple of the fundamental cavity frequency, we obtained an optical pulse with a pulse width of ~70 ps, which is slightly shorter than the modulated pump's pulse width of ~90 ps, that gives a compression factor of ~1.2×, as shown in Figure 8.20b. When the frequency condition is matched,

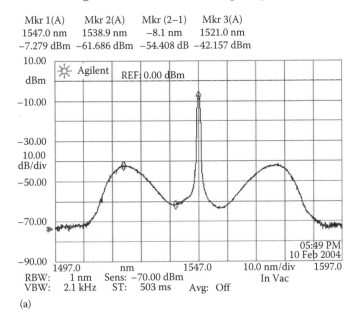

(a)

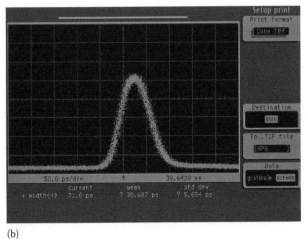

(b)

FIGURE 8.20
(a) Optical spectrum and (b) pulse shape observed when the phase condition is matched and the modulation frequency condition is not matched by a detuned frequency of 100 MHz.

we observed a strong SPM effect on the pump signal with a compression factor of ~1.5×. The results are shown in Figure 8.21.

Mode separation is not observable since the 100 MHz operation is too narrow to be seen on the optical spectrum analyzer. When the modulation frequency starts to deviate from the mode-locking frequency, some side modes start to appear, as depicted in Figures 8.22a and 8.23a. The mode separations measured are 7.7 and 10.5 nm, respectively, centered at 1547 nm.

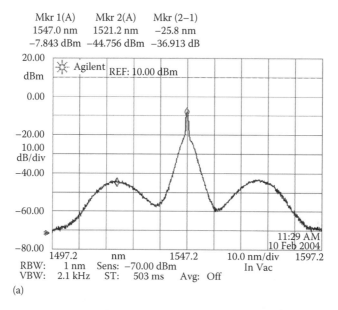

(a)

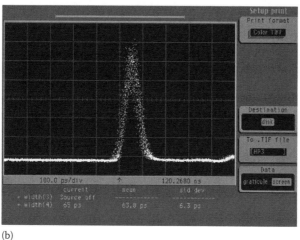

(b)

FIGURE 8.21

(a) Optical spectrum and (b) pulse shape observed when the phase and modulation frequency conditions are both matched.

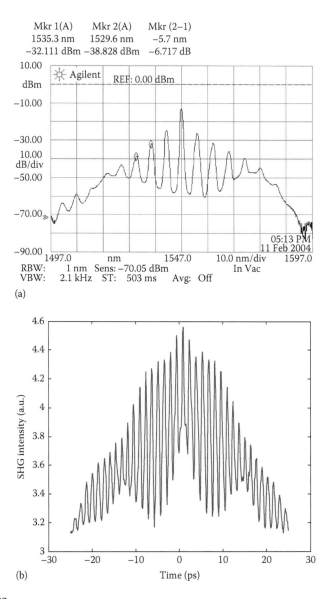

FIGURE 8.22

(a) Optical spectrum and (b) autocorrelation trace observed when the phase condition is matched and the modulation frequency is detuned for 714 GHz.

The corresponding repetition frequencies are about 714 GHz and 1.315 THz, respectively. The autocorrelation traces are shown in Figures 8.22b and 8.23b, and the estimated FWHM are about 0.6 and 0.3 ps, respectively. From the obtained results, the time-bandwidth-products (TBPs) are estimated to be 0.529 and 0.496, respectively. Hence, there is a small chirping found in the

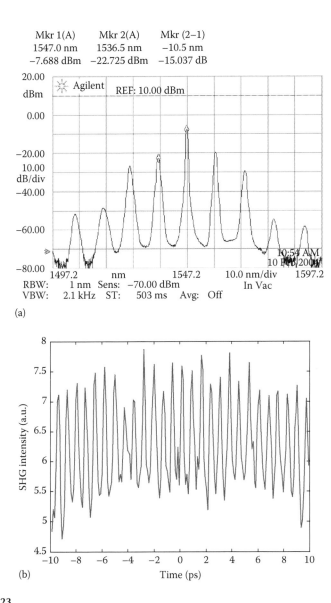

Mkr 1(A)	Mkr 2(A)	Mkr (2−1)
1547.0 nm	1536.5 nm	−10.5 nm
−7.688 dBm	−22.725 dBm	−15.037 dB

(a)

(b)

FIGURE 8.23
(a) Optical spectrum and (b) autocorrelation trace observed when the phase condition is matched and the modulation frequency is detuned for 1.315 THz.

generated pulses; however, they are close to the Gaussian transform-limited pulse. As a whole, we observed a linear relationship between the mode separation (or repetition frequency, f_r) and the detuned frequency, Δf; i.e., $f_r \, \alpha \, |\Delta f|$, as shown in Figure 8.24. We believe that the generation of these pulses is not mainly due to the frequency detuning, but also the interference between the

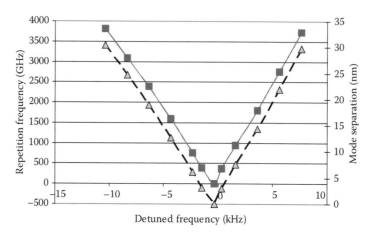

FIGURE 8.24
Repetition frequency and mode separation as a function of the detuned frequency.

pump pulses from one round trip to another, and the modulation instability due to the interaction between the nonlinear and dispersion effects within the cavity.

In summary, there are four operation regions for this PA-based mode-locked fiber ring laser system, as follows:

Region I: Phase condition is not matched, hence no gain action. The modulation frequency is not a multiple number of the fundamental cavity frequency. Noisy operation is observed in this region.

Region II: Phase condition is matched, hence parametric gain (gain loop), and new wavelength components are generated. However, the modulation frequency is not matched with the cavity frequency.

Region III: Phase condition and modulation frequency conditions are both matched and, hence, gain and mode-locking actions. Pulse shortening and spectral broadening effects are observed.

Region IV: Phase condition and modulation frequency conditions are both matched and, hence, gain and mode-locking actions, however, with some frequency detuning in the modulation frequency. Ultrahigh repetition rate optical pulse train is observed.

8.3.3.3 Ultra-Narrow Pulse Operation

As has been predicted and verified by many literatures [18], the shortest pulse width of a laser is obtainable when the cavity dispersion is minimum. Hence, we removed the SMF (dispersion) from the cavity and observed the laser behavior. The shortest pulse achieved from our laser under perfect tuning (i.e., without detuning) is shown in Figure 8.25. The obtained FWHM

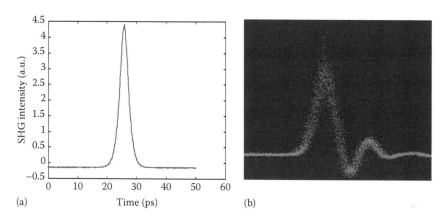

(a)   (b)

FIGURE 8.25

(a) Autocorrelation trace and (b) oscilloscope pulse shape for pulses generated by a PA-based mode-locked fiber ring laser.

is 3.36 ps; with Gaussian pulse assumption, the actual pulse width of the system is therefore 2.38 ps.

8.3.3.4 Intracavity Power

We studied the effect of the intra-cavity power in affecting the generated pulse width by inserting a variable optical attenuator into the laser cavity. The relationship between the intra-cavity power and the pulse width is shown in Figure 8.26. Shorter pulse is generated with larger cavity loss. This

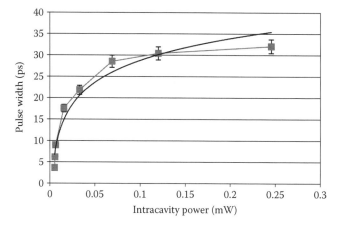

FIGURE 8.26

Generated pulse width versus the intra-cavity power.

is because the larger the cavity loss, the lesser the gain saturation effect of the system. Hence, this gives a better gain distribution to the system and the generation of a better pulse quality.

8.3.3.5 Soliton Compression

A high-peak power-pump pulse (about 33 dBm; based on an average power of 16 dBm, duty cycle of ~10% and average loss of 3 dB) with an initial pulse width of about 77 ps (through measurement) is passed through the 200 m HNLF and followed by a length of SMF. Soliton compression is observed and is shown in Figure 8.27.

The nonlinear and dispersive lengths of a system are [19]

$$L_{NL} = \frac{1}{\gamma P} \quad L_D = \frac{T_{FWHM}^2}{3.11|\beta_2|} \tag{8.23}$$

Using the experimental data, the calculated nonlinear length, L_{NL}, and dispersive length, L_D, are about 0.05 and 1.2 km, respectively. The HNLF used in our experiment is 200 m, which is much longer than the calculated nonlinear length, so the nonlinear effect will be larger in the experiment. Therefore, the corresponding dispersive length has to be increased for a balanced soliton effect. From the chart obtained, the soliton shortening effect peaks at SMF $\cong$ 2.9 km, beyond which the pulse broadening effect will start

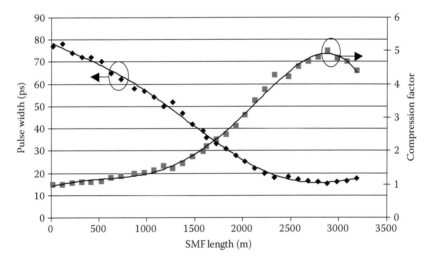

FIGURE 8.27
Soliton pulse compression: measured pulse width and compression factor as a function of the length of the SMF.

to take place. With β_2 of ~20.328 ps²/km, the T_{FWHM} is about 13.5 ps, which gives a maximum compression factor of about 5×.

8.4 Regenerative Parametric Amplifier–Based Mode-Locked Fiber Ring Laser

In order to improve the stability of the PA-based mode-locked fiber ring laser and to promote the self-locking condition, we include a regenerative feedback loop into the system. The generated output pulse train will be fed into the feedback circuit to reduce the error signal or rather the deviation between the harmonic cavity frequency and the modulating frequency that drives the modulator. The concept is similar to RMLFRL as described in Chapter 4.

8.4.1 Experimental Setup

The experimental setup of the regenerative PA-based mode-locked fiber ring laser is shown in Figure 8.14. The system is similar to the experimental setup described in the previous section, but with some modifications. The dispersive material, i.e., the 1.5 km SMF has been removed to ensure the short pulse operation. In addition, an FBG filter is inserted to select the signal wavelength and block the pump and idler wavelengths. The main add-in to the system is the regenerative part, which is made up of RFAs, phase shifter (PS), mixer, and RF filter.

Part of the output signal is taken out as the feedback signal. It is converted into an electrical signal by the PIN photo-detector (PD) and amplified by the RFA. A PS is used to adjust the electrical phase of the signal, i.e., electrical length of the signal, so that it will match with the optical frequency of the signal. The signal will be mixed with the signal generated by the RF signal generator. The mixer will then produce the error signal based on the mismatch of two signals. An RF lowpass filter is used to filter out the high frequency components of the signal so as to have a better-quality driving signal to the RF signal generator.

8.4.2 Results and Discussion

The generated feedback signal (which is derived from the output pulses) and the output mode-locked pulses are shown in Figure 8.28. Some areas of concern as regards RFAs have to be addressed so as to minimize the saturation effect in the feedback signal.

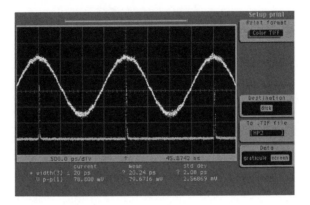

FIGURE 8.28
Output mode-locked pulses and feedback clock signal.

The generated pulse train is similar to the case described in the previous section, but with improved stability. The output pulse retains its shape without drift when it is properly locked for a period about 10 h (in laboratory condition).

8.5 Conclusions

We have developed a model for a Gaussian modulating signal in an active mode-locked laser system and studied the pulse width dependency on the duty cycle. For the Gaussian modulating signal with a duty cycle of less than 30%, one can expect to have a shorter pulse width than in the conventional case under similar system parameters. In addition, we have also showed the stable operation region for soliton pulse generation in an active mode-locked laser system using the Gaussian-like modulating signal. Similar to the previous case, a smaller duty cycle will lead to a wider stable operation region and looser control parameters.

By applying the Gaussian-like modulating signal in our active harmonically mode-locked fiber ring laser, we obtained a record-high harmonic order multiplication in the harmonically mode-locked fiber laser system, and achieved a 123 GHz operation. Stability analysis using the phase plane technique has also been performed.

Furthermore, in this chapter, parametric amplification is used in a mode-locked fiber ring laser system to achieve the ultrahigh bit-rate operation by frequency detuning and modulation instability. In addition, some other effects such as soliton compression and intracavity power influences can also be observed.

References

1. H.A. Haus, Mode-locking of lasers, *IEEE J. Select. Top. Quant. Electron.*, 6(6), 1173–1185, 2000.
2. E. Kreyszig, *Advanced Engineering Mathematics*, 8th edn., John Wiley & Sons Inc., New York, 1999, Chapter 10.
3. M. Abramowitz and I.A. Stegun, *Handbook of Mathematical Functions with Formulas, Graphs, and Mathematical Tables*, 10th Printing, John Wiley & Sons Inc., New York, 1972, Eq. 7.1.29, p. 299.
4. H.A. Haus and A. Mecozzi, Long-term storage of a bit stream of solitons, *Opt. Lett.*, 17(21), 1500–1502, 1992.
5. H.A. Haus and A. Mecozzi, Noise of mode-locked lasers, *IEEE J. Quantum Electron.*, 29(3), 983–996, 1993.
6. L.F. Mollenauer, J.P. Gordon, and M.N. Islam, Soliton propagation in long fibers with periodically compensated loss, *IEEE J. Quantum Electron.*, 22(1), 157–173, 1986.
7. C. Wu and N.K. Dutta, High repetition rate optical pulse generation using a rational harmonic mode-locked fiber laser, *IEEE J. Quantum Electron.*, 36(2), 145–150, 890, 2000.
8. G.R. Lin, Y.C. Chang, and J.R. Wu, Rational harmonic mode-locking of erbium-doped fiber laser at 40 GHz using a loss-modulated Fabry-Perot laser diode, *IEEE Photon. Technol. Lett.*, 16(8), 1810–1812, 2004.
9. G. Zhu, Q. Wang, H. Chen, H. Dong, and N.K. Dutta, High-quality optical pulse train generation at 80 Gb/s using a modified regenerative-type mode-locked fiber laser, *IEEE J. Quantum Electron.*, 40(6), 721–725, 2004.
10. E.R. Thoen, M.E. Grein, E.M. Koontz, E.P. Ippen, H.A. Haus, and L.A. Kolodziejski, Stabilization of an active harmonically mode-locked fiber laser using two-photon absorption, *Opt. Lett.*, 25(13), 948–950, 2000.
11. Y. Shiquan, L. Zhaohui, Z. Chunliu, D. Xiaoyi, Y. Shuzhong, K. Guiyun, and Z. Qida, Pulse-amplitude equalization in a rational harmonic mode-locked fiber ring laser by using modulator as both mode locker and equalizer, *IEEE Photon. Technol. Lett.*, 15(3), 389–391, 2003.
12. D. Zhao, Y. Lai, X. Shu, L. Zhang, and I. Bennion, Supermode-noise suppression using a nonlinear Fabry-Perot filter in a harmonically mode-locked fiber ring laser, *Appl. Phys. Lett.*, 81(24), 4520–4522, 2002.
13. D.R. Kirkby, A picosecond optoelectronic cross correlator using a gain modulated Avalanche photodiode for measuring the impulse response of tissue, PhD thesis, University of London, London, U.K., 1999, Chapter 5.
14. M.J. Chudobiak, New approaches for designing high voltage, high current silicon step recovery diodes for pulse sharpening applications, PhD thesis, Carleton University, Ottawa, Ontario, Canada, 1996, Chapter 2.
15. G. Zhu, H. Chen, and N. Dutta, Time domain analysis of a rational harmonic mode-locked ring fiber laser, *J. Appl. Phys.*, 90(5), 2143–2147, 2001.
16. K.K. Gupta, N. Onodera, and M. Hyodo, Technique to generate equal amplitude, higher-order optical pulses in rational harmonically modelocked fiber ring lasers, *Electron. Lett.*, 37(15), 948–950, 2001.

17. J. Hansryd, P.A. Andrekson, M. Westlund, J. Li, and P.O. Hedekvist, Fiber-based optical parametric amplifiers and their applications, *IEEE J. Select. Top. Quant. Electron.*, 8(33), 506–520, 2002.

18. H.A. Haus and W.S. Wong, Solitons in optical communications, *Rev. Mod. Phys.*, 68(2), 423–444, 1996.

19. G.P. Agrawal, *Nonlinear Fiber Optics*, 3rd edn., John Wiley & Sons Inc., Hoboken, NJ, 2001.

9

Nonlinear Fiber Ring Lasers

This chapter covers bidirectional lightwave propagations in an erbium-doped fiber ring laser and the optical bistability behavior. We exploit the bidirectional lightwave propagation effect, which is normally undesirable in many linear photonics systems, in constructing a kind of fiber laser configuration based on a nonlinear optical loop mirror (NOLM) and a nonlinear amplifying loop mirror (NALM) structure. The laser is operated in different regions of the operation, and the switching capability of the laser is identified based on its bistability behavior.

9.1 Introduction

The nonlinear phenomenon of optical bistability has been studied in nonlinear resonators since 1976 by placing a nonlinear medium inside a cavity formed by using multiple mirrors [1]. As for the fiber-based devices, a single-mode fiber was used as the nonlinear medium inside a ring cavity in 1983 [2]. Since then, the study of nonlinear phenomena in fiber resonators has remained a topic of considerable interest.

Fiber ring laser is a rich and active research field in optical communications. Many fiber ring laser configurations have been proposed and constructed to achieve different objectives. It can be designed for CW or pulse operation, linear or nonlinear operation, fast or slow repetition rate, narrow or broad pulse width, etc., for various kinds of photonic applications.

The simplest fiber laser structure shall be an optical closed loop with a gain medium and some associated optical components such as optical couplers. The gain medium used can be any rare earth element–doped fiber amplifier, such as erbium and ytterbium, semiconductor optical amplifier, parametric amplifier, etc., as long as it provides the gain requirement for lasing. Without any mode-locking mechanism, the laser shall operate in the CW regime. By inserting an active mode locker into the laser cavity, i.e., either amplitude or phase modulator, the resulting output shall be an optical pulse train operating at the modulating frequency, when the phase conditions are matched, as discussed in Chapters 2 and 3. This often results in a high-speed optical pulse train, but with a broad pulse width. There is another kind of fiber laser, which uses the nonlinear effect in generating the optical pulses, and is known as the passive mode-locking technique. The technique has been briefly mentioned in Chapter 2. Saturable absorber, stretched pulse mode locking [3], nonlinear polarization rotation [4], figure-eight fiber ring lasers [5], etc., are grouped under this category. This type of laser generates a shorter optical pulse in exchange of the repetition frequency. Hence, this is the trade-off between the active and passive mode-locked laser systems.

Although there are different types of fiber lasers with different operating regimes and operating principles, one common criterion is the unidirectionality, besides the gain- and phase-matching conditions. Unidirectional operation has been proved to offer better lasing efficiency, to be less sensitive to back reflections, and to be a good potential for single-longitudinal mode operation [6], and can be achieved by incorporating an optical isolator within the laser cavity. Shi et al. [7] have demonstrated a unidirectional inverted S-type erbium-doped fiber ring laser without the use of an optical isolator, but with optical couplers. In their laser system, the lightwaves were passed through in one direction and suppressed in the other using certain coupling ratios, achieving unidirectional operation. However, only the power difference between the two CW lightwaves was studied, and this work did not extend further into unconventional and nonlinear regions of operation.

In this chapter, we will study the bidirectional lightwave propagations in an erbium-doped fiber ring laser and observe the optical bistability behavior. This optical bistability behavior has been well reported previously by Oh and Lee [8], Mao and Lit [9], and Luo et al. [10], but only in a unidirectional manner. We exploit this commonly known as undesirable bidirectional lightwave propagation in constructing a kind of fiber laser configuration based on an NOLM and an NALM structure. This laser configuration is similar to the one demonstrated by Shi et al. [7]; however, we operate the laser in different regions of the operation. We investigate the switching capability of the laser based on its bistability behavior. Furthermore, we focus on the nonlinear dynamics of the constructed laser and observe optical bifurcation phenomenon.

9.2 Optical Bistability, Bifurcation, and Chaos

All real physical systems are nonlinear in nature. Apart from systems designed for linear signal processing, many systems have to be nonlinear by assumption, for instance, flip-flops, modulators, demodulators, amplifiers, etc. In this section, we will focus on the optical bistability, bifurcation, and chaos of a nonlinear system.

An optical bistable device is a device with two possible operation points. It will remain stable in any of the two optical states, one of high transmission and the other of low transmission, depending on the intensity of the light passing through it. In this section, we will discuss the effect of optical nonlinearity together with the proper feedback; which can give rise to optical bistability

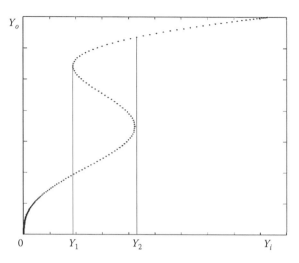

FIGURE 9.1
Bistability curve.

and hysteresis. This is expected as these two effects are also observed in nonlinear electronic circuits with feedback, such as Schmitt trigger, as well as hybrid optical devices, such as an acousto-optic device with feedback [11].

A typical bistability curve is shown in Figure 9.1, where Y_i and Y_o are the input and output parameters, respectively. For an input value between Y_1 and Y_2, there are three possible output values. The middle segment, with a negative slope, is known to be always unstable. Therefore, the output will eventually have two stable values. When two outputs are possible, which one of the outputs is eventually realized depends on the history of how the input is reached, and hence the hysteresis phenomenon. As the input value is increased from zero, the output will follow the lower branch of the curve until the input value reaches Y_2. Then it will jump up and follow the upper branch. However, if the input value is decreased from some points after the jump, the output will remain on the upper branch until the input value hits Y_1, then the output will jump down and follow the lower branch. Hence, the bistability region observed is from Y_1 to Y_2.

In an optical ring cavity, the lightwave can be separated into two components propagating in opposite directions. These two components interact with each other through an atomic medium, which leads to gain competition. Hence, in a ring laser, the lightwave may propagate in one direction or another depending on the initial configuration and is thus a running wave in general. To eliminate this randomness, a device such as an optical isolator is often added into the ring cavity to block the unwanted wave.

There are two different types of optical bistability, namely, absorptive bistability and dispersive bistability in a nonlinear ring cavity comprising a two-level gain medium as a nonlinear medium. Absorptive bistability is the case when the incident optical frequency is close to or equal to the transition frequency of the atoms from one level to another. In other words, the system is in a perfect resonance condition. In this case, the absorption coefficient becomes a nonlinear function of the incident frequency.

On the other hand, if the frequencies are far apart, the gain medium behaves like a Kerr-type material and the system exhibits what is called dispersive bistability. A nonzero atomic detuning will introduce saturable dispersion in response to the medium. In this case, the material can be modeled by an effective nonlinear refractive index, which is a nonlinear function of the optical intensity [10].

For optical bistability in a ring cavity, there is a possible instability due to the counter-propagating wave. Hence, it is interesting to know the lightwave behavior for both co- and counter-propagating components. The bidirectional operation of an optical bistable ring system has not yet been studied thoroughly. Although some studies have been done relating to the bistability properties of the ring cavities, they focused on forcing the system to support only unidirectional operation [12]. The restriction to unidirectional propagation has been quite consistent with the experimental results; however, the exceptions were noted [13]. Unfortunately, no further investigation has been carried out since then.

It was discovered that the nonlinear response of a ring resonator could initiate a period doubling (bifurcation) route to optical chaos [14]. In a dynamical system, a bifurcation is a period doubling, quadrupling, etc., that accompanies the onset of chaos. It represents a sudden appearance of a qualitatively different solution for a nonlinear system as some parameters are varied. A typical bifurcation map is shown in Figure 9.2, where g and y are the input and output system parameters, respectively. Period doubling action starts at $g = \sim 1.5$, and period quadrupling at $g = \sim 2$. The region beyond $g = \sim 2.3$ is known as chaos.

Each of the local bifurcations may give rise to distinct routes to chaos if the bifurcations appear repeatedly when changing the bifurcation parameter. These routes are important since it is difficult to conclude from the experimental data alone whether irregular behavior is due to measurement noise or chaos. The recognition of one of the typical routes to chaos in experiments is a good indication that the dynamics may be chaotic [15].

1. Period doubling route to chaos—When a cascade of successive period-doubling bifurcations occurs when changing the value of the bifurcation parameter, it is often the case that finally the system will reach chaos.

2. Intermittency route to chaos—The route to chaos cased by saddle-node bifurcations. The common feature of which is a direct transition from regular motion to chaos.

3. Torus breakdown route to chaos—The quasiperiodic route to chaos results from a sequence of Hopf bifurcations.

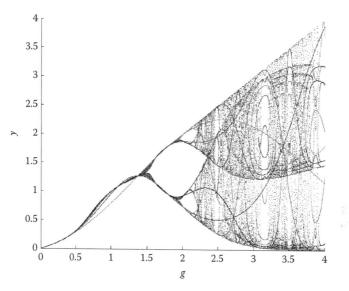

FIGURE 9.2
Typical bifurcation map.

It is well known that self-pulsing often leads to optical chaos in the laser output, following a period doubling or a quasiperiodic route. The basic idea is that the dynamics of the intra-cavity field is different from one round trip to another in a nonlinear fashion. The characteristics of a chaotic system [16] are as follows:

1. Sensitive dependence on initial conditions—it gives rise to an apparent randomness in the output of the system and the long-term unpredictability of the state. Because the chaotic system is deterministic, two trajectories that start from identical initial states will follow precisely the same paths through the state space.

2. Randomness in the time domain—in contrast to the periodic waveforms, chaotic waveform is quite irregular and does not appear to repeat itself in any observation period of finite length.

3. Broadband power spectrum—every periodic signal may be decomposed into its Fourier series, a weighted sum of sinusoids at a multiple integer of the fundamental frequency. Thus, a periodic signal appears in the frequency domain as a set of spikes at the fundamental frequency and its harmonics. The chaotic signal is qualitatively different from the periodic signal. The aperiodic nature of its time-domain waveform is reflected in the broadband noise-like power spectrum. This broadband structure of the power spectrum persists even if the spectral resolution is increased to a higher frequency.

A typical example of the chaotic system is a Lorenz attractor, as shown in Figure 9.3, or more commonly known as a butterfly attractor, with the following system description, with $a = 10$, $b = 28$, and $c = 8/3$:

$$\frac{dx}{dt} = a(y - x) \tag{9.1}$$

$$\frac{dy}{dt} = x(b - z) - y \tag{9.2}$$

$$\frac{dz}{dt} = xy - cz \tag{9.3}$$

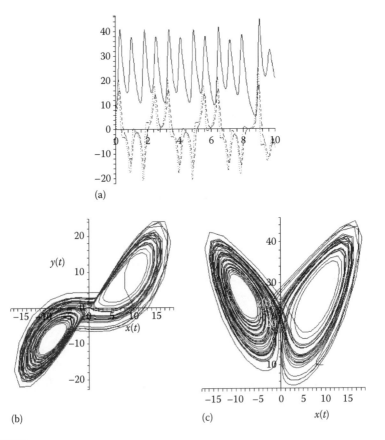

(a)

(b)

(c)

FIGURE 9.3
System response of a Lorenz attractor: (a) time traces of x, dash; y, dotted; and z, solid; (b) y–x trajectory; and (c) z–x trajectory.

9.3 Nonlinear Optical Loop Mirror

The concept of NOLM was proposed by Doran and Wood [17]. It is basically a fiber-based Sagnac interferometer that uses the nonlinear phase shift of optical fiber for optical switching. This configuration is inherently stable since the two arms of the structure reside in the same fiber and same optical path lengths for the signals propagating in both arms, however in opposite directions. There is no feedback mechanism in this structure since all light-waves entering the input port exit from the loop after a single round trip. The NOLM in its simplest form contains a fiber coupler, with two of its output ports connected together, as shown in Figure 9.4, where κ is the coupling ratio of the coupler, and E_1, E_2, E_3, and E_4 are the fields at ports 1, 2, 3, and 4, respectively. The NOLM considered here is assumed to be polarization independent.

Under lossless condition, the input–output relationships of a coupler with coupling ratio of κ are

$$E_3 = \sqrt{1-\kappa}\,E_1 + j\sqrt{\kappa}\,E_2$$
$$E_4 = j\sqrt{\kappa}\,E_1 + \sqrt{1-\kappa}\,E_2 \tag{9.4}$$

where $j^2 = -1$. If a low-intensity light is fed into port 1, i.e., no nonlinear effect, the intensity transmission, T, of the device is

$$T = \frac{P_{out}}{P_{in}} = 1 - 4\kappa(1-\kappa) \tag{9.5}$$

If $\kappa = 0.5$, all light will be reflected back to the input, hence the name loop mirror; otherwise, the counter-propagating lightwaves in the loop will have different intensities, leading to different nonlinear phase shifts. This device can be designed to transmit a high-power signal while reflecting it at low power level, thus acting as an all-optical switch.

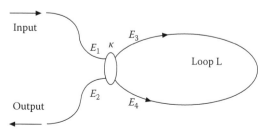

FIGURE 9.4
Nonlinear optical loop mirror.

After propagating through a distance of L, the output signals, E_{3L} and E_{4L}, become as follows with nonlinear phase shifts taken into account:

$$E_{3L} = E_3 \exp\left(j \frac{2\pi n_2 L}{\lambda} |E_3|^2 \right)$$

(9.6)

$$E_{4L} = E_4 \exp\left(j \frac{2\pi n_2 L}{\lambda} |E_4|^2 \right)$$

where
$|E_3|^2$ and $|E_4|^2$ are the light intensities at the coupler outputs
n_2 is the nonlinear coefficient
L is the loop length
λ is the operating wavelength of the signal
$E_1 = E_{01} \exp(j\omega_1 t)$ and $E_2 = E_{02} \exp(j\omega_2 t)$

$$|E_{3L}|^2 = |E_3|^2$$

$$= \kappa |E_1|^2 + (1-\kappa)|E_2|^2 + j\sqrt{\kappa(1-\kappa)}(E_1^* E_2 - E_1 E_2^*)$$

$$= \kappa |E_{01}|^2 + (1-\kappa)|E_{02}|^2 - 2\sqrt{\kappa(1-\kappa)}E_{01}E_{02} \sin[(\omega_2 - \omega_1)t]$$

$$|E_{4L}|^2 = |E_4|^2$$

(9.7)

$$= \kappa |E_2|^2 + (1-\kappa)|E_1|^2 + j\sqrt{\kappa(1-\kappa)}(E_2^* E_1 - E_2 E_1^*)$$

$$= \kappa |E_{02}|^2 + (1-\kappa)|E_{01}|^2 + 2\sqrt{\kappa(1-\kappa)}E_{01}E_{02} \sin[(\omega_2 - \omega_1)t]$$

In (9.7), "*" denotes the complex conjugate, and (E_{01}, E_{02}) and (ω_1, ω_2) are the amplitudes and frequencies of the input fields E_1 and E_2, respectively, into the inputs of the coupler.

The last term of this equation represents the interference pattern between input fields E_1 and E_2. Most literatures consider only the simple case with a single input at port 1, i.e., no input field at input port 2 (i.e., $E_2 = 0$), and hence reduce the interference effect between the signals in the loop, and the output fields at ports 1 and 2 are given as follows:

$$|E_{o1}|^2 = |E_{01}|^2 \left\{ 2\kappa(1-\kappa)\left[1 + \cos\left((1-2\kappa)\frac{2\pi n_2 |E_{01}|^2 L}{\lambda} \right) \right] \right\}$$

(9.8)

$$|E_{o2}|^2 = |E_{01}|^2 \left\{ 1 - 2\kappa(1-\kappa)\left[1 + \cos\left((1-2\kappa)\frac{2\pi n_2 |E_{01}|^2 L}{\lambda} \right) \right] \right\}$$

However, for more complete studies, we consider inputs at both input ports (i.e., $E_1 \neq 0$ and $E_2 \neq 0$); and the output fields are given by

$$|E_{o1}|^2 = \kappa|E_{4L}|^2 + (1-\kappa)|E_{3L}|^2 + j\sqrt{\kappa(1-\kappa)}(E_{4L}^*E_{3L} - E_{4L}E_{3L}^*)$$

$$= [\kappa^2 + (1-\kappa)^2]|E_{02}|^2 + 2\kappa(1-\kappa)|E_{01}|^2$$

$$-2(1-2\kappa)\sqrt{\kappa(1-\kappa)}E_{01}E_{02}\sin[(\omega_2 - \omega_1)t]$$

$$+j\sqrt{\kappa(1-\kappa)}(E_3E_4^*\exp(j\Delta\theta) - E_3^*E_4\exp(-j\Delta\theta))$$

$$= |E_{01}|^2\left[2\kappa(1-\kappa)[1+\cos(\Delta\theta)]\right] + |E_{02}|^2\left[1-2\kappa(1-\kappa)[1+\cos(\Delta\theta)]\right]$$

$$-2\sqrt{\kappa(1-\kappa)}E_{01}E_{02}\left[(1-2\kappa)\sin[(\omega_2 - \omega_1)t][1+\cos(\Delta\theta)]\right.$$

$$\left. +\cos[(\omega_2 - \omega_1)t]\sin(\Delta\theta)\right] \tag{9.9}$$

$$|E_{o2}|^2 = \kappa|E_{3L}|^2 + (1-\kappa)|E_{4L}|^2 + j\sqrt{\kappa(1-\kappa)}(E_{3L}^*E_{4L} - E_{3L}E_{4L}^*)$$

$$= [\kappa^2 + (1-\kappa)^2]|E_{01}|^2 + 2\kappa(1-\kappa)|E_{02}|^2 + 2(1-2\kappa)\sqrt{\kappa(1-\kappa)}E_{01}E_{02}\sin[(\omega_2 - \omega_1)t]$$

$$+j\sqrt{\kappa(1-\kappa)}(E_4E_3^*\exp(j\Delta\theta) - E_4^*E_3\exp(-j\Delta\theta))$$

$$= |E_{02}|^2\left[2\kappa(1-\kappa)[1+\cos(\Delta\theta)]\right] + |E_{01}|^2\left[1-2\kappa(1-\kappa)[1+\cos(\Delta\theta)]\right]$$

$$+2\sqrt{\kappa(1-\kappa)}E_{01}E_{02}\left[(1-2\kappa)\sin[(\omega_2 - \omega_1)t][1+\cos(\Delta\theta)]\right.$$

$$\left. +\cos[(\omega_2 - \omega_1)t]\sin(\Delta\theta)\right] \tag{9.10}$$

where

$$\Delta\theta = \theta_3 - \theta_4$$

$$= 2\pi n_2 L\left\{(1-2\kappa)\left[\frac{|E_{02}|^2}{\lambda_2} - \frac{|E_{01}|^2}{\lambda_1}\right] - \frac{4\sqrt{\kappa(1-\kappa)}|E_{01}||E_{02}|\sin[(\omega_2 - \omega_1)t]}{\sqrt{\lambda_1\lambda_2}}\right\} \tag{9.11}$$

and λ_1 and λ_2 are the operating wavelengths of E_1 and E_2, respectively.

For $\lambda_1 = \lambda_2$, and $E_2 = pE_1$ with p being a constant, the switching characteristics for (a) $\kappa = 0.45$ and (b) $\kappa = 0.2$ are shown in Figure 9.5. Switching occurs only for large energy difference between the two ports, i.e., $p \gg 1$. Also, as can be seen from the figure, the switching behavior is better for a coupling ratio close to 0.5.

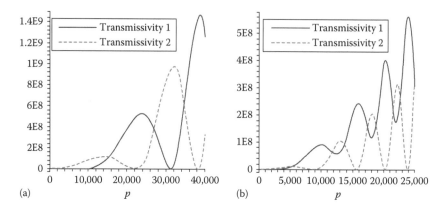

FIGURE 9.5
Output intensities or switching characteristics of the NOLM with (a) $\kappa=0.45$ and (b) $\kappa=0.2$.

9.4 Nonlinear Amplifying Loop Mirror

The structure of the NALM is somewhat similar to the NOLM structure, and it is an improved exploitation of NOLM. For the NALM configuration, a gain medium with gain coefficient, G, is added to increase the asymmetric nonlinearity within the loop [18]. The amplifier is placed at one end of the loop, closer to port 3 of the coupler, and is assumed short relative to the total loop length, as shown in Figure 9.6. One lightwave is amplified at the entrance to the loop, while the other experiences amplification just before exiting the loop. Since the intensities of the two lightwaves differ by a large amount throughout the loop, the differential phase shift can be quite large.

By following the analysis procedure stated in the previous section, we arrive at the input–output relationships as follows:

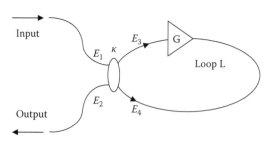

FIGURE 9.6
Nonlinear amplifying loop mirror.

$$|E_{o1}|^2 = G \left\{ \begin{array}{l} \left[|E_{01}|^2 \left[2\kappa(1-\kappa)[1+\cos(\Delta\theta)] \right] + |E_{02}|^2 \left[1 - 2\kappa(1-\kappa)[1+\cos(\Delta\theta)] \right] \right] \\ -2\sqrt{\kappa(1-\kappa)} E_{01} E_{02} \left[(1-2\kappa) \sin[(\omega_2 - \omega_1)t][1+\cos(\Delta\theta)] \right] \\ + \cos[(\omega_2 - \omega_1)t] \sin(\Delta\theta) \end{array} \right\}$$

(9.12)

$$|E_{o2}|^2 = G \left\{ \begin{array}{l} \left[|E_{02}|^2 \left[2\kappa(1-\kappa)[1+\cos(\Delta\theta)] \right] + |E_{01}|^2 \left[1 - 2\kappa(1-\kappa)[1+\cos(\Delta\theta)] \right] \right] \\ +2\sqrt{\kappa(1-\kappa)} E_{01} E_{02} \left[(1-2\kappa) \sin[(\omega_2 - \omega_1)t][1+\cos(\Delta\theta)] \right] \\ + \cos[(\omega_2 - \omega_1)t] \sin(\Delta\theta) \end{array} \right\}$$

(9.13)

$$\Delta\theta = 2\pi n_2 L \left\{ \begin{array}{l} (1-\kappa-G\kappa) \left[\dfrac{|E_{02}|^2}{\lambda_2} - \dfrac{|E_{01}|^2}{\lambda_1} \right] \\ -\dfrac{2(G+1)\sqrt{\kappa(1-\kappa)}\,|E_{01}||E_{02}|\sin[(\omega_2 - \omega_1)t]}{\sqrt{\lambda_1\lambda_2}} \end{array} \right\}$$

(9.14)

9.5 NOLM–NALM Fiber Ring Laser

9.5.1 Simulation of Laser Dynamics

The configuration of an NOLM–NALM fiber ring laser is shown in Figure 9.7. It is simply a coupled loop mirror, with NOLM (ABCEA) on one side and NALM (CDAEC) on the other side of the laser, with a common path in the middle section (AEC). One interesting feature about this configuration is the feedback mechanism of the fiber ring: one is acting as the feedback path to another, i.e., NOLM is the feeding-back part of the NALM's signal and vice versa.

As a matter of fact, complex systems tend to encounter bifurcations, which when amplified, can lead to either order or chaos. The system can transit into chaos, through period doubling, or order through a series of feedback loops. Hence, bifurcations can be considered as critical points in these system transitions.

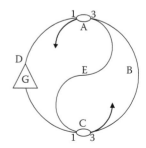

FIGURE 9.7
NOLM–NALM fiber ring laser.

In the formularization process, we ignore the nonlinear phase shift due to laser pulsation. The assumption is valid because of the saturation effect of the gain medium as well as the energy stabilization provided by the filter.

We numerically simulate the laser behavior by combining the effects described in the previous section and obtain the bifurcation maps based on the CW operation, as shown in Figures 9.8 and 9.9, with P_{o1} and P_{o2} being the output powers at points A and C, respectively, $\kappa_1 = 0.55$ and $\kappa_2 = 0.65$; L_a, L_b, and L_c are 20, 100, and 20 m, respectively; nonlinear coefficient of $3.2 \times 10^{-20}\,\mathrm{m}^2/\mathrm{W}$; gain coefficient of 0.4/m at 1550 nm with EDF length of 10 m and a saturation power of 25 dBm; and fiber-effective area of 50 μm². κ_1 and κ_2 are the coupling ratios of the couplers from port 1 to port 3 at point A and C, respectively; L_a, L_b, and L_c are the fiber lengths of left, middle, and right arm of the laser cavity. The maps are obtained with 200 iterations. Figure 9.8 shows

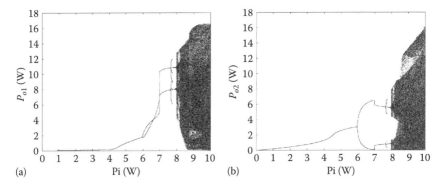

FIGURE 9.8
Bifurcation maps for (a) P_{o1} and (b) P_{o2} with $\kappa_1 = 0.55$ and $\kappa_2 = 0.65$ with SPM consideration only.

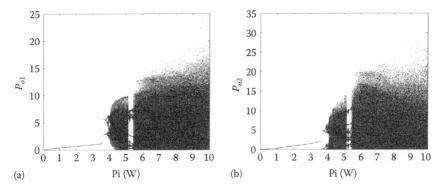

FIGURE 9.9
Bifurcation maps for (a) P_{o1} and (b) P_{o2} with $\kappa_1 = 0.55$ and $\kappa_2 = 0.65$ with both SPM and XPM considerations.

the bifurcation maps with self-phase modulation (SPM) consideration only, whereas Figure 9.9 considers both SPM and cross-phase modulation (XPM).

In constructing the bifurcation maps of the system, we separate the light-waves into clockwise and counterclockwise directions; similarly for the couplers involved in the system, i.e., four couplers are used in simulating this bifurcation behavior. Initial conditions are set to be the pump power of the EDFA. We then propagate the lightwaves in both directions within the fiber ring, with the effects of the various components taken into account, such as coupling ratios, gain, SPM, and XPM.

The outputs of the system are then served as the input to the system for the next iteration. The process is repeated for a number of times, i.e., 200 times. Each iterated set of outputs is then combined together in constructing the bifurcation maps.

There are some similarities between Figures 9.8 and 9.9. Both of them indicate different operating regions at different power levels. With both modulation (SPM and XPM) effects taken into account (Figure 9.9), the transition from one operation state to another is faster, leading to an earlier chaotic operation with a lower-pump power. This is simply because the XPM effect is twice the SPM effect. For simplicity, we discuss only the bifurcation behavior of Figure 9.8 to illustrate the transition behavior from one state to another.

From the maps obtained (Figure 9.8), under this system setting, there are three operation regions within the NOLM–NALM fiber ring laser. The first is the linear operation region $(0 < P < 6\,W)$, where there are single-value outputs, and the output power increases with the input power. The period-doubling effect starts to appear when the input power reaches ~6 W. This is the second operation region $(6 \leq P < 8\,W)$ of the laser, where double-periodic and quasiperiodic signals can be found here. When the input power goes beyond 8 W, the laser will enter into the chaotic state of the operation.

The Poincare map of the above system configuration with high pump power is shown in Figure 9.10. It shows the pattern of attractor of the system when the laser is operating in the chaotic region. The powers required for the operations can be reduced by increasing the lengths of the fibers, L_a, L_b, and L_c.

By setting $\kappa_2 = 0$, the laser will behave like an NOLM. With an input pulse to port 1 of coupler A, we observe pulse compression at its output of the ring laser of Figure 9.7, as shown in Figures 9.11 and 9.12. Figure 9.12 shows the transmission capability of the setup at port 2 for various κ_2 values, when the input is injected to port 1 and $\kappa_2 = 0$. When $\kappa_1 = 0.5$, we observe no transmission at port 2, as the entire injected signal has been reflected back to port 1, where the mirror effect takes place. By changing the coupling ratio of κ_2, we are able to change the zero transmission point away from $\kappa_1 = 0.50$. Figure 9.13 shows the transmissivities of P_{o1} and P_{o2} for various sets of κ_1 and κ_2. The figure shows the complex switching dynamics of the laser for different sets of κ_1 and κ_2 values, also with the pulse compression capability.

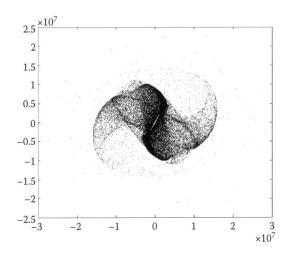

FIGURE 9.10
Poincare map of the system with high pump power with $\kappa_1 = 0.55$ and $\kappa_2 = 0.65$.

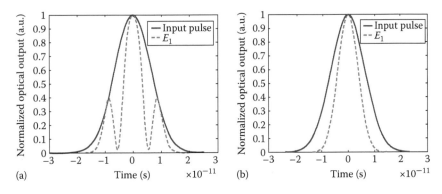

FIGURE 9.11
Comparisons of input and output pulse at port 1 and port 2 with $\kappa_2 = 0$ and (a) $\kappa_1 = 0.41$ and (b) $\kappa_1 = 0.49$. (Solid line—input pulse; dotted line—output pulse at port 1.)

9.5.2 Experiment

9.5.2.1 Bidirectional Erbium-Doped Fiber Ring Laser

We start with a simple erbium-doped fiber ring laser; an optical closed loop with EDFA and some fiber couplers. It is used to study the bidirectional lightwave propagation behavior of the laser, as shown in Figure 9.14. The EDFA is made of a 20 m EDF, which is dual pumped by 980 and 1480 nm diode lasers, with a saturation power of about 15 dBm. The slope efficiencies of the pump lasers are 0.45 mW/mA and 0.22 W/A, respectively. No isolator and filter are used in the setup to eliminate the

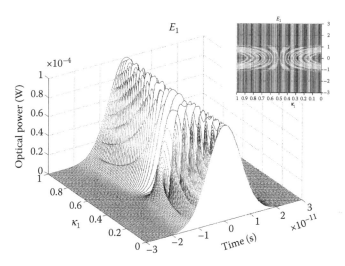

FIGURE 9.12
Output pulse behavior for various κ_1 values, when $\kappa_2 = 0$ for small input power. (Inset: top view of the pulse behavior.)

direction and spectra constraints. Two outputs are taken for examinations, i.e., Output 1 in counterclockwise direction, and Output 2 in clockwise direction. All connections within the fiber ring are spliced together to reduce the possible reflections within the system. Output 1 (counterclockwise, ccw) and Output 2 (clockwise, cw), as shown in the figure, are taken for investigations.

The ASE spectrum of the laser covers the range from 1530 to 1570 nm. By increasing both 980 and 1480 nm pumping currents to their maximum allowable values, we obtained bidirectional lasing, which is shown in Figure 9.15. The upper plot is the lightwave propagating in the clockwise direction while the other one in the opposite direction. To obtain bidirectional lasing, the losses and the gain must be balanced for the two lightwaves that propagate around the cavity in opposite directions [19]. The pump lasers used in the setup are not identical (in terms of power and pumping wavelength), and this gives rise to different lasing behaviors for both clockwise and counterclockwise directions. Due to the laser diode controllers' limitation, the maximum pumping currents for both 980 and 1480 nm laser diodes are capped at 300 and 500 mA, respectively, which correspond to 135 and 110 mW in clockwise pump and counterclockwise pump directions. This explains the domineering clockwise lasing, as shown in the figure. One thing to note is that the lasings of the laser in both directions are not very stable due to the disturbance from the opposite propagating lightwave and the ASE noise contribution due to the absence of filter. The Output 1 is mainly contributed by the back reflections from the fiber ends and connectors, as well as some backscattered noise. Since all the connections within the fiber ring are spliced

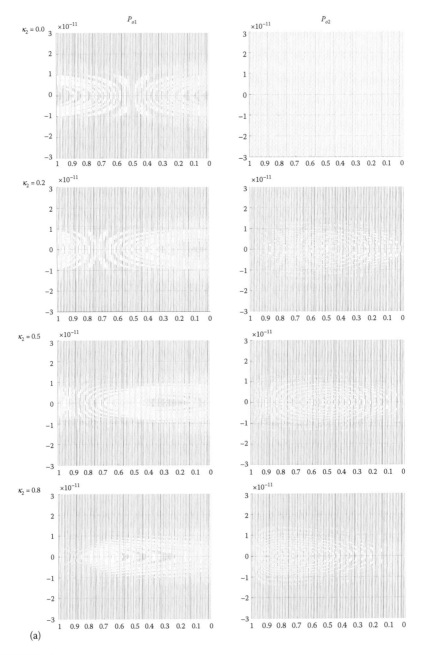

(a)

FIGURE 9.13
Intensity plots for P_{o1} and P_{o2} for different κ_2 values. (y, time axis; x, axis is the values of κ_1.)

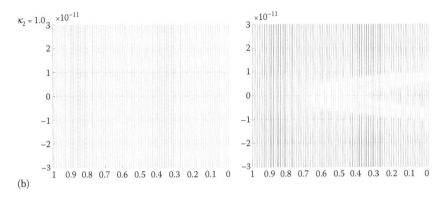

(b)

$\kappa_2 = 1.0$

FIGURE 9.13 (continued)

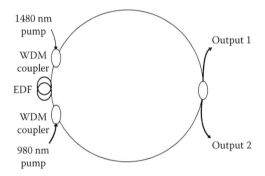

FIGURE 9.14
Bidirectional erbium-doped fiber ring laser.

together, the reflection due to the connections is minimum. However, there are still some unavoidable reflections from the fiber ends, which contribute to the lightwaves in the opposite directions. Besides that, back-scattered noise also adds to the lightwaves in the opposite directions.

We maintain the 980 nm pump current at a certain value, and adjust the 1480 nm pump current upward and then downward to examine the bistability behavior of the laser. The bistable characteristics at a lasing wavelength of about 1562.2 nm for both clockwise (cw) and counterclockwise (ccw) directions are shown in Figures 9.16 and 9.17, with log and linear scales for the vertical axes of Figures 9.16a and 9.17a, and Figures 9.16b and 9.17b, respectively. We obtain about 15 dBm difference between the two propagating lightwaves. The bistable region obtained is ~30 mA of 1480 nm pump current at a fixed value of 100 mA of 980 nm pump current. No lasing is observed in the ccw lightwave propagation. This bistable region can be further enhanced by increasing the 980 nm pump current to a higher level. By maintaining 980 nm pump current at ~175 mA, we obtain a bistable region

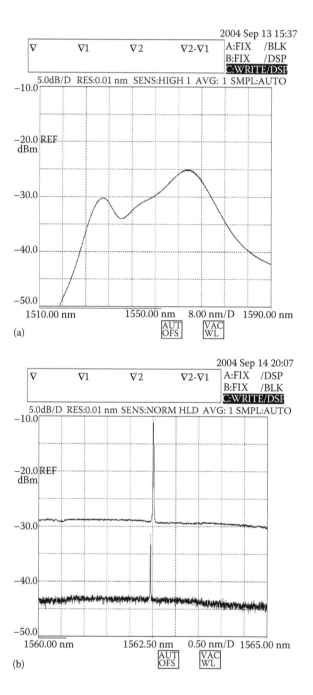

FIGURE 9.15

(a) ASE spectrum of the laser; (b) lasing characteristics in both directions. (Upper trace—Output 2; lower trace—Output 1.)

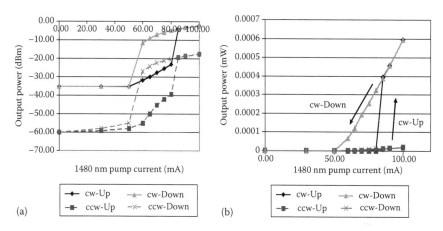

FIGURE 9.16
Hysteresis loops obtained from the EDFRL for 980 nm pump current = 100 mA: (a) log scale and (b) linear scale.

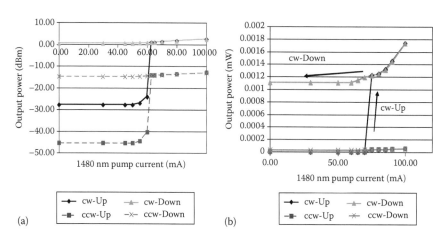

FIGURE 9.17
Hysteresis loops obtained from the EDFRL for 980 nm pump current = 200 mA: (a) log scale and (b) linear scale.

as wide as ~70 mA of the 1480 nm pump current. When the 980 nm pump current is maintained at a higher level (>175 mA), the lasing of Output 2 remains even when the 1480 nm is switched off, as shown in Figure 9.17. This bistable behavior is mainly due to the saturable absorption of the EDF.

9.5.2.2 Continuous-Wave NOLM–NALM Fiber Ring Laser

The experimental setup of the NOLM–NALM fiber ring laser under continuous-wave (CW) operation is shown in Figure 9.18. It is simply a combination

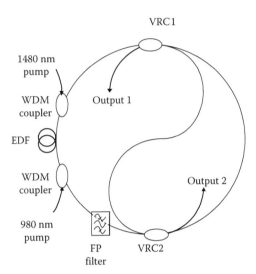

FIGURE 9.18
Experimental setup of NOLM–NALM fiber ring laser.

of NOLM on one side and NALM on the other side of the laser, with a common path in the middle section. The principal element of the laser is an optical closed loop with an optical gain medium, two variable ratio couplers (VRCs), an optical bandpass filter (BPF), optical couplers, and other associated optics. The gain medium used in our fiber laser system is the amplifier used in the preceding experiment. The two VRCs, with coupling ratios ranging from 20% to 80% and insertion loss of about 0.2 dB are added into the cavity at positions shown in the figure to adjust the coupling power within the laser. They are interconnected in such a way that the output of one VRC is the input of the other. A tunable BPF with 3 dB bandwidth of 2 nm at 1560 nm is inserted into the cavity to select the operating wavelength of the generated signal and to reduce the noise in the system. One thing to note is that the lightwave is traveling in both directions, as there is no isolator used in the laser. Output 1 and Output 2, as shown in the figure, are taken out and analyzed as the outputs of the laser.

One interesting phenomenon observed before the BPF is inserted into the cavity is the wavelength tunability. The lasing wavelength is tunable from 1530 to 1560 nm (almost the entire EDFA C-band), by changing the coupling ratios of the VRCs. We believe that this wavelength tunability is due to the change in the traveling lightwaves' intensities, which contributes to the nonlinear refractive index change, and in turn modifies the dispersion relations of the system, and hence the lasing wavelength. Therefore, the VRCs within the cavity not only determine the directionality of the lightwave propagation, but also the lasing wavelength.

For a conventional erbium-doped fiber ring laser, bistable state is not observable when the pump current is far above the threshold value, where

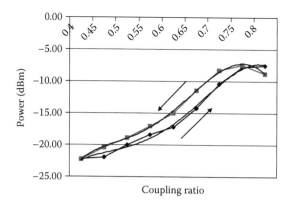

FIGURE 9.19

Hysteresis loop observed when changing the coupling ratio of one VRC while maintaining that of the other, when operating at high pump current.

saturation starts to take place. However, a small hysteresis loop has been observed in our laser setup even with high pump current, i.e., near saturation region, when changing the coupling ratio of one VRC while that of the other one remains unchanged, as shown in Figure 9.19. Changing the coupling ratio of the VRC directly alters the total power within the cavity, and hence modifying its gain and absorption behavior. As a result, a small hysteresis loop is observable even with a constant high pump power, which can be an added advantage to the existing bistable state.

The power distribution of Output 1 and Output 2 of the NOLM–NALM fiber laser obtained experimentally is depicted in Figure 9.20. We are able to obtain the switching between the outputs by tuning the coupling ratios of VRC1 and VRC2. The simulation results for transmissivities for various coupling ratios under linear operation are shown in Figure 9.21, since the available pump power of our experiment setup is insufficient to generate high power within the cavity. Both experimental and numerical results have come to some agreement, but not all, since the model developed is simple and does not consider the polarization, dispersion characteristics, etc., of the propagating lightwaves.

9.5.2.3 Amplitude-Modulated NOLM–NALM Fiber Ring Laser

The schematic of an amplitude-modulated (AM) NOLM–NALM fiber ring laser is shown in Figure 9.22. A few new photonic components are added into the laser cavity. A 10 Gbps, Ti:LiNbO$_3$ Mach–Zehnder amplitude modulator based on the asymmetric coplanar traveling-wave structure is used in the inner loop of the cavity with a half-wave voltage, $V\pi$ of 5.3 V. The modulator is DC biased near the quadrature point and not driven higher than $V\pi$ such that it operates on the linear region of its characteristic curve and to ensure minimum chirp imposed on the modulated lightwaves. It is driven by a sinusoidal signal derived from an Anritsu 68347C Synthesizer

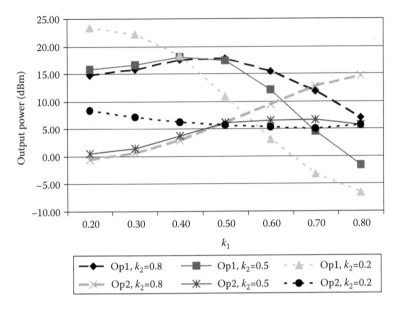

FIGURE 9.20
Experimental results for Output 1 (Op1) and Output 2 (Op2) for various coupling ratios (k_1, coupling ratio of VRC1; k_2, coupling ratio of VRC2).

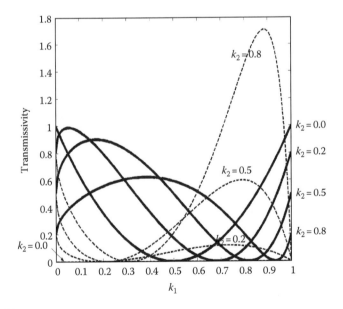

FIGURE 9.21
Simulation results for transmissivities of Output 1 (solid line) and Output 2 (dotted line) for various coupling ratios of VRC1 and VRC2.

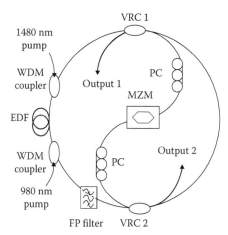

FIGURE 9.22
Experimental setup for AM NOLM–NALM fiber laser (VRC, variable ratio coupler; PC, polarization controller; MZM, Mach–Zehnder modulator).

Signal Generator. The modulator has an insertion loss of ≤7 dB. Two polarization controllers (PCs) are placed prior to the inputs of the modulator in both directions to ensure proper polarization alignment into the modulator. A wider bandwidth, i.e., 5 nm tunable FP filter is used in this case to allow more longitudinal modes within the laser for possible mode-locking process.

With the insertion of an AM modulator into the laser cavity, we are able to obtain the pulse operation from the laser, by means of an active harmonic mode-locking technique. Both propagation lightwaves are observed. By the proper adjustment of the modulation frequency, the PCs, and the VRCs, unidirectional pulsed operation at a modulating frequency of 10 GHz is obtained. However, it is highly sensitive to environmental changes. The direction of the lightwave propagation of the laser can be controlled by the VRCs. The unidirectional pulsed train propagation obtained experimentally and numerically are shown in Figure 9.23.

However, with slight deviations from the system parameters' values, either with a slight modulation frequency detuned or adjustments of the PCs, period doubling and quasiperiodic operations in the laser are observed, as shown in Figure 9.24. We believe that the effect is due the interference between the bidirectional propagation of lightwaves, which have suffered nonlinear phase shifts in each direction, since we are operating in the saturation region of the EDFA. Another factor for this formation is that the lightwave in one direction is the feedback signal for the other. Furthermore, the intensity modulation of the optical modulator is not identical for co- and counter-interactions between the lightwaves and the traveling microwaves on the surface of the optical waveguide. This would contribute to the mismatch of the locking condition of the laser.

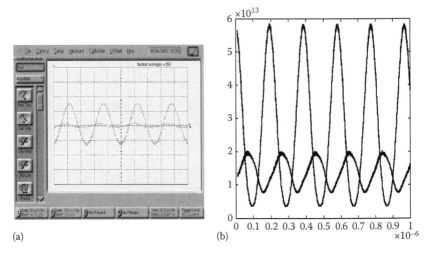

(a)   (b)

FIGURE 9.23

Unidirectional pulsed operation of the AM NOLM–NALM fiber laser: (a) Experimental results and (b) simulation results.

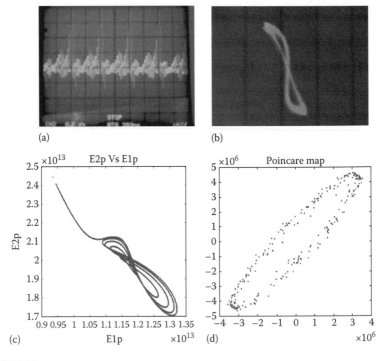

FIGURE 9.24

Quasiperiodic operation in the AM NOLM–NALM fiber laser: (a) photograph of the oscilloscope trace, (b) XY plot of Outputs 1 and 2, (c) simulated XY plot, and (d) simulated Poincaré map.

9.6 Conclusions

Bidirectional optical bistability in a dual-pumped erbium-doped fiber ring laser without an isolator has been studied. A bistable region using ~70 mA 1480 nm pump current has also been obtained. With this bidirectional bistability capability, we experimentally constructed and numerically simulated a NOLM–NALM fiber laser for its switching and bifurcation behaviors. From the simulated bifurcations maps, three basic operation regions can be identified, namely, unidirectional, period doubling, and chaos operations. The VRCs used in the setup not only control the lightwave's directionality, but also its lasing wavelength. Unidirectional lightwave propagations, without an isolator, were achieved in both CW and pulse operations by tuning the coupling ratios of the VRCs within the laser system. Bifurcation was also obtained from the AM NOLM–NALM fiber laser. However, chaotic operation was not observed experimentally due to the hardware limitation of our system, which required higher gain coefficient and input power, as predicted in our simulation.

The configuration is somehow similar to optical flip-flop concept, which can be useful in optical communication systems. The optical flip-flop concept has been used in optical packet switches and optical buffers [20–23]. This NOLM–NALM fiber laser possesses many interesting optical behaviors, which can be used in many applications, such as photonic flip-flops, optical buffer loop, photonic pulse-sampling devices, and secured optical communications.

References

1. H.M. Gibbs, S.L. McCall, and T.N.C. Venkatesan, Differential gain and bistability using a sodium-filled Fabry-Perot interferometer, *Phys. Rev. Lett.*, 36(19), 1135–1138, 1976.
2. H. Nakatsuka, S. Asaka, H. Itoh, K. Ikeda, and M. Matsuoka, Observation of bifurcation to chaos in an all-optical bistable system, *Phys. Rev. Lett.*, 50(2), 109–112, 1983.
3. K. Tamura, E.P. Ippen, H.A. Haus, and L.E. Nelson, 77-fs pulse generation from a stretched-pulse mode locked all fiber ring laser, *Opt. Lett.*, 18(13), 1080–1082, 1993.
4. G. Yandong, S. Ping, and T. Dingyuan, 298 fs passively mode locked ring fiber soliton laser, *Microw. Opt. Tech. Lett.*, 32(5), 320–333, 2002.
5. K.K. Gupta, N. Onodera, and M. Hyodo, Technique to generate equal amplitude, higher-order optical pulses in rational harmonically modelocked fiber ring lasers, *Electron. Lett.*, 37(15), 948–950, 2001.
6. A.E. Siegman, *Lasers*, University Science Books, Mill Valley, CA, 1986.

7. Y. Shi, M. Sejka, and O. Poulsen, A unidirectional Er^{3+}-doped fiber ring laser without isolator, *IEEE Photon. Technol. Lett.*, 7(3), 290–292, 1995.

8. J.M. Oh and D. Lee, Strong optical bistability in a simple L-band tunable erbium-doped fiber ring laser, *J. Quantum Electron.*, 40(4), 374–377, 2004.

9. Q. Mao and J.W.Y. Lit, L-band fiber laser with wide tuning range bases on dual-wavelength optical bistability in linear overlapping grating cavities, *IEEE J. Quantum Electron.*, 39(10), 1252–1259, 2003.

10. L. Luo, T.J. Tee, and P.L. Chu, Bistability of erbium doped fiber laser, *Opt. Commun.*, 146, 151–157, 1998.

11. P.P. Banerjee, *Nonlinear Optics—Theory, Numerical Modeling, and Applications*, Marcel Dekker Inc., New York, 2004.

12. P. Meystre, On the use of the mean-field theory in optical bistability, *Opt. Commun.*, 26(2), 277–280, 1978.

13. L.A. Orozco, H.J. Kimble, A.T. Rosenberger, L.A. Lugiato, M.L. Asquini, M. Brambilla, and L.M. Narducci, Single-mode instability in optical bistability, *Phys. Rev. A*, 39(3), 1235–1252, 1989.

14. K. Ikeda, Multiple-valued stationary state and its instability of the transmitted light by a ring cavity, *Opt. Commun.*, 30(2), 257–261, 1979.

15. M.J. Ogorzalek, Chaos and complexity in nonlinear electronic circuits, *World Sci. Ser. Nonlinear Sci., Ser. A*, (22), 1997.

16. M.P. Kennedy, Three steps to chaos—Part I: Evolution, *IEEE Trans. Circuits Syst. I, Fund. Theory Appl.*, 40(10), 640–656, 1993.

17. N.J. Doran and D. Wood, Nonlinear-optical loop mirror, *Opt. Lett.*, 13(1), 56–58, 1988.

18. M.E. Fermann, F. Haberl, M. Hofer, and H. Hochreiter, Nonlinear amplifying loop mirror, *Opt. Lett.*, 15(13), 752–754, 1990.

19. M. Mohebi, J.G. Mejia, and N. Jamasbi, Bidirectional action of a titanium-sapphire ring laser with mode-locking by a Kerr lens, *J. Opt. Technol.*, 69(5), 312–316, 2002.

20. S. Zhang, Y. Liu, D. Lenstr, M.T. Hill, H. Ju, G.D. Khoe, and H.J.S. Dorren, Ring-laser optical flip-flop memory with single active element, *IEEE J. Sel. Top. Quant. Electron.*, 10(5), 1093–1100, 2004.

21. H.J.S. Dorren, D. Lenstra, Y. Liu, M.T. Hill, and G.D. Khoe, Nonlinear polarization rotation in semiconductor optical amplifiers: Theory and application to all optical flip-flop memories, *IEEE J. Quantum Electron.*, 39(1), 141–148, 2003.

22. R. Langenhorst, M. Eiselt, W. Pieper, G. GroBkopf, R. Ludwig, L. Küller, E. Dietrich, and H.C. Weber, Fiber loop optical buffer, *J. Lightwave Technol.*, 14(3), 324–335, 1996.

23. A. Liu, C. Wu, Y. Gong, and P. Shum, Dual-loop optical buffer (DLOB) based on a 3×3 collinear fiber coupler, *IEEE Photon. Technol. Lett.*, 16(9), 2129–2131, 2004.

10

Bound Solitons by Active Phase Modulation Mode-Locked Fiber Ring Lasers

In Chapters 5 and 8, we discussed that a single pulse sequence in the form of a soliton shape can be generated from MLFLs under active modulation of the pulse energy in a ring resonator; this chapter describes the generation of bound solitons in active MLLs. This forms new photonic sources of ultrashort pulse pairs and multiple pulse pairs or groups of solitons for optical transmission or logic operations. In this chapter, we present the experimental generation of stable bound states of multiple solitons in an active MLFL using continuous phase modulation for wideband phase matching of the optical field circulating in the fiber ring cavity. Not only can dual-bound soliton states be established, but triple- and quadruple-soliton pulses can also be established. The simulations of the generated solitons are demonstrated and agree with the experimental results. We also prove by simulation that the experimental relative phase difference and chirping caused by the phase modulation of an LiNbO$_3$ modulator in the fiber loop significantly influences the interaction between the solitons and hence their stability as they circulate in the anomalous path-averaged dispersion fiber loop. Furthermore, we investigate the evolution of solitons to bound soliton states using bi-spectra techniques, i.e., two-dimensional spectral distribution of the amplitude and phases of the complex amplitude optical signals are used to distinguish the solitonic states.

10.1 Introduction

Mode-locked fiber lasers are considered as important laser sources for generating ultrashort soliton pulses. Recently, soliton fiber lasers have attracted significant research interests with the experimental demonstration of the bound states of solitons as predicted in some theoretical works [1,2]. However these bound soliton states have been mostly observed in passive MLFLs [3–6]. There are, however, few reports on bound solitons in active MLFLs. The active mode-locking mechanism offers significant advantage in the control of the repetition rate that would be critical for optical transmission systems. The observation of bound soliton pairs was first reported in a hybrid FM MLFL [7], in which a regime of bound soliton pair harmonically mode locked at 10 GHz could be established. There are, however, no report on multiple bound soliton states in such lasers. Depending on the strength of the soliton interaction, the bound solitons can be classified into two types: loosely bound solitons and tightly bound solitons, depending on the relative phase difference between the adjacent solitons. The phase difference may take the value of π or $\pi/2$ or any value, depending on the fiber laser structure and mode-locking conditions.

In Chapters 5 and 8 of this book, a single pulse sequence in the form of soliton shape was generated from MLFLs under active modulation of the pulse energy in a ring resonator, this chapter describes the generation of bound solitons in active MLLs. We describe the bound states of multiple solitons in an active MLFL using continuous phase modulation (PM) or FM mechanism in the laser ring cavity. By tuning the phase matching of the lightwaves circulating in the fiber loop, we can not only observe the dual-bound soliton state but also the triple- and quadruple-soliton bound states. Relative phase difference and chirping caused by the phase modulation of an $LiNbO_3$ modulator in the fiber loop significantly influences the interaction between the solitons and, hence, their stability as they circulate in the anomalous path-averaged dispersion fiber loop.

This chapter is, thus, organized as follows: Section 10.2 describes the formation of the bound soliton states in the active MLFL. Section 10.3 outlines the simulation of the formulation of the locking mechanism and the active control of the generation of the bound solitons in the active cavity. Section 10.4 describes the bound soliton states. Section 10.5 gives further simulations of the dynamics of the bound solitons and comments on the experimental and simulated results.

10.2 Formation of Bound States in an FM Mode-Locked Fiber Ring Laser

Although the formation of the stable bound soliton states determined by the Kerr effect and anomalous-averaged dispersion regime has been discussed in a number of passive MLFLs [4–6], it can be quite comparatively distinct in our active MLFL with the contribution of the phase modulation of the $LiNbO_3$ modulator incorporated in the ring cavity in order to establish the phase-matching conditions of the lightwaves circulating and the resonant condition of the ring cavity. Furthermore, this phase-matching mechanism contributes to the stabilization of bound states. The formation of bound soliton states in an FM MLL can, thus, be experienced through two phases, first, a process of pulse splitting and second, the stabilization of multi-soliton bound states in the presence of a phase modulator in the cavity of an MLL.

In the initial phase, the splitting of a single pulse into multi pulses is formed when the power in the fiber loop reaches above a specific mode-locking threshold [7,8]. Under a higher optical energy circulating in the ring, normally above the saturation level of optical amplification of the ring, higher-order solitons can exit. Furthermore, the accumulated nonlinear phase shift in the loop is sufficiently high for a single pulse to break up into many pulses [9]. The number of split pulses depends on the optical power in the loop, so there is a specific range of power for each splitting level. The fluctuation of pulses may occur at the transitional region of power where the transient occurs from the lower splitting to the higher level. Moreover, the chirping caused by a phase modulator in the loop makes the process of pulse conversion from a chirped single pulse into multi-pulses more easy [10,11].

After splitting into multi pulses, the multi-pulse bound states are stabilized subsequently through the balance of the repulsive and attractive forces between neighboring pulses during circulation in a fiber loop of anomalous-averaged dispersion. The repulsive force comes from direct soliton interaction subject to the relative phase difference between neighboring pulses [12]. Furthermore, the effectively attractive force comes from the variation of group velocity of soliton pulse caused by frequency chirping [13,14]. Thus, in an anomalous-averaged dispersion regime, the locked pulses can be located symmetrically around the extreme of positive phase modulation half-cycle. In other words, the bound soliton pulses acquire an up chirping when passing through the phase modulator. Besides the optical power level and dispersion of the fiber cavity, the modulator-induced chirp or the phase modulation index determine not only the pulse width but also the time separation of bound soliton pulses at which the interaction mechanisms cancel out each other.

The presence of a phase modulator in the cavity to balance the effective interactions among the bound soliton pulses is similar to the use of this device in a long-haul soliton transmission system to reduce the timing jitter. Thus, a simple perturbation theory can be applied to model the role of phase modulation on the mechanisms of bound solitons formation. The envelope of the pulses of a multi-soliton bound state can be written as

$$u_{bs} = \sum_{i=1}^{N} u_i(z,t) \tag{10.1}$$

with

$$u_i = A_i \mathrm{sec}\, h\left\{ A_i\left[\frac{(t - T_i)}{T_0} \right] \right\} \exp(j\theta_i - j\omega_i t) \tag{10.2}$$

where
   N is the number of solitons in the bound state
   T_0 is the pulse width of the soliton
   A_i, T_i, θ_i, ω_i represent the amplitude, position, phase, and frequency of the
      soliton, respectively

In the simplest case of a multi-soliton bound state, N is equal 2 or the dual-
bound soliton state with the identical amplitude of pulse and the phase dif-
ference of π value ($\Delta\theta = \theta_{i+1} - \theta_i = \pi$), the ordinary differential equations for the
frequency difference and the pulse separation can be derived by using the
perturbation method [13,14]

$$\frac{d\omega}{dz} = -\frac{4\beta_2}{T_0^3}\exp\left[-\frac{\Delta T}{T_0}\right] - 2\alpha_m\Delta T \tag{10.3}$$

$$\frac{d\Delta T}{dz} = \beta_2\omega \tag{10.4}$$

where
   β_2 is the averaged group-velocity dispersion of the fiber loop
   ΔT is the pulse separation between the two adjacent solitons ($T_{i+1} - T_i = \Delta T$)
      and $\alpha_m = m\omega_m^2/(2L_{cav})$
   L_{cav} is the total length of the loop
   m is the phase modulation index
   ω_m is the modulating frequency

Equations 10.3 and 10.4 show the evolution of frequency difference and position of bound solitons in the fiber loop in which the first term on the right
hand side represents the accumulated frequency difference of two adjacent
pulses during a round trip of the fiber loop and the second one represents
the relative frequency difference of these pulses when passing through the
phase modulator. In the steady state, the pulse separation is constant and the
induced frequency differences are nullified. Thus, by setting Equation 10.3
to zero, we have

$$-\frac{4\beta_2}{T_0^3}\exp\left[-\frac{\Delta T}{T_0}\right] - 2\alpha_m\Delta T = 0 \tag{10.5}$$

or

$$\Delta T\exp\left[\frac{\Delta T}{T_0}\right] = -\frac{4\beta_2}{T_0^3}\frac{L_{cav}}{m\omega_m^2} \tag{10.6}$$

Thus, through (10.6), we can see the effect of phase modulation on the pulse separation; in addition, β_2 and α_m must have opposite signs indicating that in an anomalous dispersion fiber loop with negative value of β_2, the pulses are up chirped. Under a specific operating condition of the FM fiber laser, when the magnitude of chirping is increased, the bound-pulse separation will decrease subsequently. The pulse width would also reduce correspondingly to the increase in the phase modulation index and the modulation frequency, so that the change of the ratio $\Delta T/T_0$ can be infinitesimal. Thus, the binding of solitons in the FM MLFL is assisted by the phase modulator incorporated in the fiber ring. Bound solitons in the loop periodically experience the frequency shift and, hence, their velocity in response to changes in their temporal positions by the interactive forces of the equilibrium state.

10.3 Experimental Technique

Figure 10.1 shows the experimental setup of an FM MLFL. Two EDFAs pumped at 980 nm are used in the fiber loop to moderate the optical power in the loop for mode locking. Both amplifiers operate in the saturation region. The phase modulator driven in the region of 1 GHz frequency assumes the role as a mode locker and controls the states of locking in the fiber ring. At the input of the phase modulator, a polarization controller (PC) consisting of two quarter-wave plates and one half-wave plate is used to control the polarization of light, which is related to the nonlinear polarization evolution and influences to the multi-pulse operation in formation of the bound soliton states.

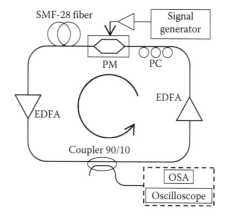

FIGURE 10.1

Experimental setup of the FM MLFL. (PM, phase modulator; PC, polarization controller; OSA, optical spectrum analyzer.)

A 50 m Corning SMF-28 fiber can be inserted after the phase modulator to ensure that the average dispersion in the loop is anomalous. The fundamental frequency of the fiber loop is 1.7827 MHz that is equivalent to the 114 m total loop length. The outputs of the MLL from the 90:10 coupler are monitored by an optical spectrum analyzer (HP 70952B) and an oscilloscope (Agilent DCA-J 86100C) with an optical bandwidth of 65 GHz.

Under normal dispersion conditions, the single-pulse mode-locking operation is achieved at an average optical power of 5 dBm and a modulation frequency of 998.315 MHz (i.e., harmonic mode locking at the 560th order), as shown in Figure 10.2. Pulse train of 8–14 ps width, depending on the RF driving power of the phase modulator was observed on a sampling oscilloscope. The measured pulse spectrum has a spectral shape of a soliton rather than a Gaussian pulse.

By adjusting the polarization states of the PC's wave plates at higher optical power, dual-bound solitons or bound soliton pairs can be observed at an average optical power circulating inside the fiber loop of about 10 dBm. Figure 10.3a shows a typical time-domain waveform and, Figure 10.3b the

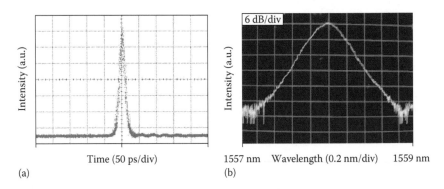

FIGURE 10.2
(a) The oscilloscope trace and (b) optical spectrum of a single soliton.

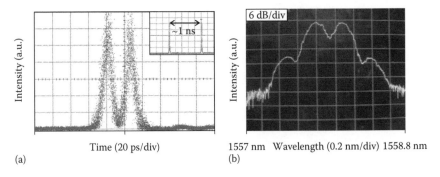

FIGURE 10.3
(a) The oscilloscope trace (inset: the periodic bound soliton pairs in the time domain) and (b) optical spectrum of a dual-bound soliton state.

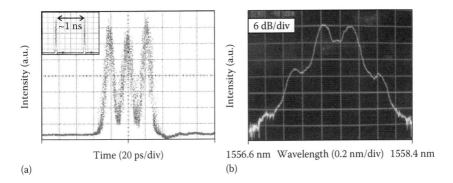

FIGURE 10.4
(a) The oscilloscope trace (inset: periodic groups of a triple-bound soliton in the time domain) and (b) optical spectrum of a triple-bound soliton state.

corresponding spectrum of the dual-bound soliton state. The estimated FWHM pulse width is about 9.5 ps and the temporal separation between two bound pulses is 24.5 ps, which are correlated exactly to the distance between two spectral main lobes of 0.32 nm of the observed spectrum. When the average power inside the loop is increased to about 11.3 dBm and a slight adjustment of the polarization controller is performed, the triple-bound soliton state occurs as shown in Figure 10.4 where the FWHM pulse width and the temporal separation of two close pulses are slightly less than 9.5 and 22.5 ps, respectively. Insets in Figures 10.3a and 10.4a show the periodic sequence of bound solitons at a repetition rate exactly equal to the modulation frequency. This feature is quite different from that in a passive MLFL in which the positions of bound solitons is not stable and the direct soliton interaction causes a random movement and phase shift of bound soliton pairs [4]. Therefore, it is advantageous to perform a stable periodic bound soliton sequence in an FM MLFL.

The symmetrical shapes of the optical modulated spectra in Figures 10.3b and 10.4b clearly indicate that the relative phase difference between two adjacent bound solitons is a π value. At the center of the spectrum, there is a dip due to the suppression of the carrier with a π phase difference in the case of a dual-bound soliton state. While, there is a small hump in the case of a triple-bound soliton state because three soliton pulses are bound together with the first and the last pulses in phase and out of phase with the middle pulse. The spectral shape will change, which can be symmetrical or asymmetrical when this phase relationship varies. We believe that the bound state with the relative phase relationship of π between solitons is the most stable in our experimental setup and this observation also agrees with the theoretical prediction of stability of bound soliton pairs relating to photon-number fluctuation in different regimes of phase difference [15].

Similar to the case of a single soliton, the phase coherence of bound solitons is still maintained as a bound unit when propagating through a dispersive

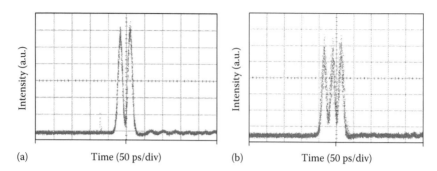

FIGURE 10.5
The oscilloscope traces of (a) the dual-bound soliton state and (b) the triple-bound soliton state after propagating through 1 km of SMF.

medium. Figure 10.5 shows the dual-bound soliton state and triple-soliton state waveforms on the oscilloscope after propagating through 1 km of SMF. There is no difference in the observed spectral shapes for both cases.

To prove that the multi-soliton bound operation can be formed in an FM MLFL by operating at a critical optical power level and using a phase modulator–induced chirp in a specific fiber loop of anomalous average dispersion, we increased the average optical power in the loop to a maximum of 12.6 dBm and decreased the RF driving power of 15 dBm. It is really amazing that the quadruple-soliton state was generated, as observed in Figure 10.6. The bound state occurring at a lower phase modulation index is due to the maintenance of a sufficiently small frequency shifting in a wider temporal duration of bound solitons to hold the balance between interactions of the group of four solitons in the fiber loop. However, the optical power is still not sufficient to stabilize the bound state, as well as the effects of the amplified stimulated emission noise from EDFAs is larger at a higher pump power. Therefore the time-domain waveform of the quadruple-soliton bound state has more noise and its spectrum can only be seen clearly where two main

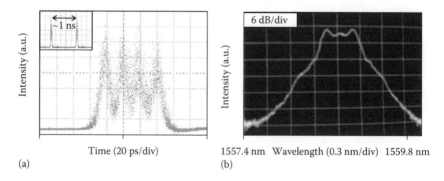

FIGURE 10.6
(a) The oscilloscope trace and (b) optical spectrum of a quadruple-soliton bound state.

lobes are inversely proportional to the temporal separation of pulses of about 20.5 ps. The pulse separation reduces when the number of pulses in the bound states increases.

Thus, through the experimental results, both the phase modulation index and the cavity's optical power influence the existence of the multi-soliton bound states in the FM MLFL. The optical power determines the splitting of a single mode-locked pulse into a number of pulses and maintains the pulse shape in the loop. Observations from the experiment indicate that there is a threshold of optical power for each bound level. At this threshold value, the bound solitons show strong fluctuations in amplitude and oscillations in position considered as a transient state between different bound levels. Furthermore, even the collisions of the adjacent pulses occur, as shown in Figure 10.7. The phase difference between the adjacent pulses can also change in these unstable states, indicating that the neighboring pulses are no longer out of phase but in phase, as observed through the measured spectrum shown in Figure 10.8. Although the decrease in the phase modulation

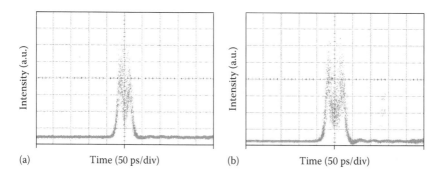

(a) Time (50 ps/div)   (b) Time (50 ps/div)

FIGURE 10.7
The oscilloscope traces of (a) the dual-bound soliton state at an optical power threshold of 8 dBm and (b) the triple-bound soliton state at 10.7 dBm.

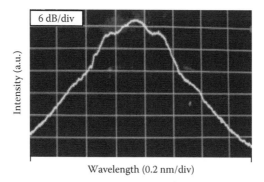

Wavelength (0.2 nm/div)

FIGURE 10.8
The spectrum of the bound soliton state with in-phase pulses.

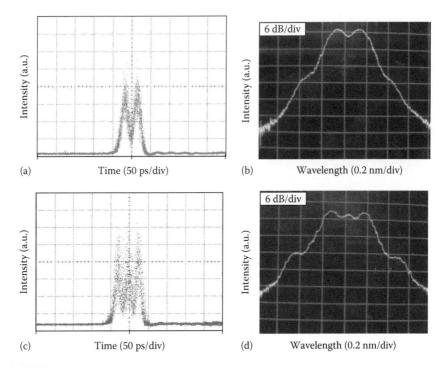

FIGURE 10.9
The oscilloscope traces of (a) the dual-bound soliton state and (c) the triple-bound soliton state and (b) and (d) their spectra, respectively, at low RF-driving power of 11 dBm.

index is required to maintain the stability at a higher bound soliton level, it increases the pulse width and reduces the peak power. So the observed waveform has much more noise and is more sensitive to the ambient environment and its spectrum is not strongly modulated, as shown in Figure 10.9.

10.4 Dynamics of Bound States in an FM Mode-Locked Fiber Ring Laser

10.4.1 Numerical Model of an FM Mode-Locked Fiber Ring Laser

To understand the dynamics of an FM bound soliton fiber laser, a simulation technique is developed to observe the evolution of the pulses and bound soliton states in the FM MLL loop. A recirculation model of the fiber loop is used to simulate the propagation of the bound solitons in the fiber cavity. The model consists of basic components of an FM MLFL, as shown in Figure 10.10. In other words, the cavity of the FM MLFL is modeled as a sequence

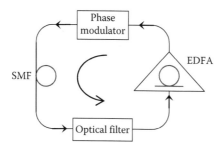

FIGURE 10.10
A circulating model for simulating the FM MLFRL. (SMF, standard single-mode fiber; EDFA, erbium-doped fiber amplifier.)

of differential elements. The optical filter has a Gaussian transfer function with a 3 dB spectral width of 2.4 nm. The time-domain pulse response of the phase modulator to a sinusoidal input is $u_{out} = u_{in} \exp[jm\,\cos(\omega_m t)]$ where $j = \sqrt{-1}$, m is the phase modulation index and $\omega_m = 2\pi f_m$ is the angular modulation frequency. It is assumed that the response follows a harmonic of the fundamental frequency of the fiber loop. The pulse propagation through the optical fiber is governed by the NLSE [12]:

$$\frac{\partial u}{\partial z} + j\frac{\beta_2}{2}\frac{\partial^2 u}{\partial T^2} - \frac{\beta_3}{6}\frac{\partial^3 u}{\partial T^3} + \frac{\alpha}{2}u = j\gamma |u|^2 u \tag{10.7}$$

where
  u is the complex envelop of an optical pulse sequence
  β_2 and β_3 account for the second- and third-order fiber dispersions
  α and γ are the loss and nonlinear parameters of the fiber, respectively

The amplification of signal including the saturation of the EDFA can be represented as [12]

$$u_{out} = \sqrt{G}u_{in} \tag{10.8}$$

with

$$G = G_0 \exp\left(-\frac{G-1}{G}\frac{P_{out}}{P_s}\right) \tag{10.9}$$

where
  G is the optical amplification factor
  G_0 is unsaturated amplifier gain
  P_{out} and P_{sat} are output power and saturation power, respectively

TABLE 10.1

Parameter Values Used in the Simulations

$\beta_2^{SMF} = -21 \text{ ps}^2/\text{km}$	$\beta_2^{ErF} = 6.43 \text{ ps}^2/\text{km}$	$\Delta\lambda_{filter} = 2.4 \text{ nm}$
$\gamma^{SMF} = 0.0019/\text{W/m}$	$\gamma^{ErF} = 0.003/\text{W/m}$	$f_m \approx 1 \text{ GHz}$
$\alpha^{SMF} = 0.2 \text{ dB/km}$	$P_{sat} = 7 \div 13 \text{ dBm}$	$m = 0.1\pi \div 1\pi$
$L_{cav} = 115 \text{ m}$	$NF = 6 \text{ dB}$	$\lambda = 1558 \text{ nm}$

SMF, standard single-mode fiber; ErF, erbium-doped fiber; NF, noise figure of the EDFA.

The difference of the dispersion between the SMF and the Er-doped fiber in the cavity results in a certain dispersion map. The fiber loop gets a positive net dispersion, an anomalous-averaged dispersion, which is important in the formation of "soliton-like" pulses in an FM fiber laser. Basic parameter values used in the simulations are listed in Table 10.1.

10.4.2 The Formation of the Bound Soliton States

First, we simulate the formation process of bound states in the FM MLFL whose parameters are those employed in the experiments described above. The lengths of the Er-doped fiber and the SMF are chosen to obtain the cavity's average dispersion $\bar{\beta}_2 = -10.7 \text{ ps}^2/\text{km}$. The Schrödinger equation is applied with initial conditions that the random optical noises are the initial amplitudes of the waves, which circulate around the ring via this propagation equation. The optical power is built up with the phase-matching conditions and, hence, the formation of the solitons. Figure 10.11 shows a simulated dual-bound soliton state building up from the initial Gaussian-distributed noise as an input seed over the first 2000 round trips with P_{sat} value of 10 dBm

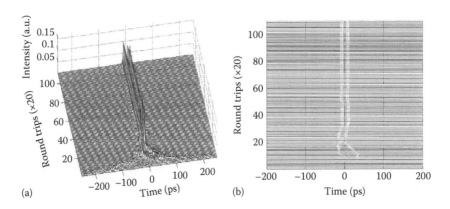

(a)   (b)

FIGURE 10.11
(a) Numerically simulated dual-bound soliton state formation from noise and (b) the evolution of the formation process in contour plot view.

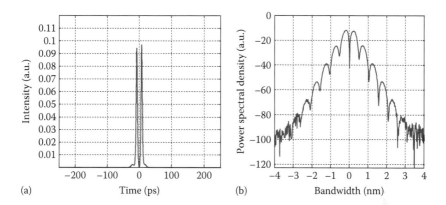

FIGURE 10.12
(a) The waveform and (b) the corresponding spectrum of the simulated dual-bound soliton state at the 2000th round trip.

and G_0 of 16 dB and m of 0.6π. The built-up pulse experiences transitions with large fluctuations of intensity, position, and pulse width during the first 1000 round trips before reaching the stable bound steady state. Figure 10.12 shows the time-domain waveform and the spectrum of the output signal at the 2000th round trip.

The bound states with higher number of pulses can be formed at higher gain of the cavity, hence, when the gain G_0 is increased to 18 dB, which is enhancing the average optical power in the loop, the triple-bound soliton steady-state is formed from the noise seeded via simulation, as shown in Figure 10.13. In the case of higher optical power, the fluctuation of signal at initial transitions is stronger and it needs more round trips to reach a more stable bound state of the three pulses. The waveform and spectrum of the

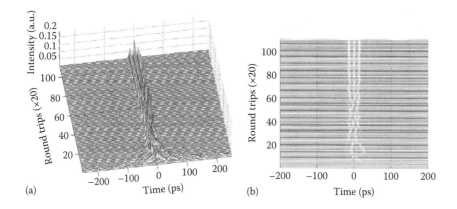

FIGURE 10.13
(a) Numerically simulated triple-bound soliton state formation from noise and (b) the evolution of the formation process in contour plot view.

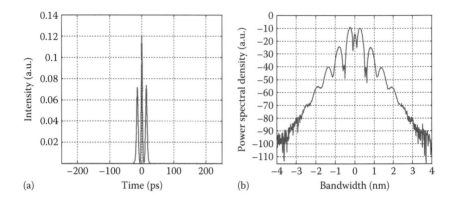

FIGURE 10.14

(a) The waveform and (b) the corresponding spectrum of the simulated triple-bound soliton state at the 2000th round trip.

output signal from the FM mode-locked fiber laser at the 2000th round trip are shown in Figure 10.14a and b, respectively.

Although the amplitude of pulses is not equal, indicating the bound state can require a larger number of round trips before the effects in the loop balance, the phase difference of pulses accumulated during circulation in the fiber loop is approximately of π value that is indicated by strongly modulated spectra. In particular from the simulation result, the phase difference between adjacent pulses is 0.98π in the case of the dual-pulse bound state and 0.89π in case of the triple-pulse bound state. These simulation results agree with the experimental results (shown in Figures 10.3b and 10.4b) discussed above to confirm the existence of multi-soliton bound states in an FM MLFL.

10.4.3 Evolution of the Bound Soliton States in the FM Fiber Loop

The stability of bound states in the FM MLFL strongly depends on the parameters of the fiber loop, which also determine the formation of these states. As stated above, besides the phase modulation and the GVD, the optical power circulating in the ring cavity also influences the existence of the bound states. The effects of active phase modulation and optical power can be simulated to observe the dynamics of bound solitons. Instead of the noise seed, the multi-soliton waveform following Equations 10.1 and 10.2 are used as an initial seed for simplifying our simulation processes. Initial bound solitons are assumed to be identical with the phase difference between adjacent pulses of π value.

First, the effect of the phase modulation index on the stability of the bound states is simulated through a typical example of the evolution of a dual-bound soliton state over 2000 round trips in the loop at different phase modulation indexes with the same saturation optical power of 9 dBm, as shown in Figure 10.15. The simulation results also indicate that the pulse separation

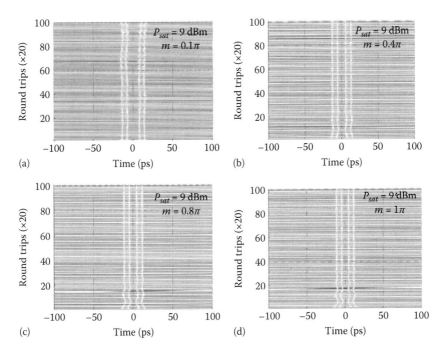

FIGURE 10.15
Simulated evolution of a dual-bound soliton state over 2000 round trips in the fiber loop at different phase modulation indexes: (a) $m = 0.1\pi$, (b) $m = 0.4\pi$, (c) $m = 0.8\pi$, and (d) $m = 1\pi$.

decreases due to the increase in the modulation index. In the initial round trips, there is a periodic oscillation of bound solitons that is considered as a transition of solitons to adjust their own parameters to match the parameters of the cavity before reaching a finally stable state. Simulations in other multi-soliton bound states also manifest this similar tendency. The periodic phase modulation in the fiber loop is not only to balance the interactive forces between solitons but also to maintain the phase difference of π between them. At a minute modulation index, the phase difference changes or reduces slightly after many round trips or, in other words, the phase coherence is no longer held leading to the amplitude oscillation. This is due to an alternative periodic exchange of energy between solitons, as shown in Figure 10.16. The higher the number of solitons in bound state, the more sensitive they are to the change in the phase modulation index. The modulation index determines time separation between the bound pulses, yet, at too high modulation index, it is more difficult to balance the effectively attractive forces between solitons, especially when the number of solitons in the bound state is larger. The enhancement of chirping leads to faster variation in the group velocity of pulses passing through the phase modulator, and this creates the periodic oscillation of the temporal pulse position. Figure 10.17 shows the evolution of the triple-bound soliton state at a multiple value

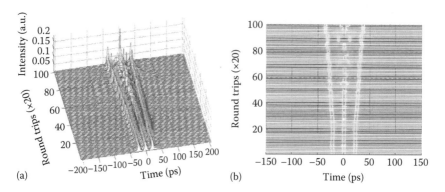

FIGURE 10.16
The simulated evolution of a triple-bound soliton state over 2000 round trips using a small
phase modulation index $m = 0.1\pi$. (a) Three-dimensional plot and (b) contour views.

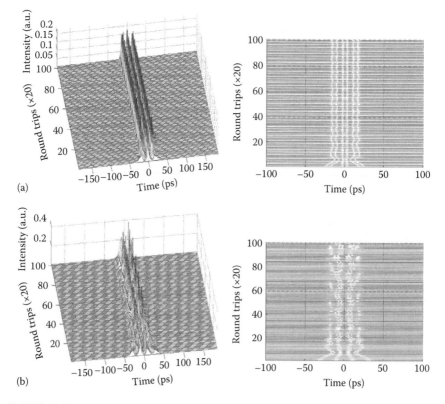

FIGURE 10.17
Simulated evolutions over 2000 round trips in the fiber loop of (a) triple-bound soliton state at
$m = 1\pi$ and (b) quadruple-soliton bound state at $m = 0.7\pi$, three-dimensional plot and contour
views.

of 1π and the quadruple-soliton bound state at the m value of 0.7π. In case of the quadruple-soliton bound state, solitons oscillate strongly and tend to collide with each other.

Another factor is the optical power of the fiber loop, it plays an important role not only in the determination of multi-pulse bound states as shown in the above simulation but also in the stabilization of the bound states circulating in the loop. As mentioned in Section 10.3, each bound state has a specific range of operational optical power. In our simulations, dual-, triple-, and quadruple-soliton bound states are in the stable evolution in the loop when the saturated power P_{sat} is about 9, 11, and 12 dBm, respectively. When the optical power of the loop is not within these ranges, the bound states become unstable and they are more sensitive to the changes in the phase modulation index. At a power lower than the threshold, the bound states are out of bound and switched to lower level of bound state. In contrast, at high power region, the generated pulses are broken into random pulse trains or decay into radiation. Figure 10.18 shows the unstable evolution of the triple-bound

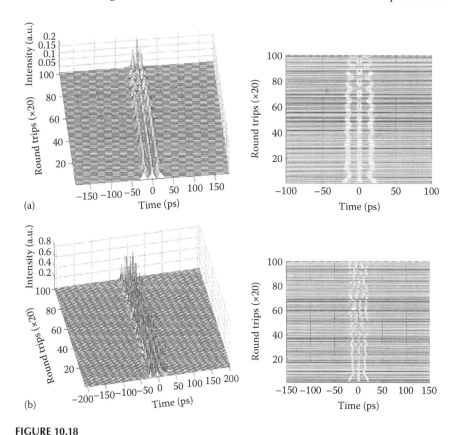

FIGURE 10.18
The simulated behavior of a triple-bound soliton state at non-optimized power (a) $P_{sat} = 9$ dBm, $m = 0.7\pi$ and (b) $P_{sat} = 13$ dBm and $m = 0.6\pi$, three-dimensional plot and contour views.

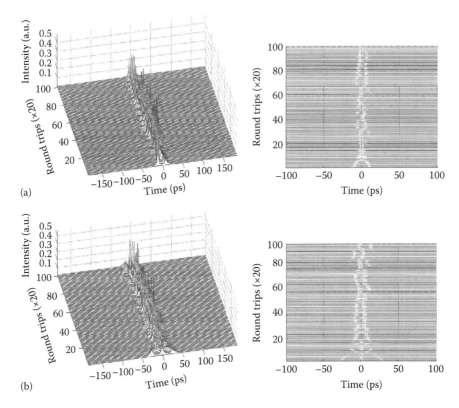

FIGURE 10.19
The time-domain behavior of (a) dual- and (b) triple-bound soliton states, respectively, with in-phase pulses. Three-dimensional plot and contour views.

soliton states under non-optimized operating conditions of the loop as an example.

Different operating conditions of the phase difference between bound solitons are also simulated. The states of non-phase difference often behave unstably and can easily be destroyed, as displayed in Figure 10.19. The simulation results agree well with the experimental observations as discussed above.

10.5 Multi-Bound Soliton Propagation in Optical Fiber

Multi-bound soliton states are generated using an active FM MLFL shown in Figure 10.20, which is similar to the setup shown in Figure 10.1. However, in this chapter, the generator employs optical amplifiers (EDFAs), of high output

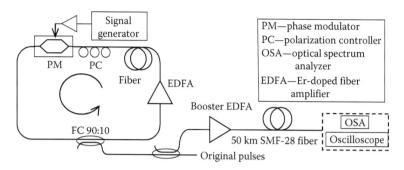

FIGURE 10.20
Experimental setup of the FM MLFL.

saturated power of 16 dBm. By optimizing the locking conditions of the fiber ring, various multi-bound solitons with ultra-high stability from dual- to quintuple-states can be generated. These states then propagate through a standard single-mode optical fiber in order to investigate their propagation dynamics. The estimated FWHM of an individual pulse of the bound soliton pairs (BSP), triple-bound solitons (TBS) and quadruple-bound solitons (QBS) are 7.9, 6.9, and 6.0 ps, respectively. The time separation between the two adjacent pulses is about three times of FWHM pulse width. The repetition rate of these multi-bound soliton sequences is about 1 ns. As shown in Figure 10.20, the sequences are then optically amplified by a booster EDFA at the output of a 90:10 coupler before launching into a spool of 50 km Corning SMF-28 fiber. The fiber propagation time is equivalent to about 50 soliton periods, which is related to the dispersion length L_D. The launched power of the multi-bound solitons can be adjusted from 2 to 16.5 dBm by tuning the pump power of the EDFA. The states of the multi-bound solitons at the output of the fiber length are monitored using an optical spectrum analyzer and the optical port of a 70 GHz Agilent signal analyzer.

It is obvious but worth mentioning that the binding property of the bound solitons in the fiber ring of Figure 10.20 is different under the case when they are propagating through the optical fiber. In the propagation, they would be interacting with each other and they are no longer supported by the periodic phase modulation when circulating in the ring laser. When propagating through a dispersive fiber, the optical carrier under the envelope of the generated multi-bound solitons is influenced by the chirping effects; hence, overlapping between the soliton pulses. When the accumulated phase difference of the adjacent pulses is π, the solitons will repel each other. In other words, solitons in multi-bound states acquire a down-chirp during propagation through the SMF due to the GVD-induced phase shift. Hence, within the group of multi-bound solitons, the front-end soliton would travel with a higher positive frequency shift; thus a higher group velocity than those at back-end of the group is influenced with a

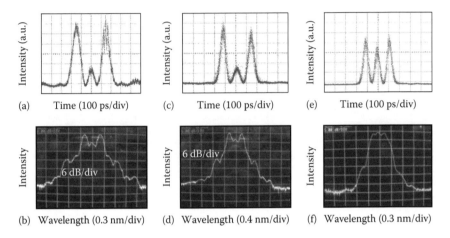

FIGURE 10.21
The oscilloscope traces and corresponding spectra of a triple-bound soliton at launching powers of (a), (b) 4.5 dBm, (c), (d) 10.5 dBm, and (e), (f) 14.5 dBm, respectively.

lower negative frequency shift. Therefore, the time separation between the adjacent pulses varies with the propagation distance. The variation of time separation between pulses depends on their relative frequency difference. Figure 10.21 shows the waveforms and their corresponding spectra of triple-bound solitons after propagating over 50 km fiber at different launching powers.

In all cases, the time separation significantly increases compared to that of the initial state due to the repulsion effect. However, propagation in a real optical fiber is considerably influenced by perturbations such as the fiber loss and initial launching powers (P_l) of solitons. Figure 10.22a and b show the dependence of the pulse width and time separation of various multi-bound solitons with the launched power as a parameter. Similar dynamics are observed for dual, quadruple, and quintuple groups of solitons. In general, an increase in the launching power leads to a shortening of the pulse width and a reduction of the pulse's temporal separation due to the enhancement of the nonlinear self-phase-modulation phase shift. The variation of the temporal separation with the launched power is shown in Figure 10.22. Two distinct areas of the curve can be observed as separated at point A. The intersection point A of these two distinct curves corresponds to the average soliton power (P_{sol}) of a bound state. For transmission over a standard SMF, it is observed that with our experimental parameters of the multi-bound solitons, the estimated average soliton powers for BSP, TBS, and QBS are 11, 13.5, and 15.2 dBm, respectively. For P_l lower than P_{sol} and in addition to the fiber attenuation, the power of multi-bound soliton is not sufficient to balance the GVD effect. As a result, the pulses are rapidly broadened by the self-adjustment of the multi-bound solitons due to the perturbations accompanied by partly shading their energy in

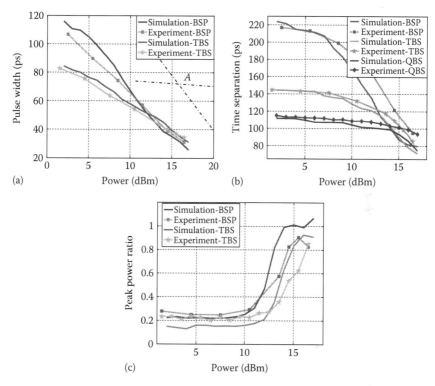

FIGURE 10.22
The launching power–dependent variation of pulse width (a) and time separation (b) and (c) peak power ratio of different multi-soliton bound states over 50 km propagation distance.

form of dispersive waves. The broadening of such a dispersive wave can be accumulated along the transmission fiber and form oscillations around the multi-bound solitons, as shown by the ripple of the tail of the soliton group in Figure 10.21a. Furthermore, the rate of broadening of the pulses is faster than that of the temporal separation at low powers, leading to the enhancement of the overlapping between the pulses. Therefore, the pulse envelope is consequently modulated by the interference of waves with the phase modulation effect due to the GVD. Because of the parabolic symmetry of the anomalous GVD-induced phase-shift profile around multi-bound solitons, the energy of inner pulses is shifted to outer pulses, resulting in a decrease in the amplitude of inner pulses. In contrast, at a higher P_l, the SPM phase shift is increased to balance the GVD effect. The pulse width is narrower and the ripple of the pulse envelope is lower with a higher level of the launched power. For the bound states with the number of solitons greater than two, there is a jump of the peak power ratio between the inner and the outer pulses at the average soliton power, as shown in Figure 10.22c.

To verify the experimental results as well as to identify the evolution of the multi-bound solitons propagating in the fiber, we model the multi-bound solitons as $u_{bs} = \sum_{i=1}^{N} u_i(z,t)$ where N is number of solitons in bound state, $u_i = A_i \sec h\{[t - q_0]\}\exp(j\theta_i)$, A_i and q_0 are the amplitude and time separation of the solitons, respectively, and the phase difference $\Delta\theta = \theta_{i+1} - \theta_i = \pi$. The propagation of multi-bound solitons in an SMF is governed by the NLSE 10.7 with the input parameters as those obtained in the experiment. Shown together with the experimental results in Figure 10.22a through c is the simulation of the evolution of the pulse width (solid curves) over 50 km propagation of various orders of multi-soliton bound states. The simulated results agree well with those obtained in the experiment. Shown in Figures 10.21 and 10.23 is the evolution of the parameters of the triple-bound solitons along the transmission fiber with different launching powers. The difference between lower and higher launching powers also obviously exhibit in the simulation results. When the launching power is far from the soliton

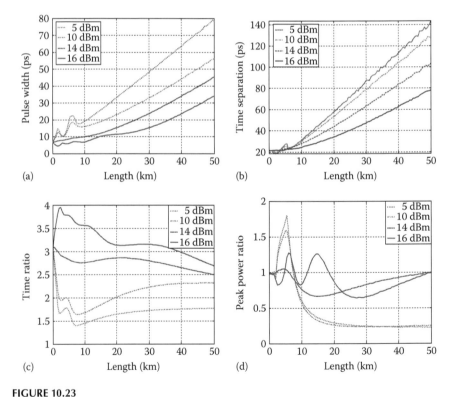

FIGURE 10.23
Evolution of a numerically simulated triple-bound soliton state after 50 km propagation distance through the SMF-28 fiber: (a) pulse width, (b) temporal separation, (c) ratio between pulse width and separation, and (d) peak power ratio between the inner pulse and the outer pulse.

power, there is oscillation or rapid variation of the parameters at an initial propagation distance due to the adjustment of multi-bound solitons to perturbations of propagation conditions as mentioned above. The pulses are compressed for P_l higher than P_{sol}, while they are rapidly broadened at P_l higher than P_{sol}. The slow variation of parameters occurs at P_l close to P_{sol}. However, the time separation of solitons remains unchanged in the propagation distance of one soliton period. We have validated this prediction in our experiment.

Another important property of multi-bound solitons is the phase difference between pulses that can be determined by the shape of the spectrum. We have monitored the optical spectrum of the multi-bound solitons at both the launched end and at the output of the fiber length (see Figure 10.24). The modulated spectrum of the multi-bound solitons is symmetrical with the carrier suppression due to a relative phase relationship of π between the adjacent solitons. At low P_l, the spectrum of the multi-bound solitons over 50 km propagation is nearly same as that at the launched end (see Figure 10.21b). The modulation of the spectrum is modified with an increase in P_l. At sufficiently high P_l, the nonlinear phase shift–induced chirp is increased at the edges of the pulses. Although, the nonlinear phase shift reduces the GVD effect, the phase transition between adjacent pulses is changed due to the direct impact of the nonlinear phase shift. Hence the small humps in the spectra of the multi-bound states are strengthened and they may be comparable to the main lobes due to the enhancement of the far interaction between the pulses, as shown in Figure 10.21f.

This section has experimentally and numerically investigated the propagation behavior of the multi-bound solitons in optical fibers. Multi-bound solitons have been under the influence of a strong repulsive interaction that is varied along the propagation distance. The changes in the soliton parameters (e.g., pulse width, soliton separation, and spectrum) depend on the launching powers, which can be lower or higher than the soliton power.

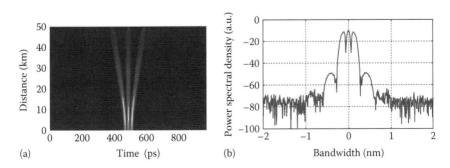

(a)   Time (ps)   (b)   Bandwidth (nm)

FIGURE 10.24
(a) The oscilloscope trace and (b) optical spectrum of a quadruple-soliton bound state.

10.6 Bi-Spectra of Multi-Bound Solitons

As identified in Sections 10.4 and 10.5 and in Chapters 5 and 8 on ultrashort pulses, the generation of such short pulses depends on the phase states of the lightwave in the ring. Even more important is the evolution of the phases of the solitons during its transition to multi-bound soliton states. This section gives a brief introduction to the use of the bi-spectrum concept for the identification of the phase transition states of multi-bound solitons.

Bi-spectrum is the Fourier transform of a triple correlation function, which represents the soliton pulses. Unlike the power spectrum studied and available in several textbooks, the bi-spectrum preserves the phase information and uniquely represents a given process in the frequency domain. The bi-spectral properties of the soliton pulses' continuous time domain are studied. This makes the bi-spectrum a powerful tool for differentiating various nonlinear system responses.

The power spectrum is a common representation of observations from linear to nonlinear dynamical systems. The power spectrum shows the distribution of power for each frequency component, while the bi-spectrum is a complex measure representing the frequency components in two dimensions. It contains information regarding deviations from the Gaussian distribution and the presence of nonlinearities [16]. The bi-coherence spectrum, which is a normalized bi-spectrum, has been successfully applied in the study of wave phenomena in plasma [17], fluctuation of water depth in the ocean [18,19], and pattern recognition [20]. The generation of bi-spectra is computationally intensive and the interpretation of results is complex.

10.6.1 Definition

The bi-spectrum $B(f_1, f_2)$ is defined as the Fourier transform of a triple correlation function $c_{xxx}(\tau_1, \tau_2)$ where

$$c_{xxx} = E\{x(t)x(t+\tau_1)x(t+\tau_2)\} \tag{10.10}$$

represents the triple correlation function and

$$B(f_1, f_2) = \int_{-\infty}^{\infty} \int_{-\infty}^{\infty} c_{xxx}(\tau_1, \tau_2) e^{-2\pi j(f_1\tau_1 + f_2\tau_2)} \, d\tau_1 \, d\tau_2 \tag{10.11}$$

defines the bi-spectrum where $j = (-1)^{1/2}$ and $E\{\cdot\}$ is the expectation operator. The discrete bi-spectrum is similarly defined as

$$B(f_1, f_2) = \sum_{p=-\infty}^{\infty} \sum_{p=-\infty}^{\infty} c_{xxx}(\tau_1, \tau_2) e^{-2\pi j(f_1 p\tau + pf_2\tau)} \tag{10.12}$$

Various symmetries are evident:

$$B(f_1, f_2) = B(f_2, f_1) = B^*(-f_2, -f_1) = B^*(-f_1, -f_2)$$

$$= B^*(-f_1 - f_2, -f_2) = B^*(f_1, -f_1 - f_2)$$

$$= B^*(-f_1 - f_2, f_1) = B^*(f_2, -f_1 - f_2) \tag{10.13}$$

where * represents the complex conjugation. Therefore the bi-spectrum needs to be calculated only for the region $0 \le f_2 \le f_1 \le f_1 + f_2 \le (1/2\tau)$, where τ is the sampling interval of the time series, as shown in Figure 10.25.

A typical single-soliton pulse and its classical power spectrum are shown in Figures 10.26 and 10.27, while Figures 10.28 and 10.29 show the amplitude and phase distribution in the frequency planes f_1 and f_2. The inner region indicates a consistency of the classical power spectrum shown in Figure 10.27. While, the outer spectral regions of the bi-spectrum indicate the noise and interactions of the frequency components of the power spectral components and noises generated in the soliton profile. The evolution of the single soliton toward the double- and triple-bound solitons in terms of amplitude and phase of the bi-spectra is shown in Figures 10.30 through 10.34. The pulse is shortened before reaching a pair of bounded twin pulses, as shown in Figure 10.30. This shows the evolution of the phases of a single soliton pulse to the bounding of the two solitons. Furthermore, the phasor spectral amplitude and the phase of this complex pulse are given in Figures 10.31 and 10.32. Comparing Figures 10.29 and 10.32, there is a periodic contour near the central region.

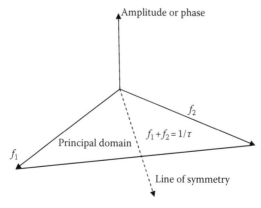

FIGURE 10.25
Principal domain of the bi-spectrum.

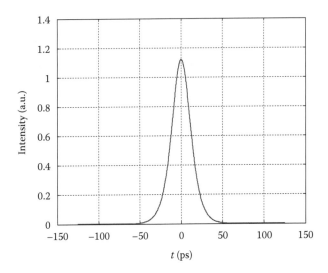

FIGURE 10.26
A typical short pulse of soliton shape.

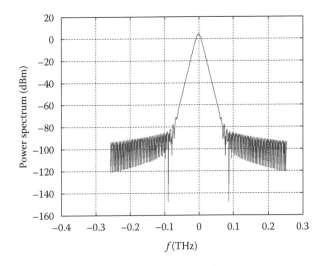

FIGURE 10.27
Classical spectrum of a typical very short soliton pulse sequence.

10.6.2 The Phasor Optical Spectral Analyzers

The schematic diagram of a phase spectral analyzer is proposed, as shown in Figure 10.35. The two couplers, No. 1 and No. 2, are used to couple lightwave-modulated signals into and out of two time-sliding devices, which can be slide by steps of very small time interval. At every slide, the output light-waves are monitored to give the amplitude and phase. This is repeated till

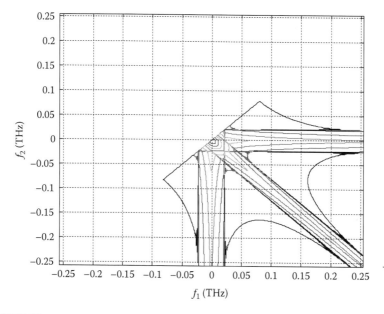

FIGURE 10.28
Output amplitude of the phasor spectrum analyzer of the short pulse envelope modulating the lightwave shown in Figure 10.26.

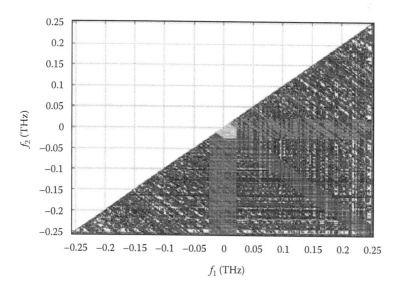

FIGURE 10.29
Output phase diagram of the phasor spectrum analyzer of the short pulse envelope modulating the lightwave shown in Figure 10.26. Frequencies f_1 and f_2 are taking both sides of the central frequency of the lightwaves.

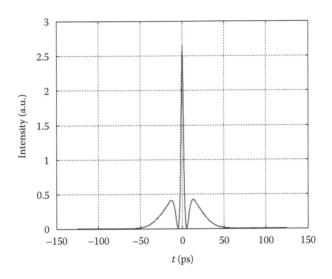

FIGURE 10.30
A short pulse near the breaking stage.

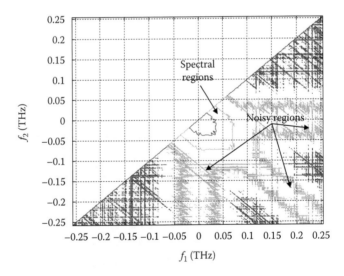

FIGURE 10.31
Output amplitude of the phasor spectrum analyzer of the short pulse envelope modulating the lightwave shown in Figure 10.30.

passing over the whole pulse temporal distance. The output amplitude and phase of the phasor spectral analyzer can be displayed, as those shown in Figures 10.31 and 10.32. It is clear that the analyzer presented here is much more detailed than the one obtained by the normal method based on the power spectrum by Wiener–Khinchin theorem.

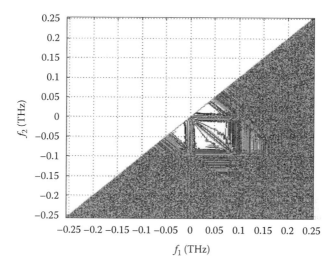

FIGURE 10.32
Output phase of the phasor spectrum analyzer of the short pulse envelope modulating the lightwave shown in Figure 10.30.

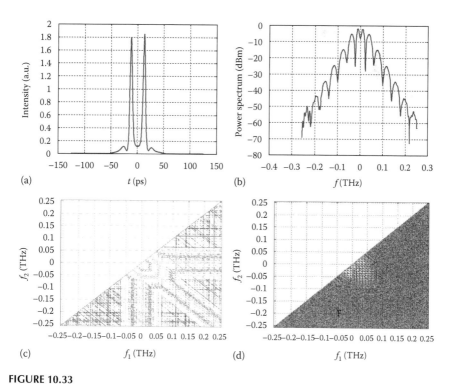

FIGURE 10.33
Bound soliton pair: (a) temporal waveform, (b) classical power spectrum, (c) amplitude distribution, and (d) phase distribution.

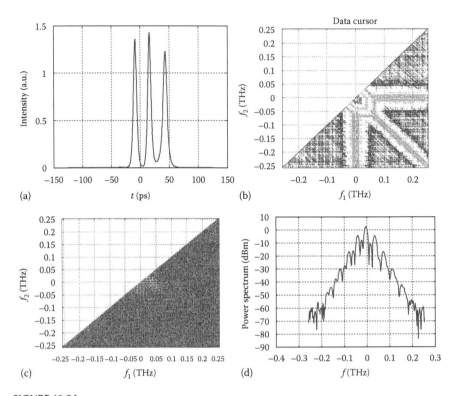

FIGURE 10.34
Triple-bound soliton group: (a) time domain, (b) three-dimensional distribution of amplitude, (c) phase distribution, and (d) classical power spectrum.

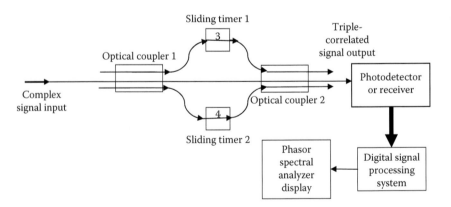

FIGURE 10.35
Schematic of the phasor spectral analyzer. Two couplers 1 and 2 couple the lightwave-modulated signals into and out of two time delay devices, which can be slide by steps of very small time interval. At every slide, the amplitude and phase of the output pulse are measured.

10.6.3 Bi-Spectrum of Duffing Chaotic Systems

Typical amplitude and phase distributions as well as the temporal and phase evolution of a Duffing nonlinear system described in Chapter 9 are shown in Figure 10.36.

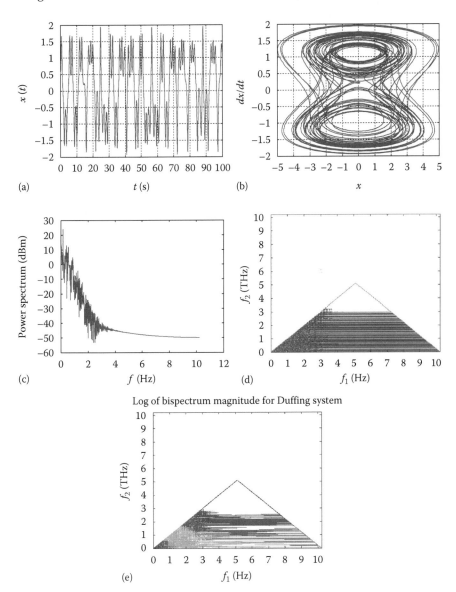

FIGURE 10.36
Duff systems: (a) temporal domain waveform, (b) classical phase evolution plane, (c) classical power spectrum, (d) bi-spectrum phase distribution, and (e) amplitude bi-spectral distribution.

10.7 Conclusions

We have experimentally demonstrated and simulated stable multi-soliton bound states that can be generated in an active MLFL under phase matching via optical phase modulation. It is believed that this stable existence of multi-soliton bound states is effectively supported by the phase modulation in an anomalous-dispersed fiber loop. Simulation results have confirmed the existence of multi-soliton bound states in the FM MLFL. Generated bound states can be easily harmonically mode locked to create a periodically multi-soliton bound sequence at a high repetition rate in this type of fiber laser, which is much more prominent than those by passive types.

The bi-spectral analyzer offers the spectral distribution of both phase and amplitude of the bound solitons states. Further works are required to identify the bound states, especially the phase distribution in the frequency domain.

References

1. B.A. Malomed, Bound solitons in coupled nonlinear Schrodinger equation, *J. Phys. Rev. A*, 45, R8321–R8323, 1991.
2. B.A. Malomed, Bound solitons in the nonlinear Schrodinger-Ginzburg-Landau equation, *J. Phys. Rev. A*, 44, 6954–6957, 1991.
3. D.Y. Tang, B. Zhao, D.Y. Shen, and C. Lu, Bound-soliton fiber laser, *J. Phys. Rev. A*, 66, 033806, 2002.
4. Y.D. Gong, D.Y. Tang, P. Shum, C. Lu, T.H. Cheng, W.S. Man, and H.Y. Tam, Mechanism of bound soliton pulse formation in a passively mode locked fiber ring laser, *Opt. Eng.*, 41(11), 2778–2782, 2002.
5. P. Grelu, F. Belhache, and F. Gutty, Relative phase locking of pulses in a passively mode-locked fiber laser, *J. Opt. Soc. Am. B*, 20, 863–870, 2003.
6. L.M. Zhao, D.Y. Tang, T.H. Cheng, H.Y. Tam, and C. Lu, Bound states of dispersion-managed solitons in a fiber laser at near zero dispersion, *Appl. Opt.*, 46, 4768–4773, 2007.
7. W.W. Hsiang, C.Y. Lin, and Y. Lai, Stable new bound soliton pairs in a 10 GHz hybrid frequency modulation mode locked Er-fiber laser, *Opt. Lett.*, 31, 1627–1629, 2006.
8. C.R. Doerr, H.A. Hauss, E.P. Ippen, M. Shirasaki, and K. Tamura, Additive-pulse limiting, *Opt. Lett.*, 19, 31–33, 1994.
9. R. Davey, N. Langford, and A. Ferguson, Interacting solitons in erbium fiber laser, *Electron. Lett.*, 27, 1257–1259, 1991.
10. D. Krylov, L. Leng, K. Bergman, J.C. Bronski, and J.N. Kutz, Observation of the breakup of a prechirped N-soliton in an optical fiber, *Opt. Lett.*, 24, 1191–1193, 1999.

11. J.E. Prilepsky, S.A. Derevyanko, and S.K. Turitsyn, Conversion of a chirped Gaussian pulse to a soliton or a bound multisoliton state in quasi-lossless and lossy optical fiber spans, *J. Opt. Soc. Am. B*, 24, 1254–1261, 2007.

12. G.P. Agrawal, *Nonlinear Fiber Optics*, Academic Press, San Diego, CA, 2001.

13. T. Georges and F. Favre, Modulation, filtering, and initial phase control of interacting solitons, *J. Opt. Soc. Am. B*, 10, 1880–1889, 1993.

14. N.J. Smith, W.J. Firth, K.J. Blow, and K. Smith, Suppression of soliton interactions by periodic phase modulation, *Opt. Lett.*, 19, 16–18, 1994.

15. R.K. Lee, Y. Lai, and B.A. Malomed, Photon-number fluctuation and correlation of bound soliton pairs in mode-locked fiber lasers, *Opt. Lett.*, 30, 3084–3086, 2005.

16. C.L. Nikias and M.R. Raghuveer, Bispectrum estimation: A digital signal processing framework, *Proc. IEEE*, 75(7), 869–891, 1987.

17. Y.C. Kim and E.J. Powers, Digital bispectral analysis and its application to nonlinear wave interactions, *IEEE Trans. Plasma Sci.*, 7(2), 120–131, 1979.

18. K. Hasselman, W. Munk, and G. MacDonald, Bispectra of ocean waves, in *Proceedings of the Symposium on Time Series Analysis*, Wiley, New York, 1963.

19. M. Hinich, Detecting a transient signals by spectral analysis, *IEEE Trans. Acoust. Speech Signal Process.*, 38(7), 1277–1283, 1990.

20. V. Chadran and S.L. Elgar, Pattern recognition using invariants defined from higher order spectra-one dimensional inputs, *IEEE Trans. Signal Process.*, 41(1), 205–212, 1993.

11

Actively Mode-Locked Multiwavelength Erbium-Doped Fiber Lasers

This chapter presents the simulation and experimental results of a particular type of an actively mode-locked multiwavelength erbium-doped fiber laser (AMLM-EDFL) operating in the 1550 nm communication window. Section 11.1 provides an overview of the various important applications of the AMLM-EDFLs and their unique performance and cost advantages over the competing technologies. Section 11.2 discusses the numerical model used in the analysis of the AMLM-EDFL through the use of the NLSE, which includes the loss, and dispersive and nonlinear effects of the fiber laser cavity. In Section 11.3, the effectiveness of the numerical model presented in Section 11.2 is demonstrated by simulating two lasing wavelengths at 10 Gbps. Section 11.4 describes the experimental results of a dual-wavelength fiber laser to verify the simulation results presented in Section 11.2. The main results drawn from the chapter are given in Section 11.5 together with some suggestions for the future.

11.1 Introduction

A single, compact, and reliable optical pulsed source that can provide simultaneous generation of high-bit-rate (>10 Gbps) ultrashort pulses at multiple wavelengths in the 1550 nm communication window is economically attractive for such applications as in the next generation of broadband optical networks combining a hybrid of optical time division multiplexing (OTDM) technology and wavelength division multiplexing (WDM) technology [1–4], optical sensing systems [5], spectroscopy [6], photonic digital-to-analog conversion [7], optical instrumentation, optical signal processing, and microwave photonic systems. There are two main types of optical sources that can meet these requirements, namely, *actively* mode-locked multiwavelength optical pulsed sources and *passively* mode-locked multiwavelength optical sources, and we shall focus our discussion on the former because they are more popular than the latter due to their unique advantages such as ultrashort and high-speed pulses [8]. Mode locking is a process in which a large number of oscillating longitudinal modes in a laser have a definite phase relation among the modes such as two consecutive modes having a constant phase difference between them. Active mode locking involves the use of a mode-locking element driven by an external source such as an amplitude modulator [9–12]. Passive mode locking does not need an external source to be a mode-locking element and could make use of some nonlinear optical effects to induce internal mode locking by exploiting, for example, the saturation of a saturable absorber or a nonlinear change in the refractive index in a suitable material (see Chapter 2). Active mode locking using an amplitude modulator is a popular method for the generation of ultrashort high-power transform-limited optical pulses through the amplitude modulation of the optical signal field for each round-trip circulation through the laser cavity, and is especially essential when a synchronization between the modulating electrical signal and the optical signal is required (see Chapter 2). The laser output consists of a train of actively mode-locked ultrashort pulses where the consecutive pulses are separated by a time interval equal to the cavity-round trip time. Although mode-locked multiwavelength semiconductor-based fiber lasers [13–16] are more compact and can overcome the homogeneous line-broadening effect present in mode-locked multiwavelength erbium-doped-based fiber lasers at room temperature and can hence suppress the gain competition among the various wavelength channels to achieve lasing stability, they lack such important features as fiber compatibility with the fiber networks, low noise, low insertion loss, uniform and broad gain spectrum, and high pulse energy.

11.2 Numerical Model of an Actively Mode-Locked Multiwavelength Erbium-Doped Fiber Laser

Most of the works reporting on AMLM-EDFLs were mainly experimental demonstrations and no comprehensive numerical models were used in the analysis of the laser dynamics [17–26]. This drawback was overcome by [27] who have come up with an elegant numerical model for the analysis of the laser dynamics and the model was verified experimentally. Thus we present here an AMLM-EDFL based mostly on the work of [27].

Figure 11.1 shows the schematic diagram of a typical AMLM-EDFL. An EDFA provides the necessary gain in the laser cavity to induce various nonlinear optical effects (e.g., four wave mixing (FWM), self-phase modulation (SPM), and cross-phase modulation (XPM)) from a highly nonlinear fiber (HNLF) while an SMF and a dispersion compensating fiber (DCF) contribute to most of the chromatic dispersion. The right amount of nonlinearity from the HNLF and the chromatic dispersion (including its sign) from the SMF/DCF is critically important to minimize the gain

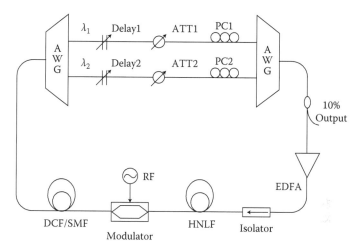

FIGURE 11.1

Schematic diagram of an AMLM-EDFL. Here the AMLM-EDFL has two lasing wavelengths λ_1 and λ_2. Modulator: LiNbO$_3$ MZIM; RF: radio frequency synthesizer; EDFA: erbium-doped fiber amplifier; HNLF: highly nonlinearly fiber; DCF: dispersion compensating fiber; SMF: single-mode fiber; AWG: array waveguide grating; Delay1, Delay2: adjustable delay lines; ATT1, ATT2: variable optical attenuators. (From Pan, S. and Lou, C., *IEEE J. Quantum Electron.*, 44, 245, 2008. With permission.)

competition between the various wavelength channels in order to generate stable and high-quality output pulses, and this is the main issue in the design and analysis of the laser. An LiNbO$_3$ Mach–Zehnder intensity modulator is used to modulate the electrically modulated radio-frequency (RF) signal as well as to achieve active mode locking of the laser. The number of adjustable delay lines (Delay1, Delay2), variable optical attenuators (ATT1, ATT2), and polarization controllers (PC1, PC2) between the two array waveguide gratings (AWGs) is determined by the number of lasing wavelengths (the figure only shows two of such elements for two wavelength channels). The adjustable delay lines (Delays) are used to ensure that all the wavelength channels experience the same delay in the cavity in order to minimize timing jitter of the multiwavelength output pulses. The variable optical attenuators (ATTs) are used to ensure that all the channels have the same optical power level in the cavity because even a small difference in the power levels of the channels could prevent effective mode locking due to the gain competition effect. The polarization controllers (PCs) are used to ensure that all the channels have the same state of polarization in the cavity. The left AWG is used as a demultiplexer while the right AWG functions as a multiplexer. The number of ports of the AWGs depends on the number of lasing wavelengths. The isolator is to ensure unidirectional propagation in the cavity. The 10/90 fiber coupler is used to extract the output lasing wavelengths.

In the AMLM-EDFL, the electrical field of the jth channel of an N-channel fiber laser system can be described by

$$E_j(z,t) = A_j(z,t)\exp\left[-i\left(\omega_j - \omega_0\right)t\right]; \quad j = 1,2,\ldots,N. \tag{11.1}$$

where

$i = \sqrt{-1}$
ω_j is the optical carrier frequency of the jth channel
ω_0 is the reference or operating frequency
z is the propagating distance along the cavity
t is the actual time
$A_j(z, t)$ is the slowly varying complex envelope of the electrical field of the jth channel

The signal gain provided by the EDFA to the jth channel, $g_j(\omega_j, z)$, by taking the cross-saturation and self-saturation effects into account, is given by

$$g_j(\omega_j, z) = \xi \frac{g_0(\omega_j)}{1 + \left(P_j / P_{sat}\right)} + (1 - \xi)\frac{g_0(\omega_j)}{1 + \displaystyle\sum_{j=1}^{N} P_j / P_{sat}} \tag{11.2}$$

where

$g_0(\omega_j)$ is the small-signal gain of the jth channel
P_j is the input power of the jth channel
P_{sat} is the saturation power
$0 < \xi < 1$

$\xi = 0$ corresponds to a homogenous broadening gain medium (e.g., an EDFA) while $\xi = 1$ represents an inhomogenous broadening gain medium (e.g., a semiconductor optical amplifier (SOA)).

A multiwavelength signal amplitude, $A(0,T)$, with N pulses at the initial condition of $z = 0$ can be described by

$$A(0,T) = \sum_{j=1}^{N} A_j(0,T)\exp\left[i\left(\omega_j - \omega_0\right)T\right] \tag{11.3}$$

where T is the time in the moving frame, which is related to the real time t through $T = t - z/v_g \equiv t - \beta_1 z$, where v_g is the group velocity of the light in the fiber and $\beta_1 = 1/v_g$.

By substituting Equation 11.3 into the well-known nonlinear Schrödinger equation [28] and separating the different frequency components, the evolution of a multiwavelength (or N channels) signal in the AMLM-EDFL can be described by a set of coupled equations

$$\frac{\partial A_j}{\partial z} = -\frac{1}{2}i\beta_2(\omega_j)\frac{\partial^2 A_j}{\partial^2 T} + \frac{1}{6}\beta_3(\omega_j)\frac{\partial^3 A_j}{\partial^3 T} - \frac{1}{2}\alpha A_j$$

$$+ i\gamma|A_j|^2 A_j + 2i\gamma\sum_{k\neq j}|A_k|^2 A_j + i\gamma\sum_{l,m,n}s_{lmn}A_lA_mA_n^*\exp(-i\Delta K_{lmn}z) \quad (11.4)$$

where * denotes the complex conjugation. The second-order and third-order dispersions of the fiber are denoted by $\beta_2(\omega_j)$ and $\beta_3(\omega_j)$, respectively, α is the attenuation of the fiber, and γ is the fiber nonlinearity (e.g., FWM, SPM, and XPM).

Note that the stimulated Brillouin scattering (SBS) effect, which only occurs in the backward direction, is negligibly small due to the unidirectional signal propagation provided by the optical isolator and the fact that the generated pulses have a very small pulse width of typically less than 100 ps [28]. Furthermore, the stimulated Raman scattering (SRS) effect is also very small and can thus be neglected because the typical pump power level used in the laser is much smaller than the required threshold pump power of typically more than 600 mW. The last term on the right side of Equation 11.4 represents the FWM effect, where s_{lmn} is a degenerate factor (if $l=m$, $s_{lmn}=1$, otherwise $s_{lmn}=2$), and $\omega_l+\omega_m=\omega_n+\omega_j$, $l\neq j$ and $m\neq j$.

The phase mismatch between the channels is given by

$$\Delta K_{lmn} = \frac{1}{2}\left[\beta_2(\omega_n)\Delta\omega_n^2 + \beta_2(\omega_j)\Delta\omega_j^2 - \beta_2(\omega_l)\Delta\omega_l^2 - \beta_2(\omega_m)\Delta\omega_m^2\right]$$

$$-\frac{\beta_3}{6}\left[\Delta\omega_n^3 + \Delta\omega_j^3 - \Delta\omega_l^3 - \Delta\omega_m^3\right] \quad (11.5)$$

where $\Delta\omega_{l,m,n}=\omega_{l,m,n}-\omega_0$. In the AMLM-EDFL, the cavity length is typically less than 2 km and the peak power of the signal is less than 100 mW so the FWM effect can be treated as a perturbation to the signal. The FWM term in Equation 11.4 can thus be considered separately as

$$\Delta A_j = i\gamma\sum_{l,m,n}A_lA_mA_n^*\int_z^{z+h}\exp(-i\Delta K_{lmn}z')dz'$$

$$= i\gamma\sum_{l,m,n}A_lA_mA_n^*\frac{\exp(-i\Delta K_{lmn}z)\left[\exp(-i\Delta K_{lmn}h)-1\right]}{-i\Delta K_{lmn}} \quad (11.6)$$

where h is the step size used in the numerical simulation. Thus, the FWM term can be easily obtained analytically from Equation 11.6, and thus the coupled equations (without the FWM term) in Equation 11.4 can be easily solved numerically using the split-step method [28]. This AMLM-EDFL model has two advantages. First, the choice of the step size h is not that critical because the FWM term can be obtained analytically according to Equation 11.6. Second, the effects due to the FWM effect and other factors can be isolated and analyzed separately because they can be considered as separate perturbations to the signal in Equation 11.4.

11.3 Simulation Results of an Actively Mode-Locked Multiwavelength Erbium-Doped Fiber Laser

Using the numerical model presented in Section 11.2, we analyze the mechanisms of gain competition suppression (GCS) provided by the chromatic dispersion of the SMF/DCF and the various nonlinear effects in the HNLF in the laser cavity. The main parameter values of the components used in the simulation are listed in Table 11.1. It is assumed in the simulation that the various signal wavelengths injected into the HNLF have the same polarization state. The EDF is an absolutely homogeneous broadening medium ($\xi = 0$), which is responsible for introducing the gain competition among the various wavelength channels. The nonlinearities of the HNLF together with the chromatic dispersion of the SMF/DCF are responsible for suppressing the gain competition among the signal wavelengths. It is assumed that the different wavelength channels experience the same small-signal gain ($g_0 = 26$ dB) in the EDFA, which is a valid reason because the small difference in the signal gain can be easily compensated for by the cavity loss in the experiment (e.g., the attenuators ATT1 and ATT2 in Figure 11.1 are used to ensure that the two wavelength channels have the same power level). To demonstrate the effectiveness of the numerical model, two wavelength channels at 1555.6 and 1557.2 nm (channel separation is 1.6 nm) are used as input signals into the cavity.

11.3.1 Effects of Small Positive Dispersion Cavity and Nonlinear Effects on Gain Competition Suppression Using a Highly Nonlinear Fiber

First, we consider the case of a small *positive* net cavity dispersion of 0.8 ps/nm at 1550 nm. Figure 11.2 shows the time-domain power evolution of the 1555.6 and 1557.2 nm signals over a large number of cycles inside the cavity. In Figure 11.2a and b where the FWM, SPM, and XPM are all included, although the loss difference between the two signal wavelengths is only 0.1 dB, the 1557.2 nm channel disappears quickly due to the serious gain competition from the 1555.6 nm signal because the cavity has a very small nonlinearity (the HNLF has $\gamma_H = 1/W/km$). In Figure 11.2c and d where the FWM, SPM,

TABLE 11.1

Parameter Values of the Components of the AMLM-EDFL (Figure 11.1) Used in the Simulation

Component	Symbol	Parameter	Value and Unit
EDF	g_0	Small signal gain	26 dB
	P_{sat}	Saturation power	0.1 mW
	ξ	Absolute homogeneous broadening gain	0
HNLF	α_H	Fiber loss	0.83 dB/km
	L_H	Fiber length	1 km
	γ_H	Nonlinear coefficient	10/W/km
	$\lambda_0 H$	Zero-dispersion wavelength	1555.5 nm
	S	Dispersion slope	0.018 ps/nm²/km
SMF	α	Fiber loss	0.2 dB/km
	L	Fiber length	50 m
	γ	Nonlinear coefficient	1/W/km
	D	Chromatic dispersion	17 ps/nm/km
Modulator	f	RF frequency	10 GHz
	α_M	Insertion loss	2.2 dB
AWG	B	3 dB passband width	40 GHz
	α_F	Insertion loss	0.8 dB
Others	$\alpha_L \cdot (= \alpha_M + \alpha_F)$	Insertion loss of the laser system	3.0 dB

and XPM effects are also included, even though the loss difference between the two signal wavelengths is slightly larger (i.e., 0.3 dB), both signal wavelengths can still oscillate nicely in the cavity due to the gain competition suppression (GCS) provided by a much stronger nonlinearity in the cavity (the HNLF has $\gamma_H = 10$/W/km). The question here is to find out whether the FWM effect alone or the combined effect of SPM and XPM contributes to the GCS. The answer can be found in Figure 11.2e and f where the FWM effect is included and the SPM and XPM effects are excluded, and the loss difference between the two signal wavelengths of 0.3 dB and the HNLF with $\gamma_H = 10$/W/km are the same as those in Figure 11.2c and d. Figure 11.2e and f show that the two signals are still nicely mode locked and their pulse shapes are pretty much maintained over a large number of cycles in the cavity. It can thus be deduced that the FWM effect alone (and not the SPM and XPM effects) mainly contributes to GCS. This fact can be further verified in Figure 11.2g and h where the 1557.2 nm pulse stops oscillating quickly because the FWM effect is excluded. The two output pulses shown in Figure 11.2e and f are superimposed after 120 cycles and are shown in Figure 11.3 in order to examine the stability of the laser. Figure 11.3b shows that the two output pulses (especially the 1557.2 nm pulse) suffer large amplitude and timing jitters because they have significant chirpings, resulting in asymmetric pulse shapes, and thus the laser cannot reach its steady state. The nonlinear chirps

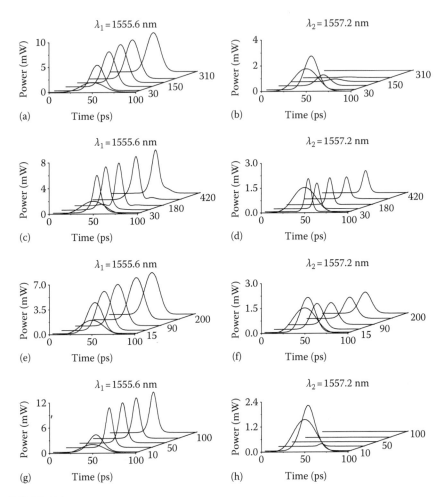

FIGURE 11.2
Simulation results for the case of a small *positive* net cavity dispersion of 0.8 ps/nm. Time-domain power evolution of the 1555.6 and 1557.2 nm pulses over a large number of cycles in the cavity. (a) and (b) The HNLF has $\gamma_H = 1/W/km$, the loss imbalance of the two signal wavelengths is 0.1 dB, and the FWM, SPM, and XPM terms are included in Equation 11.4. (c) and (d) The HNLF has $\gamma_H = 10/W/km$, the loss imbalance of the two signal wavelengths is 0.3 dB, and the FWM, SPM, and XPM effects are included. (e) and (f) The HNLF has $\gamma_H = 10/W/km$, the loss imbalance of the two signal wavelengths is 0.3 dB, and the FWM effect is included while the SPM and XPM effects are ignored. (g) and (h) The HNLF has $\gamma_H = 10/W/km$, the loss imbalance of the two signal wavelengths is 0.1 dB, and the FWM effect is ignored while the SPM and XPM effects are considered. (From Pan, S. and Lou, C., *IEEE J. Quantum Electron.*, 44, 245, 2008. With permission.)

and asymmetric shapes of the pulses would redistribute the power of the pulses in the next cycle, making the output pulses unstable after 120 cycles.

The influences of the FWM effect and the SPM and XPM effects [29] are investigated separately in greater detail to single out the effect that suppresses the gain competition between the signal wavelengths and causes the

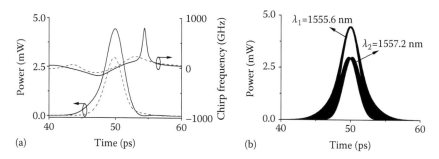

FIGURE 11.3

Simulation results for the case of a small *positive* net cavity dispersion of 0.8 ps/nm. The HNLF has $\gamma_H = 10/W/km$, the loss imbalance between the two signal wavelengths is 0.3 dB, and only the FWM effect is considered while the SPM and XPM effects are ignored. The conditions here are exactly the same as those in Figure 11.2e and f. (a) Solid line: 1555.6 nm pulse and dashed line: 1557.2 nm pulse after 120 cycles. (b) Superimposed output pulses after 120 cycles. (From Pan, S. and Lou, C., *IEEE J. Quantum Electron.*, 44, 245, 2008. With permission.)

asymmetry in the output pulse shapes. Figure 11.2e and f show the power evolution of the two pulses when the FWM term is included in Equation 11.4 while the SPM and XPM terms are removed from Equation 11.4. The good quality of the evolving pulses suggests that the FWM effect alone is mainly responsible for the GCS and does not contribute to the distortion of the pulse shapes. Figure 11.2g and h show that, when the SPM and XPM terms are included while the FWM term is excluded in Equation 11.4, the 1557.2 nm channel, which has only 0.1 dB higher loss than the other channel, disappears quickly after only tens of cycles due to the fact that the SPM and XPM effects do not contribute much to the GCS. Furthermore, the SPM and XPM effects introduce nonlinear chirps to the pulses, distorting the pulse shapes because different parts of the pulse experience different amount of phase shift or frequency chirping. In Figure 11.3a, the stronger pulse (solid line) experiences more frequency chirping than the weaker pulse (dashed line) due to the fact that the SPM effect is more pronounced for a higher-intensity pulse. We consider a numerical example to demonstrate the GCS effect due to the FWM alone. Figure 11.4a shows that the peak powers of the 1557.2 and 1555.6 nm channels at steady state before the HNLF are 36.1 and 17.9 mW, while those after the HNLF are 27 and 14.9 mW (Figure 11.4b), respectively. The power ratio of the two channels before the HNLF is thus 2.02 (36.1 mW/17.9 mW) while that after the HNLF is 1.81 (27 mW/14.9 mW), showing that power is transferred from the higher-power 1557.2 nm channel to the lower-power 1555.6 nm channel through the FWM effect. Thus the higher-power 1557.2 nm channel offers some gain to the lower-power 1555.6 nm channel, preventing the latter from vanishing through gain competition. This is a well-known FWM effect, which is described in detail by [29]. The analysis here is also consistent with the experimental demonstrations reported by [17,18].

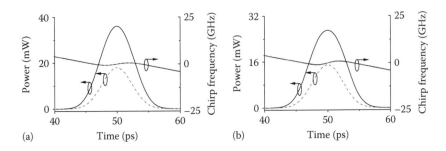

FIGURE 11.4
Simulation results for the case of a small *positive* net cavity dispersion of 0.8 ps/nm. The HNLF has $\gamma_H = 10/\text{W/km}$ and only the FWM effect is considered by excluding the SPM and XPM effects in Equation 11.4. Solid line: 1557.2 nm pulse and dashed line: 1555.6 nm pulse at steady state. The conditions here are exactly the same as those in Figures 11.2 and 11.3. (a) Before HNLF and (b) after HNLF. (From Pan, S. and Lou, C., *IEEE J. Quantum Electron.*, 44, 245, 2008. With permission.)

11.3.2 Effects of a Large Positive Dispersion and Nonlinear Effects Using a Highly Nonlinear Fiber in the Cavity on Gain Competition Suppression

Using the same procedure as in Section 11.3.1 in which a small positive net cavity dispersion of 0.8 ps/nm at 1550 nm has been considered, we now analyze the effects of a large positive net cavity dispersion of 85 ps/nm, and the SPM, XPM, and FWM effects using an HNLF on the GCS at 1550 nm. The findings here are consistent with those in Section 11.3.1 in that the FWM is mainly responsible for the GCS while the SPM and XPM create asymmetry in the pulse shapes due to the nonlinear chirping effects. In Figure 11.5a and b where the FWM, SPM, and XPM effects are all included, even though the loss difference between the two signal wavelengths is very small (only 0.1 dB), because the cavity nonlinearity is very small (the HNLF has $\gamma_H = 1/\text{W/}$ km), the GCS cannot be achieved due to the very small FWM effect, and thus the 1557.2 nm pulse (Figure 11.5b) quickly vanishes while only the 1555.6 nm pulse (Figure 11.5a) is nicely mode locked and sustained over a large number of cycles in the cavity.

However, when the cavity nonlinearity is much larger (the HNLF has $\gamma_H = 10/\text{W/km}$) and the FWM, SPM, and XPM effects are all included, Figure 11.5c and d show that the two pulses are stably mode locked and can be propagated over a large number of cycles in the cavity because the FWM effect can significantly suppress the gain competition. This finding is further validated by considering the FWM effect alone while ignoring the SPM and XPM effects in Figure 11.5e and f where the two pulses are also nicely mode locked and their qualities are maintained throughout the cavity.

Furthermore, this finding is validated again by ignoring the FWM effect while considering only the SPM and XPM effects, the 1557.2 nm pulse

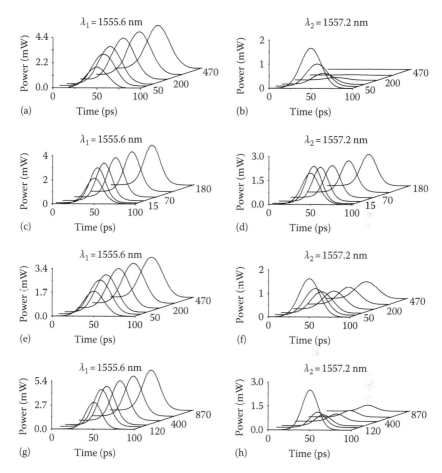

FIGURE 11.5

Simulation results for the case of a large positive net cavity dispersion of 85 ps/nm. Compared with the conditions in Figures 11.2 through 11.4 where the cavity has a very small positive dispersion of 0.8 ps/nm, the main difference here is that the cavity has a very large dispersion of 85 ps/nm. The loss imbalance between the 1555.6 and 1557.2 nm pulses is 0.1 dB. Time-domain power evolution of the two pulses over a large number of cycles in the cavity. (a) and (b) The HNLF has $\gamma_H = 1/W/km$ and the FWM, SPM, and XPM effects are considered. (c) and (d) The HNLF has $\gamma_H = 10/W/km$ and the FWM, SPM, and XPM effects are considered. (e) and (f) The HNLF has $\gamma_H = 10/W/km$ and the FWM effect is considered while the SPM and XPM effects are ignored. (g) and (h) The HNLF has $\gamma_H = 10/W/km$ and the FWM effect is excluded while the SPM and XPM are considered. (From Pan, S. and Lou, C., *IEEE J. Quantum Electron.*, 44, 245, 2008. With permission.)

(Figure 11.5h) cannot be mode locked and quickly vanishes due to the gain competition from the 1555.6 nm pulse (Figure 11.5g). The findings here (just like those in Section 11.3.1) suggest that for both small (0.8 ps/nm in Section 11.3.1) and large (85 ps/nm in this section) positive net cavity dispersions, the FWM effect plays an extremely important role to allow the propagation of the pulses over a large number of cycles by suppressing the gain competition between the multiwavelength pulses even though the power difference between the channels is as small as 0.1 dB. To suppress the gain competition, it is thus critically important that the attenuators (i.e., ATT1 and ATT2 in Figure 11.1) are properly adjusted to minimize the power difference between the multiwavelength pulses as much as possible and that an HNLF with a reasonably large nonlinearity of, for example, $\gamma_H = 10/W/km$ is used to induce a large FWM effect.

In addition to the findings described above, two new phenomena occur in the laser with a large anomalous cavity dispersion of +85 ps/nm. First, the laser can quickly reach its steady state after only 185 cycles (see Figure 11.5c and d) due mainly to the well-balanced interplay between the nonlinear chirps induced by the SPM and XPM effects and the large anomalous dispersion, resulting in little effect on the pulse shapes. We now exclude the FWM effect to validate the fact that the SPM and XPM effects can reduce the gain competition, and this is illustrated in Figure 11.6 which shows the evolution of the steady-state output pulses along the different components in the cavity. Figure 11.6a shows the amplified pulses by the EDFA, and the chirps at the center of the 1555.6 and 1557.2 nm channels are −0.46 and −0.43, respectively. These negative chirp values change to the positive values of 0.21

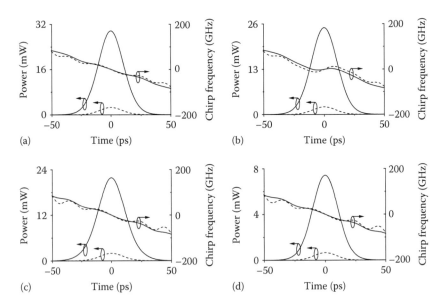

FIGURE 11.6

Simulation results of the steady-state output pulses after (a) EDF, (b) HNLF, (c) SMF, and (d) modulator (see Figure 11.1) when FWM is excluded and the SPM and XPM effects are included. Solid line: 1555.6 nm pulse and dashed line: 1557.2 nm pulse. The loss imbalance between the 1555.6 and 1557.2 nm pulses is 0.1 dB. The anomalous dispersion of the cavity is 85 ps/nm. The HNLF has $\gamma_H = 10/W/km$. (From Pan, S. and Lou, C., *IEEE J. Quantum Electron.*, 44, 245, 2008. With permission.)

and 0.49, respectively, after the HNLF (Figure 11.6b) due to the positive chirp induced by the SPM and XPM effects. The higher-power 1555.6 nm channel has a much less chirp than that of the lower-power 1557.2 nm channel due to the fact that the phase shift induced by the XPM effect is approximately two times of that induced by the SPM effect. Unlike the positive chirps induced by the SPM and XPM effects, the anomalous dispersion of the SMF has a negative linear chirp that would first narrow and then broaden the pulses. Thus, the 1555.6 nm pulse, which has a smaller positive chirp before passing through the SMF, should be broadened after going through the SMF due to the negative chirp present in the SMF. As expected, Figure 11.6c shows that the 1555.6 nm pulse has a pulse width of 26.7 ps which is larger than the 23.2 ps pulse width of the 1557.2 nm pulse. The 1555.6 nm pulse will experience a slightly larger loss when passing through the modulator due to its slightly larger pulsewidth and will thus obtain a smaller gain after passing through one cycle of the cavity. It can thus be deduced that higher-power channels can partly suppress the gain competition in a highly nonlinear cavity (where the SPM and XPM effects are included while the FWM is excluded) with a reasonably large anomalous dispersion.

11.3.3 Effects of a Large Negative Dispersion and Nonlinear Effects Using a Highly Nonlinear Fiber in the Cavity on Gain Competition Suppression

In Section 11.3.2, we have found that the combined effects of the large positive cavity dispersion of 85 ps/nm and the large cavity nonlinearity (the HNLF has $\gamma_H = 10/W/km$) can significantly suppress the gain competition between the two wavelength signals with a loss difference of 0.1 dB (see Figure 11.5c and d). The conditions here are the same as those in Figure 11.5c and d, except that here the cavity has a large negative dispersion of −85 ps/ nm (which can be obtained using appropriate lengths of the SMF and the DCF, see Figure 11.1). The laser could hardly achieve stable mode locking even though the loss imbalance between the two wavelengths was only 0.1 dB. This is because both SPM and XPM effects and the large negative dispersion in the cavity introduce large positive chirps and broaden the pulses significantly, and this further weakens the FWM effect, preventing the GCS to be achieved.

11.3.4 Effects of Cavity Dispersion and a Hybrid Broadening Gain Medium on the Tolerable Loss Imbalance between the Wavelengths

We first describe the effect of the positive and negative cavity dispersion on the tolerable loss imbalance between the wavelengths to achieve the GCS. Figure 11.7a (upper curve) shows the theoretical tolerable loss imbalance

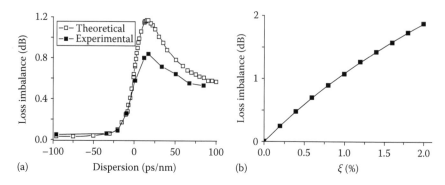

FIGURE 11.7
Loss imbalance between the 1555.6 and 1557.2 nm pulses versus (a) net cavity dispersion (the HNLF has $\gamma_H = 10/W/km$) and (b) ξ, the hybrid broadening gain medium factor (no HNLF is used). (From Pan, S. and Lou, C., *IEEE J. Quantum Electron.*, 44, 245, 2008. With permission.)

between the wavelengths versus the net cavity dispersion. In the normal (or negative) dispersion regime where the net cavity dispersion is less than −25 ps/nm, the tolerable loss imbalance can be less than 0.1 dB due to the fact that the positive chirps of both dispersion and the nonlinear effects broaden the pulses. The powers of the pulses reduce accordingly and weaken the nonlinearly effects, leading to ineffective GCS by the FWM effect. However, in the anomalous (or positive) dispersion regime where the net dispersion is greater than 0 ps/nm, different phenomena were observed. First, there was an optimum dispersion value at which the tolerable loss imbalance (~0.8 dB) was largest. At this point, the GCS can be achieved because the dispersion and the nonlinear effects are well balanced and thus the pulses have the narrowest pulsewidths and the largest powers, which enhance the nonlinear effects. Beyond this optimum dispersion point, the loss imbalance is still reasonably large and, for example, it is about 0.6 dB when the net cavity dispersion is about 100 ps/nm.

We now describe the effect of a hybrid broadening gain medium (i.e., $\xi \neq 0$ in Equation 11.2) on the tolerable loss imbalance between the wavelengths to achieve the GCS. The system configuration here is exactly the same as that in Figure 11.1, except that no HNLF is used here. Figure 11.7b shows that the theoretical tolerable loss imbalance increases linearly with ξ with a slope of 0.93 dB/percentage, which is quite close to the ideal slope of 1.0 dB/percentage. This is because an increase in ξ (or gain in the cavity) results in an increase in the nonlinear effects in the SMF/DCF, and this improves the GCS and thus increases the tolerable loss imbalance. The case of $\xi = 1.2\%$ with a loss imbalance of 1.2 dB corresponds to the above-mentioned case of using the HNLF with a loss imbalance of 1.2 dB (Figure 11.7a). The tolerable loss imbalance can be increased further by increasing ξ through the use of a higher pump power to provide large nonlinear effects.

11.4 Experimental Validation and Discussion on an Actively Mode-Locked Multiwavelength Erbium-Doped Fiber Laser

This section presents the experimental validation of the simulation results described in Section 11.3. The experimental setup of the actively AMLM-EDFL is also shown in Figure 11.1. In Section 11.2, we have described the functionalities of the various components shown in Figure 11.1. Furthermore, the EDFA provides the necessary gain required to generate the various non-linearities (i.e., FWM, SPM, and XPM) from the HNLF. The EDFA consists of a polarization-independent isolator, a section of EDF, and two 980 nm laser diodes as optical pumps. The EDFA has a saturation output power of 16.6 dBm. To achieve active mode locking, which in this case is based on loss modulation, a LiNbO$_3$ MZIM was driven by a synthesized microwave signal generator at 10 GHz. Several sections of DCFs and/or SMFs were used to introduce the required amount of chromatic dispersion in the cavity. To suppress the gain competition between the two wavelength channels, a 1 km HNLF, which was used to produce the FWM, SPM, and XPM effects, has a zero-dispersion wavelength of 1555.5 nm, a dispersion slope of 0.018 ps/nm^2/km and a nonlinear coefficient of $\gamma_H = 10$/W/km. An optical spectrum analyzer and a 40 GHz photodiode connected to a digital sampling oscilloscope were employed to observe the output pulses.

We first consider a small positive cavity dispersion of 0.8 ps/nm at 1550 nm. Figure 11.8 shows the measured optical spectra and waveforms of the individual or single-wavelength output signals at $\lambda_1 = 1555.6$ nm and $\lambda_2 = 1557.2$ nm. The single-wavelength operation was carried out by blocking the other lasing wavelength. The 1555.6 and 1557.2 nm output signals have pulse widths of 21.4 and 22.2 ps, and 3 dB bandwidths of 0.23 and 0.20 nm, respectively. The 10 GHz frequency combs can be clearly seen in the output spectra in Figure 11.8a and c. However, Figure 11.9 shows that the results are not that good when the two signals were observed simultaneously under dual-wavelength operation. Figure 11.9 shows the measured spectrum and waveform of the dual-wavelength laser with output signals at $\lambda_1 = 1555.6$ nm and $\lambda_2 = 1557.2$ nm. Figure 11.9a shows that the frequency combs (which are present in Figure 11.8a and c) disappear here due apparently to unstable mode locking, resulting in large amplitude and timing jitters, as shown in Figure 11.9b. This problem could not be solved even when the PCs and ODLs were adjusted. This drawback arises from the fact that the small net cavity dispersion of 0.8 ps/nm was not sufficient to counteract the FWM, SPM, and XPM effects from the HNLF to achieve mode locking and, hence, stable output pulses; and this phenomenon has already been explained in Section 11.3.1 (see Figure 11.2).

The experiment was then carried out with a larger net cavity dispersion of 85 ps/nm at 1550 nm. Just like Figure 11.8, Figure 11.10 shows the measured

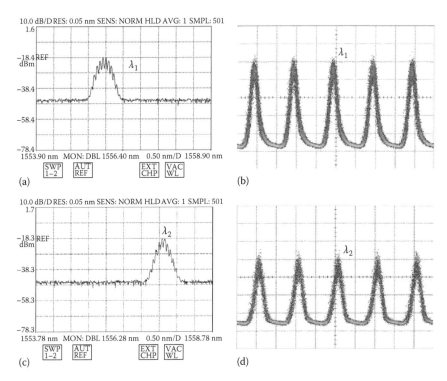

FIGURE 11.8
(a, c) Optical spectra and (c, d) waveforms of the two output signals at $\lambda_1 = 1555.6$ nm and $\lambda_2 = 1557.2$ nm. (a) and (b) Single-wavelength operation at $\lambda_1 = 1555.6$ nm. (c) and (d) Single-wavelength operation at $\lambda_2 = 1557.2$ nm. The cavity has a small positive dispersion of 0.8 ps/nm. The HNLF has $\gamma_H = 10/W/km$. (From Pan, S. and Lou, C., *IEEE J. Quantum Electron.*, 44, 245, 2008. With permission.)

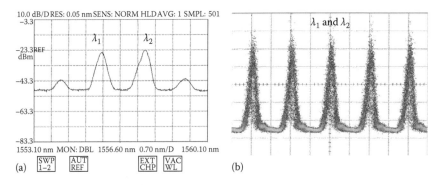

FIGURE 11.9
(a) Optical spectrum and (b) waveform of the dual-wavelength laser with output signals at $\lambda_1 = 1555.6$ nm and $\lambda_2 = 1557.2$ nm. The cavity has a small positive dispersion of 0.8 ps/nm. The HNLF has $\gamma_H = 10/W/km$. (From Pan, S. and Lou, C., *IEEE J. Quantum Electron.*, 44, 245, 2008. With permission.)

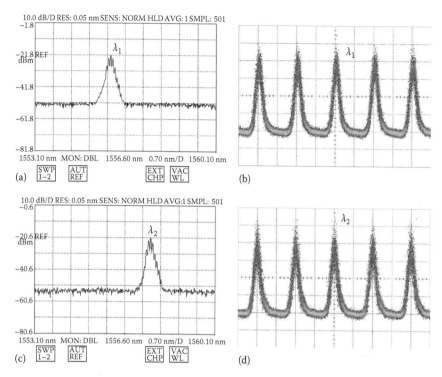

FIGURE 11.10
(a, c) Optical spectra and (c, d) waveforms of the two output signals at $\lambda_1 = 1555.6$ nm and $\lambda_2 = 1557.2$ nm. (a) and (b) Single-wavelength operation at $\lambda_1 = 1555.6$ nm. (c) and (d) Single-wavelength operation at $\lambda_2 = 1557.2$ nm. The only difference between this figure and Figure 11.8 is that the net cavity dispersion here is 85 ps/nm while it is 0.8 ps/nm in Figure 11.8. The HNLF has $\gamma_H = 10/W/km$. (From Pan, S. and Lou, C., *IEEE J. Quantum Electron.*, 44, 245, 2008. With permission.)

optical spectra and waveforms of the single-wavelength output signals at $\lambda_1 = 1555.6$ nm and $\lambda_2 = 1557.2$ nm. The two output signals have pulse widths of 22.9 and 23.5 ps, and 3 dB bandwidths of 0.19 and 0.17 nm, respectively. Figure 11.10a and c show the clear 10 GHz frequency combs. Unlike Figures 11.9a and 11.11a shows the clear frequency combs due to the good mode locking of the two wavelengths (for about half an hour without using any stabilization technique). Although the peak wavelengths were found to shift by about 0.2 nm during this time interval, the spectral shapes remained pretty much the same. As such, Figure 11.11b shows that the two output pulses overlap with each other, resulting in minimum amplitude and timing jitters. The stable mode locking is due to the well-balancing interaction between the larger positive cavity dispersion of 85 ps/nm and the HNLF's nonlinearity (FWM, SPM, and XPM) and hence sufficient GCS, resulting in high-quality output pulses. This good experimental result is consistent with the numerical results presented in Section 11.3.2 (Figure 11.5). Note in Figure 11.11a that

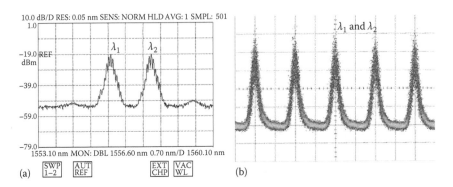

FIGURE 11.11
(a) Optical spectrum and (b) waveform of the dual-wavelength laser with output signals at $\lambda_1 = 1555.6$ nm and $\lambda_2 = 1557.2$ nm. The only difference between this figure and Figure 11.9 is that the net cavity dispersion here is 85 ps/nm while it is only 0.8 ps/nm in Figure 11.9. The HNLF has $\gamma_H = 10$/W/km. (From Pan, S. and Lou, C., *IEEE J. Quantum Electron.*, 44, 245, 2008. With permission.)

the two very small sidebands are probably due to the FWM mechanism [29]. However, when the net cavity dispersion was −96 ps/nm (large and negative), although single-wavelength operations were still stable for more than 30 min, dual-wavelength operation could only last for less than 5 s due to insufficient GCS.

The tolerable loss imbalance between the two signal wavelengths were then measured using the main parameter values shown in Table 11.1 and their experimental result is shown in Figure 11.7a which is consistent with the numerically calculated result, also shown in Figure 11.7a. In particular for the normal (or negative) cavity dispersion, the experimental and numerical results are in good agreement with each other. First, the optical attenuators were adjusted to make the power levels of the two signals to be the same. The 1557.2 nm signal was then significantly attenuated until it stopped oscillating and this signal loss was recorded as the tolerable loss imbalance. However, for the anomalous (or positive) dispersion region, the measured tolerable loss imbalance is smaller than the theoretical value. One possible reason for this discrepancy is that the two output pulses were assumed to be superimposed in the time domain while this was not really the case in the experiment due perhaps to the polarization dependence of the LiNbO$_3$ electro-optic modulator. This problem could be solved using a feedback stabilization scheme [30,31]. This is particularly for the cavity dispersion of 16 ps/nm/km at 1550 nm, where the measured pulse widths of the output pulses were less than 10 ps due perhaps to the slight walking off of the signal wavelengths, which could lead to the weakening of the nonlinear effects. The second possible reason for the discrepancy is that the polarization mode dispersion was ignored in the numerical simulation when in fact it could reduce the GCS induced by the nonlinear effects. Note that the polarization mode dispersion could be easily incorporated into the numerical calculations by splitting Equation 11.4 into two coupled sets of equations that describe pulse propagations in the two principal states of polarization [32].

11.5 Conclusions and Suggestions for Future Work

This chapter has presented a numerical model for studying the dynamic mechanism of an AMLM-EDFL operating at 10 Gbps in the 1550 nm communication window. Using the model, simulation of a dual-wavelength laser with two output pulses has been carried out by solving the nonlinear Schrödinger equation to examine the effects of the cavity dispersion and the nonlinear effects on the gain competition suppression (GCS). It has been found that the nonlinear FWM effect alone from the HNLF is mainly responsible for the GCS for both small and large positive cavity dispersions; while the SPM and XPM effects mainly contribute to the chirping and distortion of the output pulses due to the nonlinear chirping effects. Hence it is critically important that there is a good balancing interplay between the large positive cavity dispersion and the large cavity nonlinearity (with the FWM, SPM, and XPM effects) to maintain good-quality mode-locked output pulses with minimum amplitude and timing jitters. The numerical results are consistent with the experimental results of a dual-wavelength fiber laser.

For a very small positive cavity dispersion of +0.8 ps/nm, a large cavity nonlinearity of 10/W/km and the loss difference of the two signal wavelengths of 0.3 dB, it has been found that the FWM effect alone (but not the SPM and XPM effects) contributes mostly to the GCS and that the SPM and XPM effects (but not the FWM effect) create undesirable asymmetry in the pulse shapes due to the nonlinear chirping effects. However, the small net cavity dispersion of 0.8 ps/nm is not sufficient to counteract the combined effects of FWM, SPM, and XPM from the HNLF to achieve stable and mode-locked output pulses and hence minimum amplitude and timing jitters, and this phenomenon has been numerically explained in Section 11.3.1 and experimentally demonstrated in Section 11.4.

This distortion of the pulse shapes due to the SPM and XPM effects can be overcome by using a much larger positive cavity dispersion of +85 ps/nm (but not a large negative dispersion of −85 ps/nm) to achieve the GCS. This is because the well-balancing interplay between the SPM, XPM, and FWM effects (the HNLF has 10/W/km), and the large positive cavity dispersion allows the laser to quickly reach its steady state and thus good-quality mode-locked pulse shapes can be obtained. This phenomenon has been numerically explained in Section 11.3.2 and experimentally demonstrated in Section 11.4.

Although the focus and discussion here are on two signal wavelengths, the same methodology and analysis can be used for the generation of multi-wavelength signals with bit rates beyond 10 Gbps, which can be considered as future work.

Active mode locking has been achieved by loss modulation through the use of the $LiNbO_3$ MZIM that was driven by a microwave signal. However, the $LiNbO_3$ electro-optic modulator is polarization sensitive, which could result in unstable output pulses with amplitude and timing jitters. This problem may be solved using an electro-absorption modulator with reduced polarization sensitivity rather than the MZIM [26]. Similarly, due to the use of a long cavity length (more than 1 km in this case), environmental changes such as temperature change and acoustic vibration could also make the laser unstable. Thus, to make the laser practically feasible, these difficulties must be overcome using, for example, feedback stabilization circuits, which can be considered as future work [30,31].

References

1. S.C. Bigol, Y. Frignac, G. Charlet, W. Idler, S. Borne, H. Gross, R. Dischler et al., 10.2Tbit/s (256 × 42.7 Gbit/s PDM/WDM) transmission over 100 km TeraLight™ fiber with 1.28 bit/s/MHz spectral efficiency, in *Proceedings of Optical Fiber Communication Conference*, OCF 2001, Vol. 4, Anaheim, CA, 2001, pp. PD25-1–PD25-3.
2. K. Fukuchi, T. Kasamatsu, M. Morie, R. Ohhira, T. Ito, K. Sekiya, D. Ogasahara, and T. Ono, 10.92-Tb/s (273 × 40-Gb/s) triple-band/ultra-dense WDM optical-repeatered transmission experiment, in *Proceedings of Optical Fiber Communication Conference*, OCF 2001, Vol. 4, Anaheim, CA, 2001, pp. PD24-1–PD24-3.
3. J. Renaudier, G. Charlet, O. Bertran-Pardo, H. Mardoyan, P. Tran, M. Salsi, and S. Bigo, Transmission of 100 Gb/s coherent PDM-QPSK over 16 × 100 km of standard fiber with allerbium amplifiers, *Opt. Express*, 17(7), 5112–5117, March 2009.
4. G. Charlet, J. Renaudier, H. Mardoyan, P. Tran, O.B. Pardo, F. Verluise, M. Achouche et al., Transmission of 16.4-bit/s capacity over 2550 km using PDM QPSK modulation format and coherent receiver, *IEEE/OSA. J. Lightwave Technol.*, 27(3), 153–157, February 2009.
5. Z.G. Lu, F.G. Sum, G.Z. Xiao, and C.P. Grover, A tunable mutliwavelength fiber ring laser for measuring polarization-mode dispersion, *IEEE Photon. Technol. Lett.*, 16(5), 60–62, January 2007.
6. U. Sharma, C.S. Kim, and J.U. Kang, Highly stable tunable dual-wavelength Q-switched fiber laser for DIAL applications, *IEEE Photon. Technol. Lett.*, 16(5), 1277–1279, May 2004.

7. M.P. Fok, K.L. Lee, and C. Shu, 4 × 25 GHz repetitive photonic sampler for high-speed analog-to-digital conversion, *IEEE Photon. Technol. Lett.*, 16(3), 876–878, March 2004.

8. O. Svelto, *Principles of Lasers*, 4th edn., Plenum Press, New York, 1998.

9. D.J. Kuizenga and A.E. Siegman, FM and AM mode locking of the homogeneous laser—Part I: Theory, *IEEE J. Quantum Electron.*, QE-6(11), 694–708, November 1970.

10. D.J. Kuizenga and A.E. Siegman, FM and AM mode locking of the homogeneous laser—Part II: Experimental results in a NdYAG laser with internal FM modulation, *IEEE J. Quantum Electron.*, QE-6(11), 709–715, November 1970.

11. H.A. Haus, A theory of forced mode locking, *IEEE J. Quantum Electron.*, QE-11(7), 323–330, July 1975.

12. H.A. Haus, Theory of modelocking of a laser diode in an external resonator, *Jpn. J. Appl. Phys.*, 51(8), 4042–4049, August 1980.

13. K. Vlachos, C. Bintjas, N. Pleros, and H. Avramopoulos, Ultrafast semiconductor-based fiber laser sources, *IEEE J. Sel. Topics Quantum Electron.*, 10(1), 147–154, January/February 2004.

14. J. Yao, J. Yao, and Z. Deng, Multiwavelength actively mode-locked fiber ring laser with suppressed homogeneous line broadening and reduced supermode noise, *Opt. Express*, 12(19), 4529–4534, September 2004.

15. K. Vlachos, K. Zoiros, T. Houbavlis, and H. Avramopoulos, 10 × 30 GHz pulse train generation from semiconductor amplifier fiber ring laser, *IEEE Photon. Technol. Lett.*, 12(1), 25–27, January 2000.

16. W. Zhang, J. Sun, J. Wang, and L. Liu, Multiwavelength mode-locked fiber-ring laser based on reflective semiconductor optical amplifiers, *IEEE Photon. Technol. Lett.*, 19(19), 1418–1420, October 2007.

17. Y.D. Gong, M. Tang, P. Shum, C. Lu, J. Wu, and K. Xu, Dual-wavelength 10-GHz actively mode-locked erbium fiber laser incorporating highly nonlinear fibers, *IEEE Photon. Technol. Lett.*, 17(12), 2547–2549, December 2005.

18. M. Tang, X.L. Tian, P. Shum, S.N. Fu, and H. Dong, Four-wave mixing assisted self-stable 4×10 GHz actively mode-locked erbium fiber ring laser, *Opt. Express*, 14(5), 1726–1730, March 2006.

19. R. Hayashi, S. Yamashita, and T. Saida, 16-wavelength 10-GHz actively mode-locked fiber laser with demultiplexed outputs anchored on the ITU-T grid, *IEEE Photon. Technol. Lett.*, 15(12), 1692–1694, December 2003.

20. J. Vasseur, M. Hanna, J. Dudley, J.-P. Goedgebuer, J. Yu, G.-K. Chang, and J.R. Barry, Alternate multiwavelength picosecond pulse generation by use of an unbalanced Mach-Zehnder interferometer in a mode-locked fiber ring laser, *IEEE J. Quantum Electron.*, 43(1), 85–96, January 2007.

21. J. Vasseur, M. Hanna, J. Dudley, and J.-P. Goedgebuer, Alternate multiwavelength modelocked fiber laser, *IEEE Photon. Technol. Lett.*, 16(8), 1816–1818, August 2004.

22. J.-N. Maran, S. LaRochelle, and P. Besnard, Erbium-doped fiber laser simultaneously mode locked on more than 24 wavelengths at room temperature, *Opt. Lett.*, 28(21), 2082–2084, November 2003.

23. Y.L. Yang, S.C. Tjin, N.Q. Ngo, and R.T. Zheng, Double-ring cavity configuration of actively mode-locked multi-wavelength fiber laser with equally tunable wavelength spacing, *Appl. Phys. B*, 80, 445–448, 2005.

24. S. Pan and C. Lou, Multiwavelength pulse generation using an actively mode-locked erbium-doped fiber ring laser based on distributed dispersion cavity, *IEEE Photon. Technol. Lett.*, 18(4), 604–606, February 2006.

25. L.R. Chen, A.L.K. Cheng, C. Shu, S. Doucet, and S. LaRochelle, 46×2.5 GHz mode-locked erbium-doped fiber laser with 25-GHz spacing, *IEEE Photon. Technol. Lett.*, 19(23), 1871–1873, December 2007.

26. C. O'Riordan, M.J. Connelly, P.M. Anandarajah, R. Maher, and L.P. Barry, Lyot filter based multiwavelength fiber ring laser actively mode-locked at 10 GHz using an electroabsorption modulator, *Opt. Commun.*, 281, 3538–3541, 2008.

27. S. Pan and C. Lou, Analysis of gain competition suppression in multiwavelength actively mode-locked erbium-doped fiber lasers incorporating a highly nonlinear fiber, *IEEE J. Quantum Electron.*, 44(3), 245–253, March 2008.

28. G.P. Agrawal, *Nonlinear Fiber Optics*, 3rd edn., Academic Press, San Diego, CA, 2001.

29. G. Keiser, *Optical Fiber Communications*, 3rd edn., McGraw Hill, Singapore, 2000.

30. X. Shan, T. Woddowson, A.D. Ellis, and A.S. Siddiqui, Very simple method to stabilize mode-locked erbium fiber lasers, *Electron. Lett.*, 32(11), 1015–1016, May 1996.

31. H. Takara, S. Kawanishi, and M. Sarawatari, Stabilization of a mode-locked Er-doped fiber laser by suppressing the relaxation oscillation frequency component, *Electron. Lett.*, 31(4), 292–293, 1995.

32. S. Pan and C. Lou, Stable multiwavelength dispersion-tuned actively mode-locked erbium-doped fiber ring laser using nonlinear polarization rotation, *IEEE Photon. Technol. Lett.*, 18(13), 1451–1453, July 2006.

Appendix A: Er-Doped Fiber Amplifier: Optimum Length and Implementation

A.1 Determination of the Optimum Length

In order to determine the suitable EDF length for the MLFRL, the stimulated absorption and emission cross sections of the pump and signal beam in the EDF must be known first.

If no erbium ions are excited, the absorption cross section can be determined directly from an attenuation measurement using [1]

$$\sigma_{p,s}^{a} = \frac{att(\lambda_{p,s})}{10\log_{10}e \cdot 2\pi \int_{0}^{a} \rho_{Er}(r)I^{01}(\lambda_{p,s},r)rdr} \tag{A.1}$$

where

subscript p or s denotes the pump or signal wavelength
$\rho_{Er}(r)$ is the erbium concentration
$att(\lambda)$ is the attenuation in dB/m at wavelength λ
a is the core and doping radius
I^{01} is the normalized LP_{01} mode

The emission cross section can be obtained from the gain measurement, which implies that all erbium ions are excited:

$$\sigma_{p,s}^{e} = \frac{g(\lambda_{p,s})}{10\log_{10}e \cdot 2\pi \int_{0}^{a} \rho_{Er}(r)I^{01}(\lambda_{p,s},r)rdr} \tag{A.2}$$

where $g(\lambda)$ is the gain in dB/m. From the above equations, one can notice that $2\pi \int_{0}^{a} \rho_{Er}(r)I^{01}(\lambda_{p,s},r)rdr$ is in fact the overlap between the waveguide mode and the doped region. From the refractive index and erbium concentration profiles described above, we assume near full (98%) overlap between these two elements, therefore the resultant integral gives rise to ~$0.98\rho_{Er0}$. Alternatively, the emission cross sections can be obtained using the following relationship [2]:

$$\sigma^e(\lambda) = \sigma^a(\lambda)\exp\left(\frac{\varepsilon - (hc/\lambda)}{kT}\right) \tag{A.3}$$

where

ε is the net free energy required to excite one Er^{3+} ion from $^4I_{15/2}$ to $^4I_{13/2}$ state at temperature T

k is the Boltzmann's constant

It should be noted that the gain or attenuation measurement should be performed on a short fiber to avoid saturation due to ASE. In addition, the pump power must be high enough to fully invert all Er^{3+} ions in the fiber. Furthermore, the intrinsic losses of the associated optics must be considered carefully.

From the measurements, the absorption and emission cross sections for both the pump wavelength (980 nm) and signal wavelength (1550 nm) are $\sigma_p^a = 2.0192 \times 10^{-25}\,\mathrm{m}^2$, $\sigma_p^e = 4.1833 \times 10^{-33}\,\mathrm{m}^2$, $\sigma_s^a = 1.3541 \times 10^{-25}\,\mathrm{m}^2$, and $\sigma_s^e = 2.0291 \times 10^{-25}\,\mathrm{m}^2$, respectively.

The following steps are taken to determine the suitable fiber length for the MLFRL. At first, a piece of EDF with length L is chosen. R in the following equation is determined by measuring signal and pump powers at the output, P_s^{out} and P_p^{out} for given input signal and pump powers, P_s^{in} and P_p^{in} [3].

$$
\begin{aligned}
&\left[(P_s^{out} - P_s^{in}) \cdot \left(\frac{1}{I_{ss}^e} - \frac{\sigma_p^e}{\sigma_p^a I_{ss}^a}\right) + \ln\left(\frac{P_s^{out}}{P_s^{in}}\right) \cdot \frac{1}{R}\right] \frac{\sigma_p^e}{\sigma_s^a} \\
&= \left[(P_p^{in} - P_p^{out}) \cdot \left(\frac{\sigma_s^e}{\sigma_s^a I_{sp}^a} - \frac{1}{I_{sp}^e}\right) - \ln\left(\frac{P_p^{in}}{P_p^{out}}\right) \cdot \frac{1}{R}\right]
\end{aligned} \tag{A.4}
$$

$$I_{sp}^a = \frac{hc}{\lambda_p \tau \sigma_p^a} \qquad I_{sp}^e = \frac{hc}{\lambda_p \tau \sigma_p^e} \tag{A.5}$$

$$I_{ss}^a = \frac{hc}{\lambda_s \tau \sigma_s^a} \qquad I_{ss}^e = \frac{hc}{\lambda_s \tau \sigma_s^e}$$

where

h is the Planck's constant

τ is the spontaneous emission decay life time, which is about 10 ms

For maximal gain, $dP_s/dP_p = 0$, hence, the optimum output pump power will be

$$P_{p,op}^{out} = \frac{1}{R\left(\dfrac{\sigma_s^e}{\sigma_s^a I_{sp}^a} - \dfrac{1}{I_{sp}^e}\right)} \tag{A.6}$$

The maximal gain, G is calculated using numerical method for arbitrary P_s^{in} and P_p^{in}, by the following equation

$$\frac{\lambda_s}{\lambda_p}\frac{P_s^{in}}{P_{p,op}^{out}}(G-1)+\frac{\sigma_p^a}{\sigma_s^a}\ln(G)=\frac{P_p^{in}}{P_{p,op}^{out}}-1-\ln\left(\frac{P_p^{in}}{P_{p,op}^{out}}\right) \quad (A.7)$$

Subsequently, the absorption constant, α_s and intrinsic saturation power, P_s^{IS} for the signal beam are measured. Without pump power, $P_s^{out}\approx P_s^{in}\exp(-\alpha_sL)$ for small input signal power, when P_s^{in} approaches P_s^{IS}, $P_s^{out}\approx P_s^{IS}\exp(1-\alpha_sL)$. Finally, the optimum fiber length of the maximal gain is

$$L_{op}=-\frac{1}{\alpha_s}\cdot\left\{\ln(G)+\frac{hc}{\lambda_sP_s^{IS}}\left[\frac{(P_{p,op}^{out}-P_p^{in})\lambda_p}{hc}+\frac{P_s^{in}\lambda_s}{hc}(G-1)\right]\right\} \quad (A.8)$$

The maximal gain and optimum length for the EDF used in our experiment are found out to be about 30 dB and 25.9 m.

A.2 Implementation and Precautions

A typical experimental procedure can be as follows:

- Get the performance analysis of each individual optical component.
- Measure the optical loss of the ring. With all the optical components connected in an open ring structure, i.e., broken ring, monitor the wavelength and optical power of the CW signal from the 90:10 output coupler under the saturation pumping condition.
- Close the optical ring and monitor the average optical power at the output of the 90:10 fiber coupler and hence the optical power available at the photodetector is 3 dB less than the measured value after a 3 dB coupler.
- Determine if an optical amplifier is required for detecting the optical pulse train or is sufficient for opto-electronic RF feedback condition.
- Set the biasing condition for the modulator.
- Tune the synthesizer or electrical phase for the generation and locking of the optical pulse train.
- Monitor the laser output using optical oscilloscope and spectrum analyzer.

Some precaution measures must be taken, as follows:

- Optical damage threshold—excessive optical intensities can lead to dielectric breakdown and destruction of the optical fiber itself. The bulk optical damage threshold for silica is about $50\,GW/cm^2$ at 1064 nm for a single pulse focus to a 5 μm diameter. Assuming an effective mode area of $50\,\mu m^2$, the maximum peak power that can be guided in the standard optical fiber is about 25 kW, which corresponds to an energy of 250 μJ for a 10 ns pulse. The peak power–handling can be pushed farther by using fibers with a larger mode-field diameter, or multimode fibers when a diffraction-limited beam is not required.

- Maximum optical power—ensure that the input optical power to any optical detecting equipment (photodetector, optical oscilloscope, optical spectrum analyzer, etc.) is below the maximum rated power since we are operating at the nonlinear region (high optical power). An optical attenuator can be added prior to the measuring equipment if necessary.

- Electrostatic discharge—high-speed photodetector is sensitive to electrostatic discharge and can be damaged by it. Therefore, some care must be taken into account to prevent damage of equipment. Always ground cables and connectors prior to connecting them to the photodetector output. Use the wrist strap while handling the detector. Discharge the AC-coupled instruments before connecting them.

- Sensitivity of the equipment—ensure the triggering signal of the measuring equipment (e.g., optical spectrum analyzer) is above the sensitivity of the equipment.

- When laying the fiber in the fiber polarization controller, make sure that the fiber is in contact with the inside of the groove loops, but is not pulled too snug against the groove as this would cause optical losses due to induced birefringence, as the paddles are rotated with respect to each other.

References

1. B. Pedersen, A. Bjarklev, J.H. Povlsen, K. Dybdal, and C.C. Larsen, The design of erbium-doped fiber amplifiers, *J. Lightwave Technol.*, 9(9), 1105–1112, 1991.
2. W.J. Miniscalco, Erbium-doped glasses for fiber amplifiers at 1500 nm, *J. Lightwave Technol.*, 9(2), 234–250, 1991.
3. M.C. Lin and S. Chi, The gain and optimal length in the erbium-doped fiber amplifiers with 1480 nm pumping, *IEEE Photon. Technol. Lett.*, 4(4), 354–356, 1992.

Appendix B: MATLAB® Programs for Simulation

Generation of Mode-Locked Laser Pulses

```
% MLLsimple.m
% simple model of mode locked laser
% written by Lam Quoc Huy

global Ts;            % sampling period
global Fcar;          % carrier frequency (optical frequency)
c_const = 3e8;        % speed of light

lamda = 1550e-9;      % m
Fcar = c_const/lamda;

Ts = 1e-13;
N = 1024;             % number of samples in a block: Tblk = N *
Ts = 204.8 ps

% Amplifier parameters:
  GssdB = 20;         % (dB)
  PoutsatdB = 10;     % (dBm)
  NF = 8;             % (dB)

% filter bandwidth
  lamda3dB = 1e-9;    % m
  f3dB = lamda3dB*(1e11/0.8e-9);

% modulator parameters
  alpha = -0.07;
  epsilon = 40;       % (dB) extinction ratio

% modulation parameters
  m = 0.5;            % modulation index
  fm = 1e10;          % modulation frequency

% Loss
  loss = 10;          % dB
  atten = 1/10^(loss/20);

% generate an initial block of signal Ein
% Ein = 1e-3*gausswin(N,2);
Ein = wgn(N,1,-40,'complex');

Eout = Ein;
Eo = Ein;
for ii = 1:500
  [Eo,G] = amp_simp(Eo,GssdB,PoutsatdB,NF);
```

```
% [Eo,G] = AmpSimpNonoise(Eo,GssdB,PoutsatdB);  % no noise
  Eo = fft(Eo,N*4);
% Eo = filter_bessel(Eo,f3dB,G);
% Eo = filter_gaus1(Eo,f3dB);
  Eo = filter_gaus(Eo,f3dB,1);
  Eo = ifft(Eo);
  Eo = modInt(Eo(1:N),alpha,epsilon,m,fm,0.5);
% Eo = modPhase(Eo(1:N),m,fm);
  Eo = Eo*atten;
  if mod(ii,10)==0
    Eout = [Eout , Eo];
  end
end
Eout = Eout/atten;
close all
%mesh (abs(Eout'),'edgecolor','black','meshstyle','row',
'facecolor','none');
Iout = Eout.*conj(Eout);
mesh (Iout','meshstyle','row','facecolor','none');
axis tight;
% set(gca,'XTick',tt_mark);
% set(gca,'XTickLabel',tt_tick);
% set(gca,'XDir','reverse');
Xlabel('T (0.1ps)');
% set(gca,'YTick',yy_mark);
% set(gca,'YTickLabel',yy_tick);
Ylabel('Pass number');
Zlabel('intensity (W)');

N1 = size(Eout,2);
dPhi = angle(Eout(2:N,N1)) - angle(Eout(1:N-1,N1));
% figure (2);
% plot(dPhi);

Tp = fwhm(Iout(:,N1))*Ts;
pulse_alpha = 2*log(2)/(Tp^2);
pulse_beta = (dPhi(N/2+100) - dPhi(N/2-100))/200/Ts/Ts;
chirp = pulse_beta/pulse_alpha

Kmag = 8;
Nplot = 100;
Eoutfreq = fft(Eout(:,N1),N*Kmag);
Ioutfreq = Eoutfreq.*conj(Eoutfreq)/(N*Kmag)^2;
figure(2);
ind = (- Nplot/2 : Nplot/2)';
freq = ind/Ts/N/Kmag;
ind = mod((ind + N*Kmag),N*Kmag)+1;
plot(freq,Ioutfreq(ind));

n=1;
n = 2*n;
```

```
Tfil = exp(-log(2)*(2/f3dB*freq).^n);    % n order gaussian
filter VPI
Tfil = Tfil *max(Ioutfreq(ind));
hold on
plot(freq,Tfil,'r');

% plot the gaussian fit curve
% gaussFit(Iout(:,N1));

pulseBW = fwhm(Ioutfreq(ind))/Ts/N/Kmag
Tp = fwhm(Iout(:,N1))*Ts
TBP = pulseBW*Tp

% MLLsimpleDetune.m
% simple model of mode locked laser with detuning
% written by Lam Quoc Huy

clear all
close all

global Ts;            % sampling period
global Fcar;          % carrier frequency (optical frequency)
c_const = 3e8;        % speed of light

lamda = 1550e-9;      % m
Fcar = c_const/lamda;

Ts = 0.1e-12;
N = 1/2*2048;         % number of samples in a block: Tblk = N*
Ts = 102.4 ps

% Amplifier parameters:
  GssdB = 20;         % (dB)
  PoutsatdB = 10;     % (dBm)
  NF = 8;             % (dB)

% filter bandwidth
  lamda3dB = 1.2e-9;  % m
  f3dB = lamda3dB*(1e11/0.8e-9);
% modulator parameters
  alpha = -0.007;
  epsilon = 40;       % (dB) extinction ratio

% modulation parameters
  m = 0.5;            % modulation index
  fm = 10e9;          % modulation frequency
  NHar = 1000;        % harmonic order
  Ts = 1/fm/N;        % recalculate Ts so that Tm = N*Ts
  f_detune = 0.5e5;   % detuned frequency
  delay_per_pass = round(NHar*f_detune/fm^2/Ts);

% Loss
  loss = 10;          % dB
  atten = 1/10^(loss/20);
```

```
% calculate the number of samples at the ends of the window
must be
  % attenuated
  p = floor( (N*Ts*fm -1)/2);
  if (p<0)
     Ntrunc = 1;
  else
     Ntrunc = round( (N - (2*p+1)/fm/Ts)/2);
  end

% generate an initial block of signal Ein
% Ein = 1e-13*gausswin(N,2);
Ein = wgn(N,1,-40,'complex');

Eout = Ein;
Eo = Ein;
N_pass = 500;
for ii = 1:N_pass
  %[Eo,G] = amp_simp(Eo,GssdB,PoutsatdB,NF);
  [Eo,G] = AmpSimpNonoise(Eo,GssdB,PoutsatdB); % no noise
  Eo = fft(Eo);
  Eo = filter_gaus(Eo,f3dB,1);
  Eo = ifft(Eo);

  if (delay_per_pass>0)
     Et = Eo(N-delay_per_pass+1:N);
     Eo(delay_per_pass+1:N) = Eo(1:N-delay_per_pass);
     Eo(1:delay_per_pass) = Et;
  else
     Et = Eo(1:-delay_per_pass);
     Eo(1:N + delay_per_pass) = Eo(1 -delay_per_pass:N);
     Eo(N+delay_per_pass+1:N) = Et;
  end

% Eo = modInt(Eo(1:N),alpha,epsilon,m,fm,0.5); % MZI modulator
  Eo = modInt_theory(Eo(1:N),m,fm);             % Gaussian
approximation modulator
  Eo = Eo*atten;

  % ----------------------------
  % attenuate the samples at the two ends
  % comment out those codes for faster execution in case N*Ts
    = (2p+1)Tm

% temp = exp( ( (1:Ntrunc)' - Ntrunc)/5);
% Eo(1:Ntrunc) = Eo(1:Ntrunc).*temp;
% temp = exp( (N - Ntrunc - (N-Ntrunc:N)')/5);
% Eo(N - Ntrunc:N) = Eo(N - Ntrunc:N).*temp;
  % --------- End ---- attenuate the samples ----------
  % --------- output the peak power of the pulse for every
    round-
```

```
   [PeakVal MaxInd] = max(Eo.*conj(Eo));
   PulsePeak(ii) = PeakVal;
   PulsePeakPos(ii) = MaxInd;
   GOut(ii) = G;

   % ----- End -- output the peak power
   if mod(ii,N_pass/50)==0
      Eout = [Eout , Eo];
   end
   if (ii>N_pass - 10)
      Eout = [Eout , Eo];
   end
end
% Eout = Eout/atten;
% -------------- Display the results ---------
% mesh (abs(Eout'),'edgecolor','black','meshstyle','row',
'facecolor','none');
Iout = Eout.*conj(Eout);
mesh (Iout','meshstyle','row','facecolor','none');
axis tight;
% set(gca,'XTick',tt_mark);
% set(gca,'XTickLabel',tt_tick);
% set(gca,'XDir','reverse');
xlabel('T (0.1fs)');
% set(gca,'YTick',yy_mark);
% set(gca,'YTickLabel',yy_tick);
ylabel('Pass number');
zlabel('intensity (W)');

N1 = size(Eout,2);
dPhi = angle(Eout(2:N,N1)) - angle(Eout(1:N-1,N1));
% figure (2);
% plot(dPhi);

Tp = fwhm(Iout(:,N1))*Ts
pulse_alpha = 2*log(2)/(Tp^2);
[temp PeakPos] = max(Iout(:,N1));
pulse_beta = (dPhi(PeakPos+100) - dPhi(PeakPos-100))/200/Ts/Ts;
% chirp = pulse_beta/pulse_alpha
% ----- plot spectrum of the lastest pulse -----------
% Kmag = 8;
% Nplot = 100;
% Eoutfreq = fft(Eout(:,N1),N*Kmag);
% Ioutfreq = Eoutfreq.*conj(Eoutfreq)/(N*Kmag)^2;
% figure(2);
% ind = (- Nplot/2 : Nplot/2)';
% freq = ind/Ts/N/Kmag;
% ind = mod((ind + N*Kmag),N*Kmag)+1;
% plot(freq,Ioutfreq(ind));
```

```
% n = 1;
% n = 2*n;
% Tfil = exp(-log(2)*(2/f3dB*freq).^n); % n order gaussian
filter VPI
% Tfil = Tfil *max(Ioutfreq(ind));
% hold on
% plot(freq,Tfil,'r');
% ------ End ---------- plot spectrum
% plot the gaussian fit curve
% gaussFit(Iout(:,N1));
pulse_position = (find(Iout(:,N1)==max(Iout(:,N1))) - N/2)*Ts
% calculation using the theoretical equations
wd = (2*pi*f3dB);
be = 2*log(2)/wd^2;
wm = 2*pi* fm;
al = 1/8*wm*(m*wm + sqrt(m^2*wm^2 + 2*m*wd^2/log(2)));
TpCal = sqrt(2*log(2)/al);
sprintf('Calculation using Li 2000 with corrected m: Tp = %e',
TpCal)
td = NHar*f_detune/fm^2
ts = td/4/al/be
x = fm/f3dB;
delta_max_Li = sqrt(m*2*log(2))*x/4/NHar /
(1+x*sqrt(m*2*log(2)))*fm
Gss = 10^(GssdB/10);
% delta_max_Go = sqrt(2*log(2)*log(sqrt(Gss)*atten/
sqrt(1+4*al*be))) * x/pi/NHar*fm
% delta_max_2Tp = (0.589*sqrt(m)*x/NHar -
0.575*sqrt(sqrt(m))*x^1.5/NHar )*fm
delta_max_Go = sqrt(4*be*log(sqrt(Gss)*atten/
sqrt(1+4*al*be)))/NHar *fm
delta_max_2Tp = 2*al*be/NHar/(1+4*al*be)

% plot the peak
figure(2)
subplot(2,1,1)
plot(PulsePeak)
subplot(2,1,2)
plot(PulsePeakPos - N/2)
% PulsePeak(N_pass)

% store the summarised results in a file
% FileName = 'temp.txt';
% FId = fopen(FileName,'r');
% if FId ==-1
%    FId = fopen(FileName,'w');
%    fprintf(FId,'Calculation:');
%    fprintf(FId,'\n\tTs \t\t\t Tp \t\t\t PulsePoss \t\t td');
%    fprintf(FId,'\n\t%e \t %e \t %e \t %e',Ts,TpCal,ts,td);
%    fprintf(FId,'\nSimulation:');
```

```
%    fprintf(FId,'\n\tTs \t\t N \t\t N_pass \t\t Tp \t\t
     PulsePoss tt Riple');
%    fclose(FId);
% else
%    fclose(FId);
% end
% FId = fopen(FileName,'a');
% temp = PulsePeak(3*N_pass/4:N_pass);
% Riple = (max(temp) - min(temp))/mean(temp);
% fprintf(FId,'\n%e \t %d \t %d \t %e \t %e \t %e',Ts,N,N_
pass,Tp,pulse_position,Riple);
% fclose(FId);


function Eout = modInt(Ein,alpha,epsilon,m,fm,bias)
% modIntQP(Ein,alpha,esilon,m,fm)
% intensity modulator model
% written by Lam Quoc Huy

% modulator parameters
%    chirp alpha factor: alpha
%    extinction ratio: esilon (dB)
% modulation parameters
%    modulation index: m
%    modulation frequency: fm
%    bias: 0:1

global Ts;

N = size(Ein,1);
k = (1:N)';
Vm = m*cos(2*pi*fm*Ts*(k-N/2));
ext = 1 - 4/pi*atan(1/sqrt(10^(epsilon/10)));
%delta_phi = pi/4*(1-ext*Vm);
delta_phi = pi/4*(2- bias*2 - ext*Vm);
Eout = Ein.*cos(delta_phi).*exp(i*alpha*delta_phi);


function Eout = modInt_theory(Ein,m,fm)
% modInt_theory(Ein,m,fm)
% intensity modulator model
% written by Lam Quoc Huy

% modulator parameters
%    chirp alpha factor: alpha
%    extinction ratio: esilon (dB)
% modulation parameters
%    modulation index: m
%    modulation frequency: fm

global Ts;
```

```
N = size(Ein,1);
k = (1:N)';
Eout = Ein.*exp(-m/4*(2*pi*fm*Ts)^2*(k-N/2).*(k-N/2));
```

function [Eout,gain] = amp_simp(Ein,GssdB,PoutsatdB,NF)
```
% amp_simp(Ein,GssdB,PoutsatdB,NF)
% simple model of optical amplifier. The model includes the
  gain
% saturation and the ASE noise
% written by Lam Quoc Huy

% Amplifier parameters:
%    small signal gain: GssdB (dB)
%    output saturation power: PoutsatdB (dBm)
%    noise figure: NF (dB)
%
% Simulation parameters:
%    Gain Tolerance: tol, used as the threshold value to exit
the gain calculation loop.
%

% The input is a column vector containing block N samples of
the optical signal sampling at the
% rate 1/Ts
% The output is calculated using
%    Eout = Ein*sqrt(G) + Enoise
% where: G is the saturated gain
%          G = Gss*exp(-(G-1)Pin/Psat) (eq1)
%      Enoise is the complex noise with the noise power
%          Pase = (10^(NF/10)) * (G-1)hf/2 * BW
%          BW = 1/Ts

global Ts;    % sampling period
global Fcar;
h_plan = 6.626e-34;
```

function [Eout,gain] = AmpSimpNonoise(Ein,GssdB,PoutsatdB)
```
% amp_simp(Ein,GssdB,PoutsatdB,NF)
% simple model of optical amplifier. The model includes the
gain
% saturation without noise
% written by Lam Quoc Huy

% Amplifier parameters:
%    small signal gain: GssdB (dB)
%    output saturation power: PoutsatdB (dBm)
%
% The input is a column vector containing block N samples of
the optical signal sampling at the
% rate 1/Ts
% The output is calculated using
```

```
%    Eout = Ein*sqrt(G)
% where: G is the saturated gain
%          G = Gss*exp(-(G-1)Pin/Psat) (eq1)

Gss = 10^(GssdB/10);
Poutsat = (10^(PoutsatdB/10))/1000;
Psat = Poutsat*(Gss-2)/Gss/log(2);
%Pinsat = 2* Poutsat/Gss;

N = size(Ein,1);
Pin = (sum(Ein.*conj(Ein))/N);

% numerical calculation of G from the equation G = (Gss -
lnG)*Psat/Pin + 1
tol = 0.05;   % tolerance for G calculation
step = Gss/2;
G = Gss;
err = 10;
while (err > tol)
   G1 = Gss*exp(-(G-1)*Pin/Psat);
   err = G1   G;
   if err>0
      if step <0
         step = -step/2;
      end
   else
      if step >0
         step = -step/2;
      end
      err = -err;
   end
   G = G + step;
end
G = G - step;

Eout = sqrt(G)*Ein;
gain = G;


function Eout = filter_gaus(Ein,f3dB,n)
% filter_gaus(Ein,f3dB,n)
% filter the input signal with the n order gaussian filter
% written by Lam Quoc Huy

% T(f) = exp(-log(sqrt(2))*(2/f3dB/Ts/N)^2n*(k.^2n))

global Ts;

N = size(Ein,1);
% the k element in the Ein corresponds to the frequency of
%    (k-N/2)/Ts/N
k = (1:N) -1;
k(N/2+1:N) = k(N/2+1:N) - N;
k = k';
```

```
n = 2*n;
% Eout = Ein.*exp(-log(sqrt(2))*(2/f3dB/Ts/N)^n*(k.^n));
% n order gaussian filter
temp = log(sqrt(2))*(2/f3dB/Ts/N)^n;
Eout = Ein.*exp(-temp*(k.^n));    % n order gaussian filter VPI


function width = fwhm(x)
% return the Full Width at Half Maximum of the pulse x
% written by Lam Quoc Huy
nsize = size(x,1);
% half_peak = max(x)/2;
[peak ind] = max(x);
half_peak = peak/2;        % 3 dB
for iii=1:nsize-1
   if (x(iii)<=half_peak) && (x(iii+1)>half_peak)
      break;
   end
end
for jjj=ind:nsize-1
   if (x(jjj)>=half_peak) && (x(jjj+1)<half_peak)
      break;
   end
end
width = jjj-iii;


function gain = gain_saturated(Pin,GssdB,PoutsatdB)
% calculate the gain of the amplifier given the input power,
the small
% signal gain and saturation power
% written by Lam Quoc Huy

% where: G is the saturated gain
%         G = Gss*exp(-(G-1)Pin/Psat) (eq1)

% Amplifier parameters:
%      GssdB = 20            % (dB)
%      PoutsatdB = 10        % (dBm)
%      NF = 7                % (dB)

Gss = 10^(GssdB/10);
Poutsat = (10^(PoutsatdB/10))/1000;
Psat = Poutsat*(Gss-2)/Gss/log(2);
% Pinsat = 2* Poutsat/Gss;

% numerical calculation of G from the equation G = (Gss -
lnG)*Psat/Pin + 1
tol = 0.05;                  % tolerance for G calculation
step = Gss/2;
G = Gss;
err = 10;
```

```
while (err > tol)
   G1 = Gss*exp(-(G-1)*Pin/Psat);
   err = G1 - G;
   if err>0
      if step <0
         step = -step/2;
      end
   else
      if step >0
         step = -step/2;
      end
      err = -err;
   end
   G = G + step;
end
G = G - step;
gain = G;
```

Generation of Bound Solitons from Phase-Modulated Mode-Locked Laser Pulses

```
%%%%%%%%%%%%%%%%%%%%%%%%%%%%%%%%%%%%%%%%%%%%%%%%%%%%%%%%%%%%%%%
% S O L I T O N   O P T I C A L   C O M M U N I C A T I O N S   %
%
%
% global h0

% Solitonpairs.m
clear;
close all
clc;
global Ts;               % sampling period
global Fcar;             % carrier frequency (optical
                         frequency)
global Vw;
global w;
global tstep;
% cla reset
h0 = figure(1);

c_const = 3e8 ;
ng = 1.47;
lamda = 1.550e-6 ;
% ==========================================================
% Fiber parameters
   Length_SMF = 70;      % Length of a span in meter
```

```
  nz = Length_SMF/100;
  alpha_indB = 0.0*1e-3; % attenuation (dB/km) -> dB/m
  D_SMF = 1*17e-6;          % GVD (ps/nm.km); if anomalous
dispersion(for compensation),D is negative
  n2_SMF = 1*2.6e-20;     % nonlinear index (m^2/W)
  Aeff_SMF = 80e-12;      % effective area (m^2)
% Slope Dispersion
  S_SMF = 0.06e3;          % ps/(nm^2.km) in SI unit

% CALCULATED QUANTITIES of SMF

  alpha_loss_SMF = 1*log(10)*alpha_indB/10;   % alpha (1/m)
  beta2_SMF = -D_SMF*lamda^2/(2*pi*c_const);  % beta2
(ps^2/km);
% ----------------------------------------------------------

  % beta 3 can be calculated from the Slope Dispersion (S) as
follows:]
  beta3_SMF = (S_SMF - (4*pi*c_const/(lamda^3))*beta2_SMF)/
((2*pi*c_const/(lamda^2))^2);
% ----------------------------------------------------------
  gamma_SMF = 2*pi*n2_SMF/(lamda*Aeff_SMF);  % nonlinearity
coef (km^-1.W^-1)
% ==========================================================
% filter bandwidth
  lamda3dB = 4.8e-9;%1.2e-9;     % 1.2nm
  f3dB = lamda3dB*(1e11/0.8e-9); % Hz
% Amplifier parameters:
  GssdB = 60;               % (dB)
  PoutsatdB = 36;           % (dBm)
  NF = 8;                   % (dB)
% EDF fiber parameters
  Length_EDF = 0;           % Length of a span in meter
  alpha_indB = 0.0*1e-3; % attenuation (dB/km) -> dB/m
  D_EDF = -10e-6;           % GVD (ps/nm.km); if anomalous
dispersion(for compensation),D is negative
  n2_EDF = 1*2.6e-20;     % nonlinear index (m^2/W)
  Aeff_EDF = 45e-12;      % effective area (m^2)
% Slope Dispersion
  S_EDF = 0.06e3;           % ps/(nm^2.km) in SIunit
% CALCULATED QUANTITIES of EDF
  alpha_loss_EDF = 1*log(10)*alpha_indB/10; % alpha (1/m)
  beta2_EDF = -D_EDF*lamda^2/(2*pi*c_const)    % beta2 (ps^2/km);
  % ----------------------------------------------------------
  % beta 3 can be calculated from the Slope Dispersion (S) as
follows:]
  beta3_EDF = (S_EDF - (4*pi*c_const/(lamda^3))*beta2_EDF)/
((2*pi*c_const/(lamda^2))^2);
% ----------------------------------------------------------
  gamma_EDF = 2*pi*n2_EDF/(lamda*Aeff_EDF);  % nonlinearity
coefficient (km^-1.W^-1)
```

```
% ===========================================================
% Average parameters
% Loss
  loss = 8;                % dB
  atten = 1/10^(loss/20);

% to test dispersion let TotDisp=-2e-6, SegNum=2.5*Zo/h
% run again with U1 removed and TotDisp=2e-6
Length_total = Length_SMF+Length_EDF;
TotDisp = (D_SMF*Length_SMF+D_EDF*Length_EDF)/Length_total;
% TotDisp = 17*1e-6 ;

ro      = 3e-6 ;
n2      = 3.2*1e-20 ;
Beta2   = -(TotDisp*lamda^2)/(2*pi*c_const);
Beta3   = 0 ;
Gamma   = (gamma_SMF*Length_SMF+gamma_EDF*Length_EDF)/Length_
total ;
% Gamma = (2*pi*n2)/(lamda*pi*ro^2) ;
alphadB = 0;
Alpha=log(10)*alphadB/(10*1000); % = field loss per meter
% ===========================================================
% modulator parameters
Vpi = 6;      % Volt
Vbias = 0;    % Volt
Vm = 0;       % Volt
m = Vm/Vpi;   % modulation index
fm = 1e9;     % modulation frequency
p0 = -0*pi/2;
% ===========================================================
% Soliton pulse parameters
Norder = 1;       % Order of soliton
Tfwhm = 8.0e-12;
To = Tfwhm/1.763; % Pulsewidth of soliton
Tb = 54e-12;      % Pulses distance
teta = 0*pi/4;    % Relative phase difference between 2
solitons
r = 1;            % Amplitude ratio
Ci = 0;           % Chirping factor of pulse
qo = Tb/2/To      % Space between 2 solitons

Ld = To^2/abs(Beta2)
Zo = pi*Ld/2 ;
Po = Norder^2/(Gamma*Ld);
P0 = 200e-3;
% ===========================================================
Ns   = 2^12;
Ts   = 0.1e-12;
tstep = Ts*(-Ns/2:1:Ns/2-1);
```

```
w       = fftshift((2*pi/((Ns-1)*Ts))*(-Ns/2:1:Ns/2-1));
Usol    = sqrt(8*Po).*sech(tstep./To).*exp(-j*teta).*exp
(-j*Ci*tstep.^2./2/To.^2);
% Usol = sqrt(Po).*(sech(tstep./To-qo).*exp
(-j*teta)+r.*sech(r.*(tstep./To+qo)).*exp(j*teta)).*exp
(-j*Ci*tstep.^2./2/To.^2) ; %
% Usol = sqrt(P0).*(sech(tstep./To-qo).*exp
(-j*teta)+r.*sech(r.*(tstep./To+qo)).*exp(+j*teta)).*exp
(-j*Ci*tstep.^2./2/To.^2) ;
% Usol = sqrt(P0).*(r*sech(tstep./
To-qo).*exp(+j*teta)+sech(tstep./To-0)+r.*sech(r.*(tstep./
To+qo)).*exp(j*teta)).*exp(-j*Ci*tstep.^2./2/To.^2) ;
% Usol = sqrt(P0).*(r*sech(tstep./To-1*qo).*exp(+j*teta)+
r*sech(tstep./To-qo/3).*exp(-j*teta/1)…
% +sech(tstep./To+qo/3).*exp(+j*teta/1)+r.*sech(r.*(tstep./
To+1*qo)).*exp(-j*teta)).*exp(-j*Ci*tstep.^2./2/To.^2) ;
% Eo = Usol;
plot(tstep,abs(Usol).^2)
grid
ARR2(1,:) = abs(Usol).^2 ;
Up(1,:) = Usol;
pause(0.1)
% ==========================================================
N_pass = 1000;
for ii = 2:N_pass
% h1 = waitbar((ii-1)/N_pass,'The program is running…');
% [Usol,G] = AmpSimpNoise(Usol,GssdB,PoutsatdB,NF);
% [Usol,A3d,z]=hconst(Usol,Length_EDF,5,1*beta2_EDF,1*beta3_
EDF,alpha_loss_EDF,1*gamma_EDF,0); % Fiber section #1
% Usol = GaussLPfilt(tstep,Usol,1,f3dB);
% [Eo,A3d,z]=hconst(Eo,Length_SMF,5,1*beta2_SMF,1*beta3_
SMF,alpha_loss_SMF,1*gamma_SMF,0); % Fiber section #2
% Eo = modInt(Eo(1:N),alpha,epsilon,m,fm,0.5); % MZI modulator
Usol = pha_mod(Usol(1:Ns),Vm,Vbias,Vpi,fm,p0); % Phase
modulator
  % Usol = Usol.*exp(-j*Ci*tstep.^2./2/To.^2);
% Usol = hpha_mod(Usol(1:Ns),Vm,Vbias,Vpi,fm,0,p0);
% Usol = Usol*atten;
  [Usol,A3d,z]=hconst(Usol,Length_SMF,5,1*beta2_SMF,1*beta3_
SMF,alpha_loss_SMF,1*gamma_SMF,0);      % Fiber section #2
% ----------------------------------------------------------
% ARR2(ii,:)=abs(Usol).^2 ;
% Up = Usol.';
  ind = mod(ii,20);
  if ind==0
   in = round(ii/20) + 1;
   ARR2(in,:)=abs(Usol).^2 ;
   Up(in,:) = Usol;
  end
```

```
  if (ii>N_pass - 20)
    in = round(ii/20) + 1;
    ARR2(in,:)=abs(Usol).^2 ;
    Up(in,:) = Usol;
  end
end
% Eout = Eout/atten;
% =========================================================
figure(1);
plot(tstep,abs(Usol).^2)
colormap('default')
mesh (real(abs(Up).^2),'meshstyle','row','facecolor','none');
% waterfall(ARR2);
% waterfall(abs(Up).^2);
view(17.5,42);
axis tight;
% delete solpair.mat
save solpair5 Up tstep Po TotDisp Gamma Ld To Ci r teta Tfwhm
Tb Vm Vpi

NN = size(Up,1);
figure(2);
PP = angle(Up(1,:));
Pha_tinput = unwrap(PP);
ff = -diff(Pha_tinput)/Ts;
chrate = diff(ff)/Ts;
subplot(221);
plot (tstep*1e12,Pha_tinput);grid;
% plot (tstep(1:Ns-1)*1e12,ff);grid;
% plot (tstep(2:Ns-1)*1e12,chrate);grid;
PP = angle(Up(5,:));
Pha_tinput = unwrap(PP);
ff = -diff(Pha_tinput)/Ts;
chrate = diff(ff)/Ts;
subplot(222);
plot (tstep*1e12,Pha_tinput);grid;
% plot (tstep(1:Ns-1)*1e12,ff);grid;
% plot (tstep(2:Ns-1)*1e12,chrate);grid;
in = round(NN/2);
PP = angle(Up(in,:));
Pha_tinput = unwrap(PP);
ff = -diff(Pha_tinput)/Ts;
chrate = diff(ff)/Ts;
subplot(223);
plot (tstep*1e12,Pha_tinput);grid;
% plot (tstep(1:Ns-1)*1e12,ff);grid;
% plot (tstep(2:Ns-1)*1e12,chrate);grid;
PP = angle(Up(NN,:));
Pha_tinput = unwrap(PP);
ff = -diff(Pha_tinput)/Ts;
```

```
chrate = diff(ff)/Ts;
subplot(224);
plot (tstep*1e12,Pha_tinput);grid;
% plot (tstep(1:Ns-1)*1e12,ff);grid;
% plot (tstep(2:Ns-1)*1e12,chrate);grid;
figure(3);
title('The phase evolution of soliton pairs');
subplot(221);
plot (tstep*1e12,ARR2(1,:),tstep*1e12,Vw);grid;
subplot(222);
plot (tstep*1e12,ARR2(5,:));grid;

subplot(223);
in = round(NN/2);
plot (tstep*1e12,ARR2(in,:));grid;

subplot(224);
plot (tstep*1e12,ARR2(NN,:));grid;
figure(4);

Kmag = 1;
Nplot = 1000;
Uf = Up.';
Eoutfreq = fft(Uf,Ns);          % (:,N1)
Eoutfreq1 = fft(Uf,Ns*Kmag);  % (:,N1)
Ioutfreq = Eoutfreq1.*conj(Eoutfreq1)/(Ns*Kmag)^2;
ind = (- Nplot/2 : Nplot/2)';
freq = ind/Ts/Ns/Kmag;
ind = mod((ind + Ns*Kmag),Ns*Kmag)+1;
title('The spectrum evolution of pulse');
subplot(221)
% plot(freq,Ioutfreq(ind,1));
plot(freq,10*log10(Ioutfreq(ind,1))+30);
Xlabel('Freq (Hz)');
Ylabel('P (W)');
subplot(222)
% in = round(N1/4);
% plot(freq,Ioutfreq(ind,5));
plot(freq,10*log10(Ioutfreq(ind,5))+30);
Xlabel('Freq (Hz)');
Ylabel('P (W)');
subplot(223)
in = round(NN/2);
% plot(freq,Ioutfreq(ind,in));
plot(freq,10*log10(Ioutfreq(ind,in))+30);
Xlabel('Freq (Hz)');
Ylabel('P (W)');
subplot(224)
% plot(freq,Ioutfreq(ind,NN));
plot(freq,10*log10(Ioutfreq(ind,NN))+30);
Xlabel('Freq (Hz)');
```

```
Ylabel('P (W)');
figure(5);grid;
Pha = Vw*pi./Vpi;                % Phase change in time
dff = -diff(Pha)/Ts;
subplot(211);plot(tstep*1e12,Pha);xlabel('Time
(ps)');ylabel('Phase (rad)');axis tight;grid;
subplot(212);plot(tstep(1:Ns-1)*1e12,dff*1e-12);xlabel('Time
(ps)');ylabel('Chirping (THz)');grid;

%%%%%%%%%%%%%%%%%%%%%%%%%%%%%%%%%%%%%%%%%%%%%%%%%%%%%%%%%%%%%%
% S O L I T O N  O P T I C A L  C O M M U N I C A T I O N S  %
%
%
% global h0
% solitonpairsnoise.m
clear;
close all
clc;
global Ts;     % sampling period
global Fcar;   % carrier frequency (optical frequency)
global Vw;
global w;
global tstep;
% cla reset
h0 = figure(1);
c_const = 3e8 ;
ng = 1.47;
lamda = 1.550e-6 ;
% ============================================================
% Fiber parameters
  Length_SMF = 80;         % Length of a span in meter
  nz = Length_SMF/100;
  alpha_indB = 0.2*1e-3;   % attenuation (dB/km) --> dB/m
  D_SMF = 1*17e-6;         % GVD (ps/nm.km); if anomalous
dispersion(for compensation),D is negative
  n2_SMF = 1*2.6e-20;      % nonlinear index (m^2/W)
  Aeff_SMF = 55e-12;       % effective area (m^2)
% Slope Dispersion
  S_SMF = 0.06e3;          % ps/(nm^2.km) in SI unit

% CALCULATED QUANTITIES of SMF

  alpha_loss_SMF = 1*log(10)*alpha_indB/10;    % alpha (1/m)
  beta2_SMF = -D_SMF*lamda^2/(2*pi*c_const);   % beta2
(ps^2/km);
  % --------------------------------------------------------
  % beta 3 can be calculated from the Slope Dispersion (S) as
follows:]
  beta3_SMF = (S_SMF - (4*pi*c_const/(lamda^3))*beta2_SMF)/
((2*pi*c_const/(lamda^2))^2);
% --------------------------------------------------------
```

```
  gamma_SMF = 2*pi*n2_SMF/(lamda*Aeff_SMF);    % nonlinearity
coef (km^-1.W^-1)
% ========================================================
% filter bandwidth
  lamda3dB = 4.0e-9;%1.2e-9;                        % 1.2nm
  f3dB = lamda3dB*(1e11/0.8e-9);                    % Hz
% Amplifier parameters:
  GssdB = 21;              % (dB)
  PoutsatdB = 13.5;        % (dBm)
  NF = 5;                  % (dB)
% EDF fiber parameters
  Length_EDF = 15;         % Length of a span in meter
  alpha_indB = 0.5*1e-3;   % attenuation (dB/km) --> dB/m
  D_EDF = 15e-6;           % GVD (ps/nm.km); if anomalous
dispersion(for compensation),D is negative
  n2_EDF = 1*2.6e-20;      % nonlinear index (m^2/W)
  Aeff_EDF = 45e-12;       % effective area (m^2)
% Slope Dispersion
  S_EDF = 0.06e3;          % ps/(nm^2.km) in SI unit
% CALCULATED QUANTITIES of EDF
  alpha_loss_EDF = 1*log(10)*alpha_indB/10; % alpha (1/m)
  beta2_EDF = -D_EDF*lamda^2/(2*pi*c_const) % beta2 (ps^2/km);
% -------------------------------------------------------
  % beta 3 can be calculated from the Slope Dispersion (S) as
follows:]
  beta3_EDF = (S_EDF - (4*pi*c_const/(lamda^3))*beta2_EDF)/
((2*pi*c_const/(lamda^2))^2);
% -------------------------------------------------------
  gamma_EDF = 2*pi*n2_EDF/(lamda*Aeff_EDF); % nonlinearity
coef (km^-1.W^-1)
% ========================================================
% Average parameters
% Loss
  loss = 11.5;             % dB
  atten = 1/10^(loss/20);

% to test dispersion let TotDisp=-2e-6, SegNum=2.5*Zo/h
% run again with U1 removed and TotDisp=2e-6
Length_total = Length_SMF+Length_EDF;
TotDisp = (D_SMF*Length_SMF+D_EDF*Length_EDF)/Length_total;
% TotDisp = 17*1e-6 ;

ro = 3e-6 ;
n2 = 3.2*1e-20 ;
Beta2 = -(TotDisp*lamda^2)/(2*pi*c_const);
Beta3 = 0 ;
Gamma = (gamma_SMF*Length_SMF+gamma_EDF*Length_EDF)/Length_
total ;
% Gamma = (2*pi*n2)/(lamda*pi*ro^2) ;
alphadB = 0;
Alpha=log(10)*alphadB/(10*1000); % = field loss per meter
```

```
% ============================================================
% modulator parameters
Vpi = 6;                   % Volt
Vbias = 0;                 % Volt
Vm = 1.8;                  % Volt
m = Vm/Vpi;                % modulation index
fm = 1e9;                  % modulation frequency
p0 = 0*pi/2;
% ============================================================
% Soliton pulse parameters
Norder = 1;                % Order of soliton
Tfwhm = 6.0e-12;
To = Tfwhm/1.763;          % Pulsewidth of soliton
Tb = 58.5e-12;             % Pulses distance
teta = 2*pi/4; % Relative phase difference between 2 solitons
r = 1;                     % Amplitude ratio
Ci = 0;          % Chirping factor of pulse
qo = Tb/2/To;    % Space between 2 solitons
qr = Tb/2/To/2   % Space between 2 solitons

Ld = To^2/abs(Beta2)
Zo = pi*Ld/2 ;
Po = Norder^2/(Gamma*Ld);
P0 = 130e-3;
% Pav = 3*Po/(2*qr)
% ============================================================
Ns = 2^13;
Ts = 0.05e-12;
tstep = Ts*(-Ns/2:1:Ns/2-1);
w = fftshift((2*pi/((Ns-1)*Ts))*(-Ns/2:1:Ns/2-1));
% Usol = sqrt(Po).*sech(tstep./To).*exp(-j*teta).*exp
(-j*Ci*tstep.^2./2/To.^2); %
% Usol = sqrt(P0).*(sech(tstep./To-qo)+r.*sech(r.*(tstep./
To+qo))).*exp(j*teta)).*exp(-j*Ci*tstep.^2./2/To.^2) ; %
% Usol = sqrt(P0).*(sech(tstep./
To-qo).*exp(+j*teta)+r.*sech(r.*(tstep./To+qo)).*exp
(-j*teta)).*exp(-j*Ci*tstep.^2./2/To.^2) ;
% Usol = sqrt(Po).*(r*sech(tstep./
To-qo).*exp(+j*teta)+sech(tstep./To-0).*exp(-
j*teta)+r.*sech(r.*(tstep./To+qo)).*exp(+j*teta)).*exp
(-j*Ci*tstep.^2./2/To.^2) ;
Usol = sqrt(P0).*(r*sech(tstep./To-1*qo).*exp(+j*teta)+
r*sech(tstep./To-qo/3).*exp(-j*teta/1)...
+sech(tstep./To+qo/3).*exp(+j*teta/1)+r.*sech(r.*(tstep./
To+1*qo)).*exp(-j*teta)).*exp(-j*Ci*tstep.^2./2/To.^2) ;
% Usol = sqrt(P0).*(r*sech(tstep./To-1*qo).*exp(+j*teta)+
r*sech(tstep./To-qo/2).*exp(-j*teta/1)+ r*sech(tstep./
To-0).*exp(j*teta/1)...
% +sech(tstep./To+qo/2).*exp(-j*teta/1)+r.*sech(r.*(tstep./
To+1*qo)).*exp(j*teta)).*exp(-j*Ci*tstep.^2./2/To.^2) ;
```

```
% Eo = Usol;
plot(tstep,abs(Usol).^2)
grid
U2(1,:) = abs(Usol).^2 ;
Up(1,:) = Usol*atten;
pause(0.1)
% ==========================================================
N_pass = 200;
for ii = 2:N_pass
% h1 = waitbar((ii-1)/N_pass,'The program is running...');
  [Usol,G] = AmpSimpNoise(Usol,GssdB,PoutsatdB,NF);
[Usol,A3d,z]=hconst(Usol,Length_EDF,5,1*beta2_EDF,1*beta3_
EDF,alpha_loss_EDF,1*gamma_EDF,0); % Fiber section #1
  Usol = GaussLPfilt(tstep,Usol,1,f3dB);
[Usol,A3d,z]=hconst(Usol,Length_SMF,5,1*beta2_SMF,1*beta3_
SMF,alpha_loss_SMF,1*gamma_SMF,0); % Fiber section #2
  Usol = pha_mod(Usol(1:Ns),Vm,Vbias,Vpi,fm,p0); % Phase
modulator
  Usol = Usol*atten;
  % ---------------------------------------------------------
  ind = mod(ii,20);
  if ind==0
    in = round(ii/20) + 1;
    U2(in,:)=abs(Usol).^2 ;
    Up(in,:) = Usol;
  end
% % close(h1);
end
Pav = mean(abs(Usol).^2);
PavdB = 10*log10(Pav)
Up = Up./atten;
% ==========================================================
NN = size(Up,1);
figure(1);
plot(tstep,abs(Usol).^2)
colormap('default')
mesh (real(abs(Up).^2),'meshstyle','row','facecolor','none');
% waterfall(ARR2);
% waterfall(abs(Up).^2);
view(17.5,42);
axis tight;
% delete solpair.mat
% save soldoub9db06md Up tstep Po TotDisp Gamma Ld To Ci r
teta Tfwhm Tb Vm Vpi NF PoutsatdB GssdB
Pav = 3*max(max(U2))/(2*qr)
Pav1 = mean(U2(NN,:))

figure(2);
PP = angle(Up(1,:));
Pha_tinput = unwrap(PP);
```

```
ff = -diff(Pha_tinput)/Ts;
chrate = diff(ff)/Ts;
subplot(221);
plot (tstep*1e12,Pha_tinput);grid;
% plot (tstep(1:Ns-1)*1e12,ff);grid;
% plot (tstep(2:Ns-1)*1e12,chrate);grid;
PP = angle(Up(5,:));
Pha_tinput = unwrap(PP);
ff = -diff(Pha_tinput)/Ts;
chrate = diff(ff)/Ts;
subplot(222);
plot (tstep*1e12,Pha_tinput);grid;
% plot (tstep(1:Ns-1)*1e12,ff);grid;
% plot (tstep(2:Ns-1)*1e12,chrate);grid;
in = round(NN/2);
PP = angle(Up(in,:));
Pha_tinput = unwrap(PP);
ff = -diff(Pha_tinput)/Ts;
chrate = diff(ff)/Ts;
subplot(223);
plot (tstep*1e12,Pha_tinput);grid;
% plot (tstep(1:Ns-1)*1e12,ff);grid;
% plot (tstep(2:Ns-1)*1e12,chrate);grid;
PP = angle(Up(NN,:));
Pha_tinput = unwrap(PP);
ff = -diff(Pha_tinput)/Ts;
chrate = diff(ff)/Ts;
subplot(224);
plot (tstep*1e12,Pha_tinput);grid;
% plot (tstep(1:Ns-1)*1e12,ff);grid;
% plot (tstep(2:Ns-1)*1e12,chrate);grid;

figure(3);
title('The phase evolution of soliton pairs');
subplot(221);
plot (tstep*1e12,U2(1,:));grid;

in = round(NN/4);
subplot(222);
plot (tstep*1e12,U2(in,:));grid;

subplot(223);
in = round(NN/2);
plot (tstep*1e12,U2(in,:));grid;

subplot(224);
plot (tstep*1e12,U2(NN,:));grid;

figure(4);

Kmag = 1;
Nplot = 100;
Uf = Up.';
```

```
Eoutfreq = fft(Uf,Ns); %(:,N1)
Eoutfreq1 = fft(Uf,Ns*Kmag); %(:,N1)
Ioutfreq = Eoutfreq1.*conj(Eoutfreq1)/(Ns*Kmag)^2;
ind = (- Nplot/2 : Nplot/2)';
freq = ind/Ts/Ns/Kmag;
ind = mod((ind + Ns*Kmag),Ns*Kmag)+1;
title('The spectrum evolution of pulse');
subplot(221)
% plot(freq,Ioutfreq(ind,1));
plot(freq,10*log10(Ioutfreq(ind,1))+30);
Xlabel('Freq (Hz)');
Ylabel('P (W)');
subplot(222)
% in = round(N1/4);
% plot(freq,Ioutfreq(ind,5));
plot(freq,10*log10(Ioutfreq(ind,5))+30);
Xlabel('Freq (Hz)');
Ylabel('P (W)');
subplot(223)
in = round(NN/2);
% plot(freq,Ioutfreq(ind,in));
plot(freq,10*log10(Ioutfreq(ind,in))+30);
Xlabel('Freq (Hz)');
Ylabel('P (W)');
subplot(224)
% plot(freq,Ioutfreq(ind,NN));
plot(freq,10*log10(Ioutfreq(ind,NN))+30);
Xlabel('Freq (Hz)');
Ylabel('P (W)');
figure(5);grid;
Pha = Vw*pi./Vpi; % Phase change in time
dff = -diff(Pha)/Ts;
subplot(211);plot(tstep*1e12,Pha);xlabel('Time
(ps)');ylabel('Phase (rad)');axis tight;
subplot(212);plot(tstep(1:Ns-1)*1e12,dff*1e-12);xlabel('Time
(ps)');ylabel('Chirping (THz)');

%%%%%%%%%%%%%%%%%%%%%%%%%%%%%%%%%%%%%%%%%%%%%%%%%%%%%%%%%%%%%%%%%%%
% S O L I T O N   O P T I C A L   C O M M U N I C A T I O N S   %
%
%
% global h0

% Boundsolmix.m
clear;
close all
clc;
global Ts;   % sampling period
global Fcar; % carrier frequency (optical frequency)
```

```
global Vw;
global w;
global tstep;
% cla reset
h0 = figure(1);

c_const = 3e8 ;
ng = 1.47;
lamda = 1.550e-6 ;
% ===========================================================
% Fiber parameters
  Length_SMF = 65;          % Length of a span in meter
  nz = Length_SMF/100;
  alpha_indB = 0.0*1e-3;    % attenuation (dB/km) --> dB/m
  D_SMF = 1*17e-6;          % GVD (ps/nm.km); if anomalous
dispersion(for compensation),D is negative
  n2_SMF = 1*2.6e-20;       % nonlinear index (m^2/W)
  Aeff_SMF = 55e-12;        % effective area (m^2)
% Slope Dispersion
  S_SMF = 0.06e3;           % ps/(nm^2.km) in SI unit

% CALCULATED QUANTITIES of SMF

  alpha_loss_SMF = 1*log(10)*alpha_indB/10; % alpha (1/m)
  beta2_SMF = -D_SMF*lamda^2/(2*pi*c_const) % beta2 (ps^2/km);
  % ----------------------------------------------------------
  % beta 3 can be calculated from the Slope Dispersion (S) as
follows:]
  beta3_SMF = (S_SMF - (4*pi*c_const/(lamda^3))*beta2_SMF)/
((2*pi*c_const/(lamda^2))^2);
  % ----------------------------------------------------------
  gamma_SMF = 2*pi*n2_SMF/(lamda*Aeff_SMF); % nonlinearity
coefficient (km^-1.W^-1)
% ===========================================================
% filter bandwidth
  lamda3dB = 4.8e-9;%1.2e-9;                 % 1.2 nm
  f3dB = lamda3dB*(1e11/0.8e-9);             % Hz
% Amplifier parameters:
  GssdB = 60;               % (dB)
  PoutsatdB = 36;           % (dBm)
  NF = 8;                   % (dB)
% EDF fiber parameters
  Length_EDF = 50;          % Length of a span in meter
  alpha_indB = 0.0*1e-3;    % attenuation (dB/km) --> dB/m
  D_EDF = -5e-6;            % GVD (ps/nm.km); if anomalous
dispersion(for compensation),D is negative
  n2_EDF = 1*2.6e-20;       % nonlinear index (m^2/W)
  Aeff_EDF = 35e-12;        % effective area (m^2)
% Slope Dispersion
  S_EDF = 0.06e3;           % ps/(nm^2.km) in SI unit
```

```
% CALCULATED QUANTITIES of EDF
   alpha_loss_EDF = 1*log(10)*alpha_indB/10; % alpha (1/m)
   beta2_EDF = -D_EDF*lamda^2/(2*pi*c_const) % beta2 (ps^2/km);
   % --------------------------------------------------------
   % beta 3 can be calculated from the Slope Dispersion (S) as
follows:]
   beta3_EDF = (S_EDF - (4*pi*c_const/(lamda^3))*beta2_EDF)/
((2*pi*c_const/(lamda^2))^2);
% --------------------------------------------------------
   gamma_EDF = 2*pi*n2_EDF/(lamda*Aeff_EDF); % nonlinearity
coef (km^-1.W^-1)
% ========================================================
% Average parameters
% Loss
   loss = 8;                    % dB
   atten = 1/10^(loss/20);

% to test dispersion let TotDisp=-2e-6, SegNum=2.5*Zo/h
% run again with U1 removed and TotDisp=2e-6
Length_total = Length_SMF+Length_EDF;
TotDisp = (D_SMF*Length_SMF+D_EDF*Length_EDF)/Length_total;
% TotDisp = 17*1e-6 ;

ro = 3e-6 ;
n2 = 3.2*1e-20 ;
Beta2 = -(TotDisp*lamda^2)/(2*pi*c_const);
Beta3 = (beta3_SMF*Length_SMF+beta3_EDF*Length_EDF)/Length_
total ;
Gamma = (gamma_SMF*Length_SMF+gamma_EDF*Length_EDF)/Length_
total ;
% Gamma = (2*pi*n2)/(lamda*pi*ro^2) ;
alphadB = 0;
Alpha=log(10)*alphadB/(10*1000); % = field loss per meter
% ========================================================
% modulator parameters
Vpi = 6;                         % Volt
Vbias = 0;                       % Volt
Vm = 1.5;                        % Volt
m = Vm/Vpi;                      % modulation index
fm = 1e9;                        % modulation frequency
p0 = 0*pi/2;
% ========================================================
% Soliton pulse parameters
Norder = 1;                      % Order of soliton
Tfwhm = 6.0e-12;
To = Tfwhm/1.763;                % Pulsewidth of soliton
Tb = 50e-12;                     % Pulses distance
teta = 1*pi/2; % Relative phase difference between 2 solitons
r = 1;                           % Amplitude ratio
Ci = 0;                          % Chirping factor of pulse
qo = Tb/2/To                     % Space between 2 solitons
```

```
Ld = To^2/abs(Beta2)
Zo = pi*Ld/2 ;
Po = Norder^2/(Gamma*Ld);
P0 = 340e-3;
% =========================================================
Ns = 2^10;
Ts =0.2e-12;
tstep = Ts*(-Ns/2:1:Ns/2-1);
w = fftshift((2*pi/((Ns-1)*Ts))*(-Ns/2:1:Ns/2-1));
% Usol = sqrt(P0).*sech(tstep./To).*exp(-j*teta).*exp(-j*Ci*
tstep.^2./2/To.^2);
% Usol = sqrt(P0).*(sech(tstep./To-qo)+r.*sech(r.*(tstep./
To+qo)).*exp(j*teta)).*exp(-j*Ci*tstep.^2./2/To.^2) ; %
% Usol = sqrt(P0).*(sech(tstep./To-qo).*exp(-j*teta)+r.*
sech(r.*(tstep./To+qo)).*exp(+j*teta)).*exp(-j*Ci*tstep.^2./2/
To.^2) ;
Usol = sqrt(P0).*(r*sech(tstep./To-qo).*exp(-j*teta)+sech
(tstep./To-0).*exp(+j*teta)+r.*sech(r.*(tstep./To+qo)).*exp
(-j*teta)).*exp(-j*Ci*tstep.^2./2/To.^2) ;
% Usol = sqrt(P0).*(r*sech(tstep./To-1*qo).*exp(+j*teta)+
r*sech(tstep./To-qo/3).*exp(-j*teta/1)...
% +sech(tstep./To+qo/3).*exp(+j*teta/1)+r.*sech(r.*(tstep./To+
1*qo)).*exp(-j*teta)).*exp(-j*Ci*tstep.^2./2/To.^2) ;

% Eo = Usol;
plot(tstep,abs(Usol).^2)
grid
ARR2(1,:) = abs(Usol).^2 ;
Up(1,:) = Usol;
pause(0.1)

% =========================================================
%
N_pass = 2000;
for ii = 2:N_pass
% h1 = waitbar((ii-1)/N_pass,'The program is running...');
% [Usol,G] = AmpSimpNoise(Usol,GssdB,PoutsatdB,NF);
% [Usol,A3d,z]=hconst(Usol,Length_EDF,5,1*beta2_EDF,1*beta3_
EDF,alpha_loss_EDF,1*gamma_EDF,0); % Fiber section #1
% Usol = GaussLPfilt(tstep,Usol,1,f3dB);

[Usol,A3d,z]=hconst(Usol,Length_total,5,1*Beta2,1*Beta3,alpha_
loss_SMF,1*Gamma,0); % Averaged Fiber
% Eo = modInt(Eo(1:N),alpha,epsilon,m,fm,0.5); % MZI modulator
    if ii <= round(N_pass/2)
    Usol = pha_mod(Usol(1:Ns),Vm,Vbias,Vpi,fm,p0); % Phase
modulator

% Usol = Usol.*exp(-j*Ci*tstep.^2./2/To.^2);
% Usol = hpha_mod(Usol(1:Ns),Vm,Vbias,Vpi,fm,2,0);
  end
% Usol = Usol*atten;
```

```
% [Usol,A3d,z]=hconst(Usol,Length_SMF,5,1*beta2_SMF,1*beta3_
SMF,alpha_loss_SMF,1*gamma_SMF,0); % Fiber section #2
  % ----------------------------------------------------
% ARR2(ii,:)=abs(Usol).^2 ;
% Up = Usol.';
  ind = mod(ii,20);
  if ind==0
    in = round(ii/20) + 1;
    ARR2(in,:)=abs(Usol).^2 ;
    Up(in,:) = Usol;
  end
  if (ii>N_pass - 20)
    in = round(ii/20) + 1;
    ARR2(in,:)=abs(Usol).^2 ;
    Up(in,:) = Usol;
  end
% Up(ii,:) = Usol;
% plot1Hndl=plot(tstep*1e12, abs(Usol).^2) ;grid; xlabel('Time
(ps)');
% ylabel('P (W)');
  % ----------------------------------------------------
  % Display the number of round trips
% global b0
% str0=num2str(ii);
% b0 = uicontrol('Parent',h0,…
%                 'Style','text','BackgroundColor',[1 1 1], …
%                 'Units','points',…
%                 'ForegroundColor','blue',…
%                 'FontSize',10, …
%                 'String',['roundtrips = ',str0,'rounds'],
'Position',[500 310 100 15]);
% % --------------------------
% pause(0.01)
% % close(h1);
end
% Eout = Eout/atten;
% ========================================================
plot(tstep,abs(Usol).^2)
colormap('default')
mesh (real(abs(Up).^2),'meshstyle','row','facecolor','none');
% waterfall(ARR2);
% waterfall(abs(Up).^2);
view(17.5,42);
axis tight;
% delete solpair.mat
save solpair5 Up tstep Po TotDisp Gamma Ld To Ci r teta Tfwhm
Tb Vm Vpi

NN = size(Up,1);
figure(2);
```

```
PP = angle(Up(1,:));
Pha_tinput = unwrap(PP);
ff = -diff(Pha_tinput)/Ts;
chrate = diff(ff)/Ts;
subplot(221);
% plot (tstep*1e12,Pha_tinput);grid;
plot (tstep(1:Ns-1)*1e12,ff);grid;
% plot (tstep(2:Ns-1)*1e12,chrate);grid;
PP = angle(Up(5,:));
Pha_tinput = unwrap(PP);
ff = -diff(Pha_tinput)/Ts;
chrate = diff(ff)/Ts;
subplot(222);
% plot (tstep*1e12,Pha_tinput);grid;
plot (tstep(1:Ns-1)*1e12,ff);grid;
% plot (tstep(2:Ns-1)*1e12,chrate);grid;
in = round(NN/2);
PP = angle(Up(in,:));
Pha_tinput = unwrap(PP);
ff = -diff(Pha_tinput)/Ts;
chrate = diff(ff)/Ts;
subplot(223);
% plot (tstep*1e12,Pha_tinput);grid;
plot (tstep(1:Ns-1)*1e12,ff);grid;
% plot (tstep(2:Ns-1)*1e12,chrate);grid;
PP = angle(Up(NN,:));
Pha_tinput = unwrap(PP);
ff = -diff(Pha_tinput)/Ts;
chrate = diff(ff)/Ts;
subplot(224);
% plot (tstep*1e12,Pha_tinput);grid;
plot (tstep(1:Ns-1)*1e12,ff);grid;
% plot (tstep(2:Ns-1)*1e12,chrate);grid;
figure(3);
title('The phase evolution of soliton pairs');

subplot(221);
plot (tstep*1e12,ARR2 (1,:));grid;

subplot(222);
plot (tstep*1e12,ARR2(5,:));grid;

subplot(223);
in = round(NN/2);
plot (tstep*1e12,ARR2(in,:));grid;

subplot(224);
plot (tstep*1e12,ARR2(NN,:));grid;

figure(4);

Kmag = 1;
Nplot = 100;
```

```
Uf = Up.';
Eoutfreq = fft(Uf,Ns);           %(:,N1)
Eoutfreq1 = fft(Uf,Ns*Kmag);  %(:,N1)
Ioutfreq = Eoutfreq1.*conj(Eoutfreq1)/(Ns*Kmag)^2;
ind = (- Nplot/2 : Nplot/2)';
freq = ind/Ts/Ns/Kmag;
ind = mod((ind + Ns*Kmag),Ns*Kmag)+1;
title('The spectrum evolution of pulse');
subplot(221)
% plot(freq,Ioutfreq(ind,1));
plot(freq,10*log10(Ioutfreq(ind,1))+30);
Xlabel('Freq (Hz)');
Ylabel('P (W)');
subplot(222)
% in = round(N1/4);
% plot(freq,Ioutfreq(ind,5));
plot(freq,10*log10(Ioutfreq(ind,5))+30);
Xlabel('Freq (Hz)');
Ylabel('P (W)');
subplot(223)
in = round(NN/2);
% plot(freq,Ioutfreq(ind,in));
plot(freq,10*log10(Ioutfreq(ind,in))+30);
Xlabel('Freq (Hz)');
Ylabel('P (W)');
subplot(224)
% plot(freq,Ioutfreq(ind,NN));
plot(freq,10*log10(Ioutfreq(ind,NN))+30);
Xlabel('Freq (Hz)');
Ylabel('P (W)');
figure(5);grid;
Pha = Vw*pi./Vpi; % Phase change in time
dff = -diff(Pha)/Ts;
subplot(211);plot(tstep*1e12,Pha);xlabel('Time
(ps)');ylabel('Voltage (V)');axis tight;
subplot(212);plot(tstep(1:Ns-1)*1e12,dff);xlabel('Time
(ps)');ylabel('Phase (rad)');

%%%%%%%%%%%%%%%%%%%%%%%%%%%%%%%%%%%%%%%%%%%%%%%%%%%%%%%%%%%%%%%%%
% S O L I T O N   O P T I C A L   C O M M U N I C A T I O N S   %
%
%
% global h0

% Gaussianpairs.m
clear;
close all
clc;
global Ts;    % sampling period
global Fcar;  % carrier frequency (optical frequency)
```

```
global Vw;
global w;
global tstep;
% cla reset
h0 = figure(1);

c_const = 3e8 ;
ng = 1.47;
lamda = 1.550e-6 ;
% ==========================================================
% Fiber parameters
  Length_SMF = 65;          % Length of a span in meter
  nz = Length_SMF/100;
  alpha_indB = 0.0*1e-3;  % attenuation (dB/km) --> dB/m
  D_SMF = 1*17e-6;          % GVD (ps/nm.km); if anomalous
dispersion(for compensation),D is negative
  n2_SMF = 1*2.6e-20;      % nonlinear index (m^2/W)
  Aeff_SMF = 55e-12;                % effective area (m^2)
% Slope Dispersion
  S_SMF = 0.06e3;                        % ps/(nm^2.km) in SI unit

% CALCULATED QUANTITIES of SMF

  alpha_loss_SMF = 1*log(10)*alpha_indB/10;    % alpha (1/m)
  beta2_SMF = -D_SMF*lamda^2/(2*pi*c_const);   % beta2
(ps^2/km);
  % ---------------------------------------------------------
  % beta 3 can be calculated from the Slope Dispersion (S) as
follows:]
  beta3_SMF = (S_SMF - (4*pi*c_const/(lamda^3))*beta2_SMF)/
((2*pi*c_const/(lamda^2))^2);
% ---------------------------------------------------------
  gamma_SMF = 2*pi*n2_SMF/(lamda*Aeff_SMF);    % nonlinearity
coefficient (km^-1.W^-1)
% ==========================================================
% filter bandwidth
  lamda3dB = 4.8e-9              % 1.2e-9;      % 1.2nm
  f3dB = lamda3dB*(1e11/0.8e-9); % Hz
% Amplifier parameters:
  GssdB = 60;                    % (dB)
  PoutsatdB = 36;                % (dBm)
  NF = 8;                        % (dB)
% EDF fiber parameters
  Length_EDF = 50;                % Length of a span in meter
  alpha_indB = 0.0*1e-3;  % attenuation (dB/km) --> dB/m
  D_EDF = -7e-6;            % GVD (ps/nm.km); if anomalous
dispersion(for compensation),D is negative
  n2_EDF = 1*2.6e-20;            % nonlinear index (m^2/W)
  Aeff_EDF = 45e-12;              % effective area (m^2)
% Slope Dispersion
  S_EDF = 0.06e3;                % ps/(nm^2.km) in SIunit
```

```
% CALCULATED QUANTITIES of EDF
  alpha_loss_EDF = 1*log(10)*alpha_indB/10;     % alpha (1/m)
  beta2_EDF = -D_EDF*lamda^2/(2*pi*c_const)     % beta2
(ps^2/km);
  % ------------------------------------------------------
  % beta 3 can be calculated from the Slope Dispersion (S) as
follows:]
  beta3_EDF = (S_EDF - (4*pi*c_const/(lamda^3))*beta2_EDF)/
((2*pi*c_const/(lamda^2))^2);
% ------------------------------------------------------
  gamma_EDF = 2*pi*n2_EDF/(lamda*Aeff_EDF);     % nonlinearity
coefficient (km^-1.W^-1)
% ========================================================
% Average parameters
% Loss
  loss = 8;                          % dB
  atten = 1/10^(loss/20);

% to test dispersion let TotDisp=-2e-6, SegNum=2.5*Zo/h
% run again with U1 removed and TotDisp=2e-6
Length_total = Length_SMF+Length_EDF;
TotDisp = (D_SMF*Length_SMF+D_EDF*Length_EDF)/Length_total;
% TotDisp = 17*1e-6 ;

ro = 3e-6 ;
n2 = 3.2*1e-20 ;
Beta2 = -(TotDisp*lamda^2)/(2*pi*c_const);
Beta3 = 0 ;
Gamma = (gamma_SMF*Length_SMF+gamma_EDF*Length_EDF)/Length_
total ;
% Gamma = (2*pi*n2)/(lamda*pi*ro^2) ;
alphadB = 0;
Alpha=log(10)*alphadB/(10*1000); % = field loss per meter
% ========================================================
% modulator parameters
Vpi = 6;              % Volt
Vbias = 0;            % Volt
Vm = 4.5;             % Volt
m = Vm/Vpi;           % modulation index
fm = 1e9;             % modulation frequency
p0 = 2*pi/2;
% ========================================================
% Soliton pulse parameters
Norder = 1;           % Order of soliton
Tfwhm = 5.4e-12;
To = Tfwhm/1.665;     % Pulsewidth of gaussian
Tb = 24e-12;          % Pulses distance
teta = 1*pi/2; % Relative phase difference between 2 solitons
r = 1;                % Amplitude ratio
Ci = 0;               % Chirping factor of pulse
```

```
qo = Tb/2/To          % Space between 2 solitons
Ld = To^2/abs(Beta2)
Zo = pi*Ld/2 ;
Po = Norder^2/(Gamma*Ld);
P0 = 400e-3;
% ==========================================================
Ns = 2^10;
Ts = 0.2e-12;
tstep = Ts*(-Ns/2:1:Ns/2-1);
w = fftshift((2*pi/((Ns-1)*Ts))*(-Ns/2:1:Ns/2-1));

% Usol = sqrt(P0)*(exp(-(tstep-Tb/2).^2/2/To^2).*exp
(-j*teta)+r*exp(-(tstep+Tb/2).^2/2/To^2).*exp(+j*teta)).*exp
(-j*Ci*tstep.^2./2/To^2);%*exp(j*(2*pi*f*ti+phiC));
% Usol = sqrt(P0).*sech(tstep./To).*exp(-j*teta).*exp
(-j*Ci*tstep.^2./2/To.^2);
% Usol = sqrt(P0).*(sech(tstep./To-qo)+r.*sech(r.*(tstep./
To+qo)).*exp(j*teta)).*exp(-j*Ci*tstep.^2./2/To.^2) ; %
% Usol = sqrt(P0).*(sech(tstep./To-qo).*exp
(-j*teta)+r.*sech(r.*(tstep./To+qo)).*exp(+j*teta)).*exp
(-j*Ci*tstep.^2./2/To.^2) ;
% Usol = sqrt(P0).*(r*sech(tstep./
To-qo).*exp(+j*teta)+sech(tstep./To-0)+r.*sech(r.*(tstep./
To+qo)).*exp(+j*teta)).*exp(-j*Ci*tstep.^2./2/To.^2) ;

% Eo = Usol;
plot(tstep,abs(Usol).^2)
grid
ARR2(1,:) = abs(Usol).^2 ;
Up(1,:) = Usol;
pause(0.1)

% ----- Algorithm for Soliton Beam Propagation Method : Split
Step Fourier Model -----%

% for a = 2:1:Block
% for b = 1:BlockSize
%
% D = -j/2*Beta2.*(j.*w).^2 + 1/6*Beta3.*(j.*w).^3 - Alpha/2 ;
% N1 = j*Gamma.*(abs(Usol)).^2 ; %-2*Gamma*OpLamda./
(2*pi*c.*U1).*(j.*w.*(abs(U1).^2.*U1)) ;
% N2 = N1 ;
%
% % First half interval : dispersion operator only
% Udisp = ifft(exp(h/2.*D).* fft(Usol)) ;
%
% % Iteration process to find N2
% for c = 1:2
% Utmp = ifft( exp(h/2.*D).* fft(Usol)) ;
% UnonTmp = exp(h/2.*(N1+N2)).* Utmp ;
% U2tmp = ifft(exp(h/2.*D).* fft(UnonTmp)) ;
```

```
% N2 = j*Gamma.*(abs(U2tmp)).^2 ;
% end
%
% % Second half interval : non linear operator only (lumped)
% Unon = exp(h/2.*(N1+N2)).* Udisp ;
%
% % Whole segment propagation
% U2 = ifft(exp(h/2.*D).*fft(Unon));
% Usol=U2;
% end %b
%
% ARR2(a,:)=abs(U2).^2 ;
%
% plot1Hndl=plot(t*1e12, abs(U2).^2) ;grid; xlabel('Time
(ps)');
% ylabel('P (W)');
% zz = a*BlockSize*h/1000;
% global b0
% str0=num2str(zz);
% b0 = uicontrol('Parent',h0,…
%                   'Style','text','BackgroundColor',[1 1 1], …
%                   'Units','points',…
%                   'ForegroundColor','blue',…
%                   'FontSize',10, …
%                   'String',['z = ',str0,' km'],'Position',[500
310 100 15]);
%
% %axes(axHndl1);
% pause(0.5)
% end
% =========================================================
N_pass = 2000;
for ii = 2:N_pass
% h1 = waitbar((ii-1)/N_pass,'The program is running…');
% [Usol,G] = AmpSimpNoise(Usol,GssdB,PoutsatdB,NF);
[Usol,A3d,z]=hconst(Usol,Length_EDF,5,1*beta2_EDF,1*beta3_
EDF,alpha_loss_EDF,1*gamma_EDF,0); % Fiber section #1
% Usol = GaussLPfilt(tstep,Usol,1,f3dB);
% [Eo,A3d,z]=hconst(Eo,Length_SMF,5,1*beta2_SMF,1*beta3_
SMF,alpha_loss_SMF,1*gamma_SMF,0); % Fiber section #2
% Eo = modInt(Eo(1:N),alpha,epsilon,m,fm,0.5); % MZI modulator
        Usol = pha_mod(Usol(1:Ns),Vm,Vbias,Vpi,fm,p0); % Phase
modulator
% Usol = Usol.*exp(-j*Ci*tstep.^2./2/To.^2);
% Usol = hpha_mod(Usol(1:Ns),Vm,Vbias,Vpi,fm,3,p0);
% Usol = Usol*atten;
[Usol,A3d,z]=hconst(Usol,Length_SMF,5,1*beta2_SMF,1*beta3_
SMF,alpha_loss_SMF,1*gamma_SMF,0); % Fiber section #2
   % -------------------------------------------------------
```

```
%  ARR2(ii,:)=abs(Usol).^2 ;
%  Up = Usol.';
   ind = mod(ii,20);
   if ind==0
     in = round(ii/20) + 1;
     ARR2(in,:)=abs(Usol).^2 ;
     Up(in,:) = Usol;
   end
   if (ii>N_pass - 20)
     in = round(ii/20) + 1;
     ARR2(in,:)=abs(Usol).^2 ;
     Up(in,:) = Usol;
   end
%  Up(ii,:) = Usol;
%  plot1Hndl=plot(tstep*1e12, abs(Usol).^2) ;grid; xlabel('Time
(ps)');
%  ylabel('P (W)');
   %  -------------------------------------------------------
        %  Display the number of round trips
%  global b0
%  str0 = num2str(ii);
%  b0 = uicontrol('Parent',h0,…
%                   'Style','text','BackgroundColor',[1 1 1], …
%                   'Units','points',…
%                   'ForegroundColor','blue',…
%                   'FontSize',10, …
%                   'String',['roundtrips = ',str0,'rounds'],
'Position',[500 310 100 15]);
% %  -------------------------------------------------------
%  pause(0.01)
% %  close(h1);
end
%  Eout = Eout/atten;
%  ========================================================
figure(1);
plot(tstep,abs(Usol).^2)
colormap('default')
mesh (real(abs(Up).^2),'meshstyle','row','facecolor','none');
%  waterfall(ARR2);
%  waterfall(abs(Up).^2);
view(17.5,42);
axis tight;
%  delete solpair.mat
save solpair5 Up tstep Po TotDisp Gamma Ld To Ci r teta Tfwhm
Tb Vm Vpi

NN = size(Up,1);
figure(2);
PP = angle(Up(1,:));
Pha_tinput = unwrap(PP);
```

```
ff = -diff(Pha_tinput)/Ts;
chrate = diff(ff)/Ts;
subplot(221);
% plot (tstep*1e12,Pha_tinput);grid;
plot (tstep(1:Ns-1)*1e12,ff);grid;
% plot (tstep(2:Ns-1)*1e12,chrate);grid;
PP = angle(Up(5,:));
Pha_tinput = unwrap(PP);
ff = -diff(Pha_tinput)/Ts;
chrate = diff(ff)/Ts;
subplot(222);
% plot (tstep*1e12,Pha_tinput);grid;
plot (tstep(1:Ns-1)*1e12,ff);grid;
% plot (tstep(2:Ns-1)*1e12,chrate);grid;
in = round(NN/2);
PP = angle(Up(in,:));
Pha_tinput = unwrap(PP);
ff = -diff(Pha_tinput)/Ts;
chrate = diff(ff)/Ts;
subplot(223);
% plot (tstep*1e12,Pha_tinput);grid;
plot (tstep(1:Ns-1)*1e12,ff);grid;
% plot (tstep(2:Ns-1)*1e12,chrate);grid;
PP = angle(Up(NN,:));
Pha_tinput = unwrap(PP);
ff = -diff(Pha_tinput)/Ts;
chrate = diff(ff)/Ts;
subplot(224);
% plot (tstep*1e12,Pha_tinput);grid;
plot (tstep(1:Ns-1)*1e12,ff);grid;
% plot (tstep(2:Ns-1)*1e12,chrate);grid;
figure(3);
title('The phase evolution of soliton pairs');
subplot(221);
plot (tstep*1e12,ARR2(1,:));grid;

subplot(222);
plot (tstep*1e12,ARR2(5,:));grid;

subplot(223);
in = round(NN/2);
plot (tstep*1e12,ARR2(in,:));grid;

subplot(224);
plot (tstep*1e12,ARR2(NN,:));grid;

figure(4);
Kmag = 1;
Nplot = 100;
Uf = Up.';
Eoutfreq = fft(Uf,Ns); %(:,N1)
Eoutfreq1 = fft(Uf,Ns*Kmag); %(:,N1)
```

```
Ioutfreq = Eoutfreq1.*conj(Eoutfreq1)/(Ns*Kmag)^2;
ind = (- Nplot/2 : Nplot/2)';
freq = ind/Ts/Ns/Kmag;
ind = mod((ind + Ns*Kmag),Ns*Kmag)+1;
title('The spectrum evolution of pulse');
subplot(221)
% plot(freq,Ioutfreq(ind,1));
plot(freq,10*log10(Ioutfreq(ind,1))+30);
Xlabel('Freq (Hz)');
Ylabel('P (W)');
subplot(222)
% in = round(N1/4);
% plot(freq,Ioutfreq(ind,5));
plot(freq,10*log10(Ioutfreq(ind,5))+30);
Xlabel('Freq (Hz)');
Ylabel('P (W)');
subplot(223)
in = round(NN/2);
% plot(freq,Ioutfreq(ind,in));
plot(freq,10*log10(Ioutfreq(ind,in))+30);
Xlabel('Freq (Hz)');
Ylabel('P (W)');
subplot(224)
% plot(freq,Ioutfreq(ind,NN));
plot(freq,10*log10(Ioutfreq(ind,NN))+30);
Xlabel('Freq (Hz)');
Ylabel('P (W)');
figure(5);grid;
Pha = Vw*pi./Vpi;              % Phase change in time
dff = -diff(Pha)/Ts;
subplot(211);plot(tstep*1e12,Pha);xlabel('Time
(ps)');ylabel('Phase (rad)');axis tight;
subplot(212);plot(tstep(1:Ns-1)*1e12,dff*1e-12);xlabel('Time
(ps)');ylabel('Chirping (THz)');

%%%%%%%%%%%%%%%%%%%%%%%%%%%%%%%%%%%%%%%%%%%%%%%%%%%%%%%%%%%%%%%%%
% S O L I T O N  O P T I C A L  C O M M U N I C A T I O N S  %
%
%
% global h0

% gaussianpairsnoise.m
clear;
close all
clc;
global Ts;                     % sampling period
global Fcar; % carrier frequency (optical frequency)
global Vw;
global w;
global tstep;
```

```
% cla reset
h0 = figure(1);
c_const = 3e8 ;
ng = 1.47;
lamda = 1.550e-6 ;

% ===========================================================
% Fiber parameters
  Length_SMF = 95;              % Length of a span in meter
  nz = Length_SMF/100;
  alpha_indB = 0.0*1e-3;    % attenuation (dB/km) --> dB/m
  D_SMF = 1*17e-6;             % GVD (ps/nm.km); if anomalous
dispersion(for compensation),D is negative
  n2_SMF = 1*2.6e-20;       % nonlinear index (m^2/W)
  Aeff_SMF = 55e-12;        % effective area (m^2)
% Slope Dispersion
  S_SMF = 0.06e3;              % ps/(nm^2.km) in SI unit

% CALCULATED QUANTITIES of SMF

  alpha_loss_SMF = 1*log(10)*alpha_indB/10;    % alpha (1/m)
  beta2_SMF = -D_SMF*lamda^2/(2*pi*c_const);   % beta2
(ps^2/km);
  % -------------------------------------------------------
  % beta 3 can be calculated from the Slope Dispersion (S) as
follows:]
  beta3_SMF = (S_SMF - (4*pi*c_const/(lamda^3))*beta2_SMF)/
((2*pi*c_const/(lamda^2))^2);
% -------------------------------------------------------
  gamma_SMF = 2*pi*n2_SMF/(lamda*Aeff_SMF);    % nonlinearity
coefficient (km^-1.W^-1)
% ===========================================================
% filter bandwidth
  lamda3dB = 4.0e-9;                   %1.2e-9;      % 1.2nm
  f3dB = lamda3dB*(1e11/0.8e-9);  % Hz
% Amplifier parameters:
  GssdB = 20;                          % (dB)
   PoutsatdB = 10;                     % (dBm)
   NF = 5;                             % (dB)
% EDF fiber parameters
  Length_EDF = 20;                     % Length of a span in meter
  alpha_indB = 0.0*1e-3;    % attenuation (dB/km) --> dB/m
  D_EDF = -10e-6;                      % GVD (ps/nm.km); if
anomalous dispersion(for compensation),D is negative
  n2_EDF = 1*2.6e-20;                  % nonlinear index (m^2/W)
  Aeff_EDF = 45e-12;                   % effective area (m^2)
% Slope Dispersion
  S_EDF = 0.06e3;                      % ps/(nm^2.km) in SIunit
% CALCULATED QUANTITIES of EDF
  alpha_loss_EDF = 1*log(10)*alpha_indB/10;    % alpha (1/m)
```

```
  beta2_EDF = -D_EDF*lamda^2/(2*pi*c_const)    % beta2 (ps^2/
km);
  % ----------------------------------------------------------
  % beta 3 can be calculated from the Slope Dispersion (S) as
follows:]
  beta3_EDF = (S_EDF - (4*pi*c_const/(lamda^3))*beta2_EDF)/
((2*pi*c_const/(lamda^2))^2);
% ----------------------------------------------------------
  gamma_EDF = 2*pi*n2_EDF/(lamda*Aeff_EDF);    % nonlinearity
coefficient (km^-1.W^-1)
% ==========================================================
% Average parameters
% Loss
  loss = 8;                       % dB
  atten = 1/10^(loss/20);

% to test dispersion let TotDisp=-2e-6, SegNum=2.5*Zo/h
% run again with U1 removed and TotDisp=2e-6
Length_total = Length_SMF+Length_EDF;
TotDisp = (D_SMF*Length_SMF+D_EDF*Length_EDF)/Length_total;
% TotDisp = 17*1e-6 ;

ro = 3e-6 ;
n2 = 3.2*1e-20 ;
Beta2 = -(TotDisp*lamda^2)/(2*pi*c_const);
Beta3 = 0 ;
Gamma = (gamma_SMF*Length_SMF+gamma_EDF*Length_EDF)/Length_
total ;
% Gamma = (2*pi*n2)/(lamda*pi*ro^2) ;
alphadB = 0;
Alpha=log(10)*alphadB/(10*1000); % = field loss per meter
% ==========================================================
% modulator parameters
Vpi = 6;            % Volt
Vbias = 0;          % Volt
Vm = 4.5;           % Volt
m = Vm/Vpi;         % modulation index
fm = 1e9;           % modulation frequency
p0 = 0*pi/2;
% ==========================================================
% Soliton pulse parameters
Norder = 1;         % Order of soliton
Tfwhm = 5.0e-12;
To = Tfwhm/1.763;   % Pulsewidth of soliton
Tb = 20e-12;        % Pulses distance
teta = 1*pi/2; % Relative phase difference between 2 solitons
r = 1;              % Amplitude ratio
Ci = 0;             % Chirping factor of pulse
qo = Tb/2/To;       % Space between 2 solitons
qr = Tb/2/To/2      % Space between 2 solitons
```

```
Ld = To^2/abs(Beta2)
Zo = pi*Ld/2 ;
Po = Norder^2/(Gamma*Ld);
P0 = 4e-3;
% Pav = 3*Po/(2*qr)
% ==========================================================
Ns = 2^12;
Ts =0.05e-12;
tstep = Ts*(-Ns/2:1:Ns/2-1);
w = fftshift((2*pi/((Ns-1)*Ts))*(-Ns/2:1:Ns/2-1));

% Usol = sqrt(P0)*(exp(-(tstep-Tb/2).^2/2/To^2).*exp
(-j*teta)+r*exp(-(tstep+Tb/2).^2/2/To^2).*exp(+j*teta)).*exp
(-j*Ci*tstep.^2./2/To^2);%*exp(j*(2*pi*f*ti+phiC));
Usol = sqrt(P0)*exp(-tstep.^2/2/To^2).*exp(-j*Ci*tstep.^2./2/
To^2);%*exp(j*(2*pi*f*ti+phiC));

% Eo = Usol;
plot(tstep,abs(Usol).^2)
grid
ARR2(1,:) = abs(Usol).^2 ;
Up(1,:) = Usol;
pause(0.1)
% ==========================================================
N_pass = 2000;
for ii = 2:N_pass
% h1 = waitbar((ii-1)/N_pass,'The program is running…');
  [Usol,G] = AmpSimpNoise(Usol,GssdB,PoutsatdB,NF);

[Usol,A3d,z]=hconst(Usol,Length_EDF,5,1*beta2_EDF,1*beta3_
EDF,alpha_loss_EDF,1*gamma_EDF,0); % Fiber section #1
  Usol = GaussLPfilt(tstep,Usol,1,f3dB);
% [Eo,A3d,z]=hconst(Eo,Length_SMF,5,1*beta2_SMF,1*beta3_
SMF,alpha_loss_SMF,1*gamma_SMF,0); % Fiber section #2
% Eo = modInt(Eo(1:N),alpha,epsilon,m,fm,0.5); % MZI modulator
  Usol = pha_mod(Usol(1:Ns),Vm,Vbias,Vpi,fm,p0); % Phase
modulator
% Usol = Usol.*exp(-j*Ci*tstep.^2./2/To.^2);
% Usol = hpha_mod(Usol(1:Ns),Vm,Vbias,Vpi,fm,3,p0);
  Usol = Usol*atten;

[Usol,A3d,z]=hconst(Usol,Length_SMF,5,1*beta2_SMF,1*beta3_
SMF,alpha_loss_SMF,1*gamma_SMF,0); % Fiber section #2
  % --------------------------------------------------------
% ARR2(ii,:)=abs(Usol).^2 ;
% Up = Usol.';
   ind = mod(ii,20);
   if ind==0
   in = round(ii/20) + 1;
   ARR2(in,:)=abs(Usol).^2 ;
   Up(in,:) = Usol;
```

```
    end
    if (ii>N_pass - 20)
      in = round(ii/20) + 1;
      ARR2(in,:)=abs(Usol).^2 ;
      Up(in,:) = Usol;
    end
% % close(h1);
end
% Eout = Eout/atten;
% ============================================================
figure(1);
plot(tstep,abs(Usol).^2)
colormap('default')
mesh (real(abs(Up).^2),'meshstyle','row','facecolor','none');
% waterfall(ARR2);
% waterfall(abs(Up).^2);
view(17.5,42);
axis tight;
% delete solpair.mat
save solpair5 Up tstep Po TotDisp Gamma Ld To Ci r teta Tfwhm
Tb Vm Vpi
Pav = 3*max(max(ARR2))/(2*qr)

NN = size(Up,1);
figure(2);
PP = angle(Up(1,:));
Pha_tinput = unwrap(PP);
ff = -diff(Pha_tinput)/Ts;
chrate = diff(ff)/Ts;
subplot(221);
% plot (tstep*1e12,Pha_tinput);grid;
plot (tstep(1:Ns-1)*1e12,ff);grid;
% plot (tstep(2:Ns-1)*1e12,chrate);grid;
PP = angle(Up(5,:));
Pha_tinput = unwrap(PP);
ff = -diff(Pha_tinput)/Ts;
chrate = diff(ff)/Ts;
subplot(222);
% plot (tstep*1e12,Pha_tinput);grid;
plot (tstep(1:Ns-1)*1e12,ff);grid;
% plot (tstep(2:Ns-1)*1e12,chrate);grid;
in = round(NN/2);
PP = angle(Up(in,:));
Pha_tinput = unwrap(PP);
ff = -diff(Pha_tinput)/Ts;
chrate = diff(ff)/Ts;
subplot(223);
% plot (tstep*1e12,Pha_tinput);grid;
plot (tstep(1:Ns-1)*1e12,ff);grid;
% plot (tstep(2:Ns-1)*1e12,chrate);grid;
```

```
PP = angle(Up(NN,:));
Pha_tinput = unwrap(PP);
ff = -diff(Pha_tinput)/Ts;
chrate = diff(ff)/Ts;
subplot(224);
% plot (tstep*1e12,Pha_tinput);grid;
plot (tstep(1:Ns-1)*1e12,ff);grid;
% plot (tstep(2:Ns-1)*1e12,chrate);grid;
figure(3);
title('The phase evolution of soliton pairs');
subplot(221);
plot (tstep*1e12,ARR2(1,:));grid;

subplot(222);
plot (tstep*1e12,ARR2(5,:));grid;

subplot(223);
in = round(NN/2);
plot (tstep*1e12,ARR2(in,:));grid;

subplot(224);
plot (tstep*1e12,ARR2(NN,:));grid;

figure(4);

Kmag = 1;
Nplot = 100;
Uf = Up.';
Eoutfreq = fft(Uf,Ns);          %(:,N1)
Eoutfreq1 = fft(Uf,Ns*Kmag); %(:,N1)
Ioutfreq = Eoutfreq1.*conj(Eoutfreq1)/(Ns*Kmag)^2;
ind = (- Nplot/2 : Nplot/2)';
freq = ind/Ts/Ns/Kmag;
ind = mod((ind + Ns*Kmag),Ns*Kmag)+1;
title('The spectrum evolution of pulse');
subplot(221)
% plot(freq,Ioutfreq(ind,1));
plot(freq,10*log10(Ioutfreq(ind,1))+30);
Xlabel('Freq (Hz)');
Ylabel('P (W)');
subplot(222)
% in = round(N1/4);
% plot(freq,Ioutfreq(ind,5));
plot(freq,10*log10(Ioutfreq(ind,5))+30);
Xlabel('Freq (Hz)');
Ylabel('P (W)');
subplot(223)
in = round(NN/2);
% plot(freq,Ioutfreq(ind,in));
plot(freq,10*log10(Ioutfreq(ind,in))+30);
Xlabel('Freq (Hz)');
Ylabel('P (W)');
```

```
subplot(224)
% plot(freq,Ioutfreq(ind,NN));
plot(freq,10*log10(Ioutfreq(ind,NN))+30);
Xlabel('Freq (Hz)');
Ylabel('P (W)');
figure(5);grid;
Pha = Vw*pi./Vpi; % Phase change in time
dff = -diff(Pha)/Ts;
subplot(211);plot(tstep*1e12,Pha);xlabel('Time
(ps)');ylabel('Phase (rad)');axis tight;
subplot(212);plot(tstep(1:Ns-1)*1e12,dff*1e-12);xlabel('Time
(ps)');ylabel('Chirping (THz)');

function [Eout,gain] = AmpSimpNoise(Ein,GssdB,PoutsatdB,NF)
% amp_simp(Ein,GssdB,PoutsatdB,NF)
% simple model of optical amplifier. The model includes the
gain
% saturation without noise
% written by Lam Quoc Huy
% Amplifier parameters:
%   small signal gain: GssdB (dB)
%   output saturation power: PoutsatdB (dBm)
%
% The input is a column vector containing block N samples of
the optical signal sampling at the
% rate 1/Ts
% The output is calculated using
%   Eout = Ein*sqrt(G)
% where: G is the saturated gain
%    G = Gss*exp(-(G-1)Pin/Psat)  (eq1)
global Ts
f = 193.1e12;
hplank = 6.6261*1e-34;

Gss = 10^(GssdB/10);
Poutsat = (10^(PoutsatdB/10))*1e-3;
Psat = Poutsat*(Gss-2)/Gss/log(2);
% Pinsat = 2* Poutsat/Gss;

N = size(Ein,1);
% Pin = (sum(Ein.*conj(Ein))/N);
Pin = mean((Ein.*conj(Ein)));
% numerical calculation of G from the equation G = (Gss -
lnG)*Psat/Pin + 1
tol = 0.05;  % tolerance for G calculation
step = Gss/2;
G = Gss;
err = 10;
while (err > tol)
  G1 = Gss*exp(-(G-1)*Pin/Psat);
  err = G1 - G;
```

```
  if err>0
   if step <0
     step = -step/2;
   end
  else
   if step >0
     step = -step/2;
   end
   err = -err;
  end
  G = G + step;
end
G = G - step;

% Eout = sqrt(G)*Ein;
gain = G;
Egain = sqrt(G)*Ein;

dt = Ts;
Bsim = 1/dt;
FigNoise = 10^(NF/10);
nsp = (FigNoise*G-1)/(2*(G-1));
% Pase = hplank.*opfreq.*nsp*(OGain-1)*Bsim
Pase = hplank.*f.*nsp*(G-1)*Bsim/1000;
PasedB = 10*log10(Pase);
% afout = fft(Egain) + (randn(size(Egain))+i*randn(size
(Egain)))*sqrt(Pase)./sqrt(2);
% Eout = ifft(afout);
% Eout = Egain + (randn(size(Egain))+i*randn(size
(Egain)))*sqrt(Pase)./sqrt(2)./1;
Eout = Egain + wgn(N,1,PasedB,'complex');
% Eout = Egain;

function [Atout] = GaussLPfilt(t,At,N,f3dB)
% global BitRate

% if nargin == 3
% f3dB = 0.7*BitRate;
% end
S = length(t);
dt = abs(t(2)-t(1));
fin = fftshift(1/dt/1*(-S/2:S/2-1)/(S));
% k = (1:S)-1;
% k(S/2+1:S) = k(S/2+1:S) - S;
% % k = k';
% fin = k/dt/S;

Afin = fft(At);

T = exp(-log(sqrt(2))*(fin/f3dB).^(2*N)); % Ham truyen bo loc
fout = fin;
Afout = T.*Afin;
```

```
Atout = ifft(Afout);
% figure(4);
% plot(fin,T);

function [As_out,A3d,z] = hconst(As_in,L,h,b2,b3,a,g,p)
% Symmetrized Split-Step Fourier Method
% with constant step.
% Input:
%   L = Length of fiber
%   h = step size
%   As_in = Input field in the time domain
%
% Output:
%   As_out = Output field in the time domain
%
%   wrtten by N D Nhan - PTIT
global beta2
global beta3
global w
global alpha
global gam
global pmdmode

beta2 = b2;
beta3 = b3;
alpha = a;
gam = g;
pm = p;
pmdmode = 0;

Atemp = As_in;
M = round(L/h);        % number of steps
n = 1;
z = 0;
l = 1;
A3d = [];

for k = 1:M
    Atemp = ssfm(Atemp,h);
    z(1,n+1) = z(1,n)+h;
    n = n+1;
    % A3d(l,:) = At;
    l = l + 1;
end

As_out = Atemp;
        % ============Testing
% c = 3e8;
% nin = 1.5;
% vg = c/nin;
% Tr = 50/c/nin;
```

```
% Nc =size(As_in,1);
% Tw = Nc*1e-13;
% Tm = 1/10e9;
% Th = Tr/1000;
%
% As_out = As_out.*exp(j*2*pi*(Tm-Th)/Tm);

function Eout = hpha_mod(Ein,Vm,Vbias,Vpi,fm,st,ph0)
% phase modulator parameters
% m: phase modulation index
% bias: DC phase shift (rad)
% modulation frequency: fm
% ph0: initial phase
%% written by Nguyen Duc Nhan
global tstep;
global Ts;
global Vw;

% N = size(Ein,1);
% k = (1:N)';
% tstep = Ts*(k-N/2);
mrad = Vm/Vpi*pi;

Norder = 1;

% Vw = Vbias+Vm*cos(2*pi*fm*(tstep-st)+ph0)+0.27*Vm*cos
(2*pi*2*fm*(tstep-st)+ph0-0.6*pi/1)...
%   +0.001*Vm*cos(2*pi*3*fm*(tstep-st)+ph0-0.6*pi/1)+0.0001*Vm
*cos(2*pi*4*fm*(tstep-st)+ph0-0.6*pi/1);
Vw = Vbias+Norder*Vm*cos(2*pi*Norder*fm*(tstep-st)+ph0+
0.0*pi)+0.48*Norder*Vm*cos(2*pi*2*Norder*fm*
(tstep-st)+2*ph0-1.4*pi/1)...
+0.003*Norder*Vm*cos(2*pi*3*Norder*fm*(tstep-st)+
3*ph0-0.1*pi/1)+0.0001*Norder*Vm*cos(2*pi*4*fm*
(tstep-st)+4*ph0-0.1*pi/1);

Eout = Ein.*exp(j*Vw);

function Eout = pha_mod(Ein,Vm,Vbias,Vpi,fm,ph0)
% phase modulator parameters
% m: phase modulation index
% bias: DC phase shift (rad)
% modulation frequency: fm
% ph0: initial phase
% % written by Nguyen Duc Nhan

global tstep;
global Ts;
global Vw;

N = size(Ein,1);
% k = (1:N)';
% tstep = Ts*(k-N/2);
```

```
T0 = 1/fm;
modeph = 0;
if modeph == 0
    Vw = Vbias+Vm*cos(2*pi*fm*tstep+ph0);
elseif modeph == 1
    Vw = real(synth_sig(Vbias,Vm,fm,tstep,ph0*T0/(2*pi),2));
else
    Vw = real(synth_sig(Vbias,Vm,fm,tstep,ph0*T0/(2*pi),1));
end
% delta_phi = pi/4*(2- bias*2 - ext*Vm);
Phi = Vw*pi/Vpi;
Eout = Ein.*exp(j*Phi);
% Eout = Ein.*Vm;

% solplot.m
% plotting the soliton - soliton pairs
load solpair5
Ts = tstep(2)-tstep(1);
Ns = size(Up,2);
ARR2 = abs(Up).^2;
figure(1);
colormap('default')
mesh (real(abs(Up).^2),'meshstyle','row','facecolor','none');
%waterfall(abs(Up).^2);
view(17.5,42);
axis tight;

NN = size(Up,1);
figure(2);
PP = angle(Up(1,:));
Pha_tinput = unwrap(PP);
ff = -diff(Pha_tinput)/Ts;
chrate = diff(ff)/Ts;
subplot(221);
% plot (tstep*1e12,Pha_tinput);grid;
% plot (tstep(1:Ns-1)*1e12,ff);grid;
plot (tstep(2:Ns-1)*1e12,chrate);grid;
PP = angle(Up(5,:));
Pha_tinput = unwrap(PP);
ff = -diff(Pha_tinput)/Ts;
chrate = diff(ff)/Ts;
subplot(222);
% plot (tstep*1e12,Pha_tinput);grid;
% plot (tstep(1:Ns-1)*1e12,ff);grid;
plot (tstep(2:Ns-1)*1e12,chrate);grid;
in = round(NN/2);
PP = angle(Up(in,:));
Pha_tinput = unwrap(PP);
ff = -diff(Pha_tinput)/Ts;
chrate = diff(ff)/Ts;
```

```
subplot(223);
% plot (tstep*1e12,Pha_tinput);grid;
% plot (tstep(1:Ns-1)*1e12,ff);grid;
plot (tstep(2:Ns-1)*1e12,chrate);grid;
PP = angle(Up(NN,:));
Pha_tinput = unwrap(PP);
ff = -diff(Pha_tinput)/Ts;
chrate = diff(ff)/Ts;
subplot(224);
% plot (tstep*1e12,Pha_tinput);grid;
% plot (tstep(1:Ns-1)*1e12,ff);grid;
plot (tstep(2:Ns-1)*1e12,chrate);grid;
figure(3);
title('The phase evolution of soliton pairs');
subplot(221);
plot (tstep*1e12,ARR2(1,:));grid;

subplot(222);
plot (tstep*1e12,ARR2(5,:));grid;

subplot(223);
in = round(NN/2);
plot (tstep*1e12,ARR2(in,:));grid;

subplot(224);
plot (tstep*1e12,ARR2(NN,:));grid;

figure(4);

Kmag = 1;
Nplot = 100;
Uf = Up.';
Eoutfreq = fft(Uf,Ns);          %(:,N1)
Eoutfreq1 = fft(Uf,Ns*Kmag); %(:,N1)
Ioutfreq = Eoutfreq1.*conj(Eoutfreq1)/(Ns*Kmag)^2;
ind = (- Nplot/2 : Nplot/2)';
freq = ind/Ts/Ns/Kmag;
ind = mod((ind + Ns*Kmag),Ns*Kmag)+1;
title('The spectrum evolution of pulse');
subplot(221)
% plot(freq,Ioutfreq(ind,1));
plot(freq,10*log10(Ioutfreq(ind,1))+30);
Xlabel('Freq (Hz)');
Ylabel('P (W)');
subplot(222)
% in = round(N1/4);
% plot(freq,Ioutfreq(ind,5));
plot(freq,10*log10(Ioutfreq(ind,5))+30);
Xlabel('Freq (Hz)');
Ylabel('P (W)');
subplot(223)
```

```
in = round(NN/2);
% plot(freq,Ioutfreq(ind,in));
plot(freq,10*log10(Ioutfreq(ind,in))+30);
Xlabel('Freq (Hz)');
Ylabel('P (W)');
subplot(224)
% plot(freq,Ioutfreq(ind,NN));
plot(freq,10*log10(Ioutfreq(ind,NN))+30);
Xlabel('Freq (Hz)');
Ylabel('P (W)');


function [Ato] = ssfm(Ati,h)
% Symmetrized Split-Step Fourier Method
% used for single channel.
% Input:
% L = Length of fiber
% h = Variable simulation step
% Ati = Input field in the time domain
%
% Output:
%        Ato = Output field in the time domain
%
%        written by N D Nhan - PTIT
global beta2
global beta3
global w
global alpha
global gam

% c = 3e8;
% nin = 1.5;
% vg = c/nin;
% Tr = 50/c/nin;
% Nc =size(Ati,1);
% Tw = Nc*1e-13;

D = -i/2*beta2.*(i*w).^2+1/6*beta3.*(i*w).^3-alpha/2; % linear
operator
N1 = i*gam.*(abs(Ati).^2); % nonlinear operator
N2 = N1;
%Propagation in the first half dispersion region, z to z +h/2
At1 = ifft(exp(h/2.*D).*fft(Ati)); %
% At1 = ifft(exp(1/Tr*h/2.*D).*fft(Ati)); %

% ========================================================
% Iteration for the nonlinear phase shift (2 iterations)
% ========================================================
for m = 1:4
    At1temp = ifft(exp(h/2.*D).*fft(Ati));
    At2 = exp(h/2*(N1+N2)).*At1temp;
```

```
      At3 = ifft(exp(h/2.*D).*fft(At2));
      %At3 = At3.';
      N2 = i*gam.*(abs(At3).^2);
%     At1temp = ifft(exp(1/Tr*h/2.*D).*fft(Ati));
%     At2 = exp(1/Tr*h/2*(N1+N2)).*At1temp;
%     At3 = ifft(1/Tr*exp(h/2.*D).*fft(At2));
%     %At3 = At3.';
%     N2 = i*gam.*(abs(At3).^2);
end
At4 = exp(h/2.*(N1+N2)).*At1;
% Propagation in the second Dispersion region, z +h/2 to z + h
Ato = ifft(exp(h/2.*D).*fft(At4));
% Ato = Ato.*exp(j*Tr/Tw);


function Vout = synth_sig(Vbias,Vac,fm,t,initphase,opt)
% Synth_sig is a function to synthesize an arbitrary waveforms
% to generate the signal driving a phase modulator
% Vbias - DC voltage
% Vac - amplitude of the ac component
% fm - modulation frequency
% t - vector of times

Nharm = 38;

period = 1/fm;
y = 0;
if opt == 1
  step = 2;
  Vdc = Vbias;
  for ii = -Nharm:step:Nharm
   if ii == 0
     y = y + 0;
   else
     ai = -2/(pi*ii)^2;
     y = y + ai*exp(j*2*pi*ii*(t-initphase)/period);
   end
end
  Va = Vac*2*y;
elseif opt == 2
  step = 1;
  Vdc = Vbias + 0;%15/20;
  Va = Vac/2;
  for ii = -Nharm:step:Nharm
   if ii == 0
     y = y + 0;
   else
     ai = -2/(pi*ii)^2;
     y = y + ai*exp(j*2*pi*ii*(t-initphase)/period);
   end
end
```

```
Va = Va*2*y;
else
  % Vdc  = Vbias;
  % pp   = initphase*2*pi/period;
  % ff   = cos(2*pi*fm*t+pp);
  % Va   = Vac*exp(ff);
  Vdc    = Vbias;
  pp     = initphase*2*pi/period;
  mm     = 0.02;
  Tfwhm = mm*period;
  ti = t;
% ff = Vac*cos(2*pi*fm*t+pp);
  % ti = rem(t,period) ; % fraction of time within BitPeriod
  n = fix(t/period)+1 ; % extract input sequence number
  Va = Vac*exp(-log(2)*1/2*(2*ti./Tfwhm).^(2*1));
end
Vout = Vdc + Va;
```

Appendix C: Abbreviations

ACF	Autocorrelation function
AM	Amplitude modulation
AMLM-EDFL	Actively mode-locked multiwavelength erbium-doped fiber laser
AOM	Acousto-optic modulator
APE	Annealed proton exchange
APL	Additive pulse limiting
APM	Addictive pulse mode-locking
ASE	Amplified spontaneous emission
AWG	Arrayed waveguide grating
BER	Bit error rate
BPG	Bit pattern generator
BPF	Bandpass filter
ccw	Counter clockwise
CFBG	Chirped fiber Bragg grating
CSA	Communications signal analyzer
CSRZ	Carrier suppressed return-to-zero
cw	Clockwise
CW	Continuous-wave
DC	Direct current
DCF	Dispersion compensating fiber
DFB	Distributed feedback
DM	Dispersion management
DSF	Dispersion shifted fiber
DWDM	Dense wavelength division multiplexing
E/O	Electrical-optical conversion
EDF	Erbium-doped fiber
EDFA	Erbium-doped fiber amplifier
EDFL	Erbium-doped fiber laser
EO	Electro-optic
ESA	Excited state absorption
FBG	Fiber Bragg grating
FFP	Fabry–Perot filter
FFT	Fast Fourier transform
IFFT	Inverse fast Fourier transform
FM	Frequency modulation
FP	Fabry–Perot
FRL	Fiber ring laser
FS	Frequency shifter
FSK	Frequency shift keying

FSR	Free spectral range
FTTH	Fiber to the home
FTTx	Fiber to the x
FWHM	Full width at half maximum
FWM	Four wave mixing
GVD	Group velocity dispersion
HDTV	High definition television
HiBi	Hi-birefringence
HMLFL	Harmonic mode-locked fiber laser
HNLF	Highly nonlinear fiber
IPTV	Internet protocol television
KLM	Kerr lens mode-locking
LCFG	Linearly chirped fiber grating
LD	Laser diode
LHS	Left-hand side
LPFBG	Long period fiber Bragg grating
MBS	Multi-bound solitons
ML	Mode locked
MLFL	Mode-locked fiber laser
MLFRL	Mode-locked fiber ring laser
MLL	Mode-locked laser
MZ	Mach–Zehnder
MZIM	Mach–Zehnder intensity or interferometric modulator
NALM	Nonlinear amplifying loop mirror
NF	Noise figure
NLSE	Nonlinear Schrödinger equation
NOLM	Nonlinear optical loop mirror
NPR	Nonlinear polarization rotation
NTT	Nippon Telegraph and Telephone Company
NZ-DSF	Nonzero dispersion–shifted fiber
O/E	Optical-electrical conversion
OE	Opto-electronic
OEO	Opto-electrical oscilloscope
OPO	Optical parametric oscillator
OSA	Optical spectrum analyzer
OSC	Oscilloscope
OTDM	Optical time division multiplexing
OTDR	Optical time domain reflectometer
PA	Parametric amplifier
PC	Polarization controller
PCF	Photonic crystal fiber
PD	Photodetector
PDM	Polarization division multiplexing
PM	Phase modulation
PMD	Polarization mode dispersion

PM-EDF	Polarization-maintaining erbium-doped fiber
PMF	Polarization-maintaining fiber
PO	Parametric oscillator
Pol	Polarizer
PS	Phase shifter
PZT	Piezoelectric transducer
RF	Radio frequency
RFA	Radio frequency amplifier
RHML	Rational harmonic mode locking
RHMLFL	Rational harmonic mode-locked fiber laser
RHS	Right-hand side
RMLFRL	Regenerative mode-locked fiber ring laser
RMS	Root mean square
SAW	Surface acoustic waves
SMF	Single-mode fiber
SNR	Signal-to-noise ratio
SOA	Semiconductor optical amplifier
SPM	Self-phase modulation
SRD	Step recovery diode
SRS	Stimulated Raman scattering
SSFM	Split step Fourier method
TBP	Time-bandwidth product
TBRRM	Talbot-based repetition rate multiplication
TDM	Time division multiplexing
TPA	Two-photon absorption
VRC	Variable ratio coupler
WDM	Wavelength-division multiplexing
XPM	Cross-phase modulation

Index

Printed and bound by CPI Group (UK) Ltd, Croydon, CR0 4YY

21/10/2024

01777089-0016